Springer Proceedings in Mathematics & Statistics

Volume 310

Springer Proceedings in Mathematics & Statistics

This book series features volumes composed of selected contributions from workshops and conferences in all areas of current research in mathematics and statistics, including operation research and optimization. In addition to an overall evaluation of the interest, scientific quality, and timeliness of each proposal at the hands of the publisher, individual contributions are all refereed to the high quality standards of leading journals in the field. Thus, this series provides the research community with well-edited, authoritative reports on developments in the most exciting areas of mathematical and statistical research today.

More information about this series at http://www.springer.com/series/10533

Jin Cheng · Shuai Lu · Masahiro Yamamoto
Editors

Inverse Problems and Related Topics

Shanghai, China, October 12–14, 2018

Springer

Editors
Jin Cheng
Shanghai Key Laboratory for Contemporary Applied Mathematics, Key Laboratory of Mathematics for Nonlinear Sciences and School of Mathematical Sciences
Fudan University
Shanghai, China

Shuai Lu
Shanghai Key Laboratory for Contemporary Applied Mathematics, Key Laboratory of Mathematics for Nonlinear Sciences and School of Mathematical Sciences
Fudan University
Shanghai, China

Masahiro Yamamoto
Department of Mathematical Sciences
The University of Tokyo
Tokyo, Japan

ISSN 2194-1009 ISSN 2194-1017 (electronic)
Springer Proceedings in Mathematics & Statistics
ISBN 978-981-15-1591-0 ISBN 978-981-15-1592-7 (eBook)
https://doi.org/10.1007/978-981-15-1592-7

Mathematics Subject Classification (2010): 35R30, 35R11, 65M32, 65M12, 35Q93, 65J20

This Springer imprint is published by the registered company Springer Nature Singapore Pte Ltd.
The registered company address is: 152 Beach Road, #21-01/04 Gateway East, Singapore 189721, Singapore

Preface

Inverse problems are ubiquitous in various fields of science and engineering, and have been grown rapidly in recent decades. This volume contains thirteen articles which are extended versions of the presentations at International Conference on Inverse Problems at Fudan University, Shanghai, China from October 12–14, 2018 in honor of Masahiro Yamamoto on the occasion of his 60th anniversary.

These articles are authored by world-renowned researchers and rising young talents, and aim at updated accounts of various aspects of the researches on inverse problems. For the inverse problems, both theoretical researches and numerical approaches are indispensable. Thus all the articles are naturally classified into the following two categories, which indicate the main branches of researches on inverse problems:

- **Inverse problems and related topics for partial differential equations**: the seven papers discuss concrete inverse problems or control problems for partial differential equations focusing on the uniqueness, stability, and reconstruction schemes.
- **Regularization theory of inverse problems**: the six papers investigate the convergence properties of different regularization schemes and their related methods or applications.

We hope that this volume can provide current and convenient perspectives of the researches on the inverse problem.

Last but not least, we would like to express our sincere thanks to all the plenary speakers, who have given rise to the present volume. We would also like to thank Mrs. Wei Chen, Mrs. Wenqing Tang and the graduate students at Fudan University for their enormous efforts to make the conference successful.

Shanghai, China — Jin Cheng
Shanghai, China — Shuai Lu
Tokyo, Japan — Masahiro Yamamoto
August 2019

Acknowledgements The conference was financially supported by the A3 Foresight Program "Modeling and Computation of Applied Inverse Problems" by the National Natural Science Foundation of China (NSFC) No. 11421110002 and Japan Society for the Promotion of Science (JSPS). Other financial supports are NSFC projects (No. 11925104, 11971121, 91730304), Shanghai Key Lab of Contemporary Applied Mathematics, and Grant-in-Aid for Scientific Research (S) 15H05740 (JSPS).

Contents

Part II Regularization Theory of Inverse Problems

Part I
Inverse Problems and Related Topics for Partial Differential Equations

Chapter 1
An Inverse Conductivity Problem in Multifrequency Electric Impedance Tomography

Jin Cheng, Mourad Choulli and Shuai Lu

Dedicated to Masahiro Yamamoto for his sixtieth birthday.

Abstract We deal with the problem of determining the shape of an inclusion embedded in a homogenous background medium. The multifrequency electrical impedance tomography is used to image the inclusion. For different frequencies, a current is injected at the boundary and the resulting potential is measured. It turns out that the function describing the potential solves an elliptic equation in divergence form with discontinuous leading coefficient. For this inverse problem we aim to establish a logarithmic type stability estimate. The key point in our analysis consists in reducing the original problem to that of determining an unknown part of the inner boundary from a single boundary measurement. The stability estimate is then used to prove uniqueness results. We also provide an expansion of the solution of the BVP under consideration in the eigenfunction basis of the Neumann–Poincaré operator associated to the Neumann–Green function.

Keywords Inverse conductivity problem · Multifrequency electric impedance tomography · Stability estimate · Neumann-Poincaré operator

---

J. Cheng · S. Lu
Shanghai Key Laboratory for Contemporary Applied Mathematics, Key Laboratory of Mathematics for Nonlinear Sciences and School of Mathematical Science, Fudan University, Shanghai 200433, China
e-mail: jcheng@fudan.edu.cn

S. Lu
e-mail: slu@fudan.edu.cn

M. Choulli (✉)
Université de Lorraine, 34 cours Léopold, 54052 Nancy cedex, France
e-mail: mourad.choulli@univ-lorraine.fr

J. Cheng et al. (eds.), *Inverse Problems and Related Topics*,
Springer Proceedings in Mathematics & Statistics 310,
https://doi.org/10.1007/978-981-15-1592-7_1

1.1 Introduction

We firstly proceed with the mathematical formulation of the problem under consideration. In order to specify the BVP satisfied by the function u describing the electric potential, we consider Ω and D two bounded Lipschitz domains of $\mathbb{R}^n$, $n \geq 2$, so that $D \Subset \Omega$. Fix $k_0 > 0$ and define, for $k \in (0, \infty) \setminus \{k_0\}$, the function $\mathbf{a}_D$ on Ω by

$$\mathbf{a}_D(k) = k_0 + (k - k_0)\chi_D.$$

Here χ_D is the characteristic function of the inclusion D, k_0 is the conductivity of the background medium and k is the conductivity of the inclusion.

It is well known that in the present context u solves the BVP

$$\begin{cases} \text{div}\,(\mathbf{a}_D(k)\nabla u) = 0 \text{ in } \Omega, \\ k_0 \partial_\nu u = f \qquad\qquad \text{on } \partial\Omega, \end{cases} \tag{1.1}$$

where ∂_ν is the derivative along the unit normal vector field ν on $\partial\Omega$ pointing outward Ω.

Let

$$V = \left\{ u \in H^1(\Omega);\ \int_{\partial\Omega} u(x)d\sigma(x) = 0 \right\}.$$

Note that V is a Hilbert space when it is endowed with the scalar product

$$\mathfrak{a}(u, v) = \int_\Omega \nabla u \cdot \nabla v, \quad u, v \in V.$$

We leave to the reader to check that the norm induced by this scalar product is equivalent to the $H^1(\Omega)$-norm on V.

Pick $f \in L^2(\partial D)$ and $k \in (0, \infty) \setminus \{k_0\}$. Then it is not hard to see that, according to Lax–Milgram's lemma, the BVP (1.1) possesses a unique variational solution $u_D(k) \in V$. That is $u_D(k)$ is the unique element of V satisfying

$$\int_\Omega \mathbf{a}_D(k)\nabla u_D(k) \cdot \nabla v dx = \int_{\partial\Omega} f v d\sigma(x), \quad v \in V. \tag{1.2}$$

Recall that the trace operator $\mathfrak{t} : w \in C^\infty(\overline{\Omega}) \mapsto u_{|\partial\Omega} \in C^\infty(\partial\Omega)$ is extended to bounded operator, still denoted by $\mathfrak{t}$, from $H^1(\Omega)$ into $L^2(\partial\Omega)$.

Since we will consider conductivities varying with the frequency ω, we introduce the map $k : \omega \in (0, \infty) \mapsto k(\Omega) \in (0, \infty) \setminus \{k_0\}$. We assume in the sequel the condition

$$\lim_{\omega \to \infty} k(\omega) = \infty.$$

Let $\mathfrak{J}$ be a given subset of $(0, \infty)$. In the present work, we are mainly interested in determining the unknown subdomain D from the boundary measurements

$$\mathfrak{t}u_D(k(\omega)), \quad \omega \in \mathfrak{J}.$$

The inverse problem we discuss in the present paper is known as multifrequency electrical impedance tomography and has for instance applications in biomedical imaging. We refer to [3] and references therein for more explanations.

Prior to stating our main result, we introduce some notations and definitions. Let

$$\begin{aligned}
&\mathbb{R}^n_+ = \{x = (x', x_n) \in \mathbb{R}^n;\ x_n > 0\},\\
&Q = \{x = (x', x_n) \in \mathbb{R}^n;\ |x'| < 1 \text{ and } |x_n| < 1\},\\
&Q_+ = Q \cap \mathbb{R}^n_+,\\
&Q_0 = \{x = (x', x_n) \in \mathbb{R}^n;\ |x'| < 1 \text{ and } x_n = 0\}.
\end{aligned}$$

Fix $0 < \alpha < 1$. We say that the bounded domain $\mathcal{U}$ of $\mathbb{R}^n$ is of class $C^{2,\alpha}$ with parameters $\varrho > 0$ and $\aleph > 0$ if for any $x \in \partial\mathcal{U}$ there exists a bijective map $\phi : Q \to B := B_{\mathbb{R}^{n-1}}(x, \varrho)$, satisfying $\phi \in C^{2,\alpha}(\overline{Q})$, $\phi^{-1} \in C^{2,\alpha}(\overline{B})$ and

$$\|\phi\|_{C^{2,\alpha}(\overline{Q})} + \|\phi^{-1}\|_{C^{2,\alpha}(\overline{B})} \le \aleph,$$

so that

$$\phi(Q_+) = \mathcal{U} \cap B \quad \text{and} \quad \phi(Q_0) = B \cap \partial\mathcal{U}.$$

If we substitute in this definition $C^{2,\alpha}$ by $C^{0,1}$ then we obtain the definition of a Lipschitz domain with parameters ϱ and $\aleph$.

Let $\mathscr{D}_0(\varrho, \aleph)$ denote the set of subdomains D so that $D \Subset \Omega$, and $\Omega \setminus \overline{D}$ is of class $C^{2,\alpha}$ with parameters $\varrho > 0$ and $\aleph > 0$.

Fix $D_0 \Subset \Omega_0 \Subset \Omega$ and $\delta > 0$. Consider then $\mathscr{D}_1(\varrho, \aleph, \delta)$ the set of subdomains $D \Subset \Omega_0 \Subset \Omega$ satisfying $D_0 \Subset D$, $\Omega \setminus \overline{\Omega_0}$ and $\Omega_0 \setminus \overline{D}$ are domains of class $C^{2,\alpha}$ with parameters $\varrho > 0$ and $\aleph > 0$,

$$\text{dist}(D, \partial\Omega_0) \ge \delta,$$

and the following assumption holds

$$B(x_0, d(x_0, \overline{D})) \subset \Omega \setminus \overline{D}, \quad x_0 \in \Omega_0 \setminus \overline{D}. \tag{1.3}$$

Henceforward, we make the assumption that Ω_0, D_0 and δ are chosen in such a way that $\mathscr{D}_1(\varrho, \aleph, \delta)$ is nonempty.

Define also $\mathscr{C}_0(\varrho, \aleph, \varrho_0, \aleph_0, \delta)$ as the set of couples (D_1, D_2) so that $D_j \in \mathscr{D}_1(\varrho, \aleph, \delta)$, $j = 1, 2$, and $\Omega \setminus \overline{D_1 \cup D_2}$ is a domain of class $C^{0,1}$ with parameters $\varrho_0 > 0$ and $\aleph_0 > 0$.

Define the geometric distance d_g^U on a bounded domain U of $\mathbb{R}^n$ by

$$d_g^U(x, y) = \inf\left\{\ell(\psi);\ \psi : [0, 1] \to U \text{ is Lipschitz path joining } x \text{ to } y\right\},$$

where

$$\ell(\psi) = \int_0^1 \left|\dot{\psi}(t)\right| dt$$

is the length of ψ.

Note that, according to Rademacher's theorem, any Lipschitz continuous function $\psi : [0, 1] \to D$ is almost everywhere differentiable with $\left|\dot{\psi}(t)\right| \le L$ a.e. $t \in [0, 1]$, where L is the Lipschitz constant of ψ (see for instance [17, Theorem 4, p. 279]). Therefore, $\ell(\psi)$ is well defined.

From [14, Lemma 3.3], we know that $d_g^U \in L^\infty(U \times U)$ whenever U is of class $C^{0,1}$. Fix then $\mathfrak{b} > 0$ and define $\mathscr{C}_1(\varrho, \aleph, \varrho_0, \aleph_0, \delta, \mathfrak{b})$ as the subset of the couples $(D_1, D_2) \in \mathscr{C}_0(\varrho, \aleph, \varrho_0, \aleph_0, \delta)$ so that

$$d_g^{\Omega \setminus \overline{D_1 \cup D_2}} \le \mathfrak{b}.$$

Recall that the Hausdorff distance for compact subsets of $\mathbb{R}^n$ is given by

$$\mathbf{d}_H(\overline{D}_1, \overline{D_2}) = \max\left(\max_{x \in \overline{D}_1} d(x, \overline{D}_2), \max_{x \in \overline{D}_2} d(x, \overline{D}_1)\right)$$

and, following [1], we define the modified distance $\mathbf{d}_m$ by

$$\mathbf{d}_m(\overline{D}_1, \overline{D_2}) = \max\left(\max_{x \in \partial D_1} d(x, \overline{D}_2), \max_{x \in \partial D_2} d(x, \overline{D}_1)\right).$$

As it is pointed out in [1], $\mathbf{d}_m$ is not a distance. To be convinced of that, consider $D_1 = B(0, 1) \setminus B(0, 1/2)$ and $D_2 = B(0, 1)$. In that case simple computations show that

$$0 = \mathbf{d}_m(\overline{D}_1, \overline{D_2}) < \mathbf{d}_H(\overline{D}_1, \overline{D_2}) = \frac{1}{2}.$$

In this example $D_1 \subset D_2$, but $\overline{D}_1 \not\subset D_2$. However, we can enlarge slightly D_2 in order to satisfy $\overline{D}_1 \subset D_2$. Indeed, if $D_2 = B(0, 5/4)$ then $\overline{D}_1 \subset D_2$ and

$$\frac{1}{4} = \mathbf{d}_m(\overline{D}_1, \overline{D_2}) < \mathbf{d}_H(\overline{D}_1, \overline{D_2}) = \frac{1}{2}.$$

In dimension two, take $D_1 = \{(r, \theta);\ \eta < \theta < 2\pi - \eta, 1/2 < r < 1\}, 0 < \eta < \pi$. Smoothing the angles of D_1 we get a C^∞ simply connected domain so that, if again $D_2 = B(0, 5/4)$,

$$\frac{1}{4} \le \mathbf{d}_m(\overline{D}_1, \overline{D_2}) < \mathbf{d}_H(\overline{D}_1, \overline{D_2}) = \frac{1}{2}.$$

In all these examples a small translation in the direction of one of the coordinates axes for instance enables us to construct examples with $D_1 \setminus \overline{D}_2 \neq \emptyset$, $D_1 \setminus \overline{D}_2 \neq \emptyset$ and $\mathbf{d}_m(\overline{D}_1, \overline{D_2}) < \mathbf{d}_H(\overline{D}_1, \overline{D_2})$.

What these examples show is that it is difficult to give sufficient geometric conditions ensuring that the following equality holds

$$\mathbf{d}_m(\overline{D_1}, \overline{D_2}) = \mathbf{d}_H(\overline{D_1}, \overline{D_2}). \tag{1.4}$$

It is obvious to check that (1.4) is satisfied whenever D_1 and D_2 are balls or ellipses, but not only.

The subset of couples $(D_1, D_2) \in \mathscr{C}_2(\varrho, \aleph, \varrho_0, \aleph_0, \delta, \mathfrak{b})$ satisfying (1.4) will be denoted by $\mathscr{C}_1(\varrho, \aleph, \varrho_0, \aleph_0, \delta, \mathfrak{b})$.

We fix in all of this text $f \in C^{1,\alpha}(\partial\Omega)$ non negative and non identically equal to zero.

We aim in this paper to establish the following result.

Theorem 1.1 *Let* $\mathfrak{d} = (\varrho, \aleph, \varrho_0, \aleph_0, \delta, \mathfrak{b})$. *There exist two constants* $C = C(\mathfrak{d})$ *and* $0 < \Lambda^* = \Lambda^*(\mathfrak{d}) < e^{-e}$ *so that, for any* $(D_1, D_2) \in \mathscr{C}_2(\mathfrak{d})$ *with* $\partial D_1 \cap \partial D_2 \neq \emptyset$ *and any sequence* $(k_j) \in (0, \infty)$ *satisfying* $k_j \to \infty$, *we have*

$$\mathbf{d}_H(\overline{D_1}, \overline{D_2}) \leq C \left(\ln \ln |\ln \Lambda|\right)^{-1},$$

provided that

$$0 < \Lambda := \sup_j \|\mathfrak{t}u_{D_2}(k_j) - \mathfrak{t}u_{D_1}(k_j)\|_{L^2(\partial\Omega)} < \Lambda^*.$$

As an immediate consequence of this theorem we have the following corollary.

Corollary 1.1 *Let* $\mathfrak{d} = (\varrho, \aleph, \varrho_0, \aleph_0, \delta, \mathfrak{b})$. *There exist two constants* $C = C(\mathfrak{d})$ *and* $0 < \Lambda^* = \Lambda^*(\mathfrak{d}) < e^{-e}$ *so that, for any* $(D_1, D_2) \in \mathscr{C}_2(\mathfrak{d})$ *with* $\partial D_1 \cap \partial D_2 \neq \emptyset$ *and any sequence of frequencies* $(\omega_j) \in (0, \infty)$ *satisfying* $\omega_j \to \infty$, *we have*

$$\mathbf{d}_H(\overline{D_1}, \overline{D_2}) \leq C \left(\ln \ln |\ln \Lambda|\right)^{-1},$$

provided that

$$0 < \Lambda := \sup_j \|\mathfrak{t}u_{D_2}(k(\omega_j)) - \mathfrak{t}u_{D_1}(k(\omega_j))\|_{L^2(\partial\Omega)} < \Lambda^*.$$

The inverse problem we consider in the present paper was already studied, in the case of smooth star-shaped subdomains with respect to some fixed point, by Ammari and Triki [3]. For a similar problem with a single boundary measurement we refer to [9] where a Lipschitz stability estimate was established for a non monotone one-parameter family of unknown subdomains. The literature on the determination of an unknown part of the boundary is rich, but we just quote here [1, 8, 16] (see also the references therein). We note that another multifrequency medium problem considers observations corresponding to different wavenumbers [4, 6]. To have an overview, we recommend a recent review paper [5] which nicely summarizes the theoretical and numerical results in multifrequency inverse medium and source problems for acoustic Helmholtz equations and Maxwell equations.

The key step in our proof consists in reducing the original inverse problem to the one of determining an unknown part of the inner boundary from a single boundary measurement. For this last problem we provided in Sect. 1.2 a logarithmic stability estimate. This intermediate result is then used in Sect. 1.3 to prove Theorem 1.1. Section 1.4 contains uniqueness results obtained from Theorem 1.1.

The idea of reducing the original problem to the one of recovering the shape of an unknown inner part of the boundary was borrowed from the paper by Ammari and Triki [3].

Our analysis combines both ideas from [1, 3] together with some recent results related to quantifying the uniqueness of continuation in various situations [11–13].

This paper is completed by a last section in which we give an expansion of the solution of the BVP (1.1) in a basis of eigenfunctions of the Neumann–Poincaré operator (shortened to NP operator in the rest of this paper) related to the Neumann–Green function.

1.2 An Intermediate Estimate

Pick $D \in \mathscr{D}_0(\varrho, \aleph)$ and let $\tilde{u}_D^0 \in H^1(\Omega)$ satisfying $\tilde{u}_D^0 = 0$ in $\overline{D}$ and it is the variational solution of the BVP

$$\begin{cases} \Delta \tilde{u} = 0 & \text{in } \Omega \setminus \overline{D}, \\ \tilde{u} = 0 & \text{on } \partial D, \\ \partial_\nu \tilde{u} = f & \text{on } \partial \Omega. \end{cases}$$

As $\tilde{u}_D^0 \in C^\infty(\Omega \setminus \overline{D})$ by the usual interior regularity of harmonic functions, we can apply, for an arbitrary ω, $D \Subset \omega \Subset \Omega$, both the Hölder regularity theorem for Dirichlet BVP in $\omega \setminus \overline{D}$ [18, Theorem 6.14 in p. 107] and the Hölder regularity theorem for Neumann BVP in $\Omega \setminus \overline{\omega}$ [18, Theorem 6.31 in p. 128]. Therefore $\tilde{u}_D^0 \in C^{2,\alpha}(\overline{\Omega} \setminus D)$.

Also, note that, according to the maximum principle and Hopf's maximum principle, $\tilde{u}_D^0 \geq 0$.

Define then

$$\tilde{u}_D = \tilde{u}_D^0 - \frac{1}{|\partial\Omega|} \int_{\partial\Omega} \tilde{u}_D^0 d\sigma(x).$$

Lemma 1.1 *Let $D_1, D_2 \in \mathscr{D}_0(\varrho, \aleph)$ and set*

$$m_j = \frac{1}{|\partial\Omega|} \int_{\partial\Omega} \tilde{u}_D^0 d\sigma(x).$$

If $\partial D_1 \cap \partial D_2 \neq \emptyset$ then

$$|m_1 - m_2| \leq \|\tilde{u}_{D_1} - \tilde{u}_{D_2}\|_{L^\infty(\partial(D_1 \cup D_2))}. \tag{1.5}$$

Proof As $\partial D_1 \cap \partial D_2 \subset \partial(D_1 \cup D_2)$ and

$$\tilde{u}_{D_1} - \tilde{u}_{D_2} = m_1 - m_2 \quad \text{on } \partial D_1 \cap \partial D_2,$$

the expected inequality follows easily. □

Under the assumptions and the notations of Lemma 1.1, we have from (1.5)

$$\|\tilde{u}^0_{D_2} - \tilde{u}^0_{D_1}\|_{L^\infty(\partial(D_1\cup D_2))} \leq 2\|\tilde{u}_{D_2} - \tilde{u}_{D_1}\|_{L^\infty(\partial(D_1\cup D_2))}. \tag{1.6}$$

Let $\mathfrak{d}_1 = (\varrho, \aleph, \varrho_0, \aleph_0, \mathfrak{b})$. Then, checking carefully the result in [11, Sect. 2.4] we find that there exist three constants $C = C(\mathfrak{d}_1) > 0$, $c = c(\mathfrak{d}_1) > 0$ and $\beta = \beta(\mathfrak{d}_1)$ so that, for any $0 < \epsilon < 1$ and $D_1, D_2 \in \mathscr{C}_1(\mathfrak{d}_1)$, we have

$$\begin{aligned} C\|\tilde{u}_{D_2} - \tilde{u}_{D_1}\|_{L^\infty(\partial(D_1\cup D_2))} \\ \leq \epsilon^\beta \|\tilde{u}_{D_2} - \tilde{u}_{D_1}\|_{C^{1,\alpha}(\overline{\Omega}\setminus(D_1\cup D_2))} + e^{c/\epsilon}\|\tilde{u}_{D_2} - \tilde{u}_{D_1}\|_{H^1(\partial\Omega)}. \end{aligned} \tag{1.7}$$

Let us mention that [11, Proposition 2.30] holds for an arbitrary bounded Lipschitz domain (we refer to [7] or [12] for a detailed proof of this improvement).

Proposition 1.1 *Set $\mathfrak{d} = (\varrho, \aleph, \delta)$. There exists $C = C(\mathfrak{d})$ so that, for any $D \in \mathscr{D}_1(\mathfrak{d})$, we have*

$$\|\tilde{u}_D\|_{C^{2,\alpha}(\overline{\Omega}\setminus D)} \leq C.$$

Proof Let $0 \leq g \in C^{2,\alpha}(\partial\Omega)$ and denote by $w \in C^{2,\alpha}(\overline{\Omega} \setminus D_0)$ the solution of the BVP

$$\begin{cases} \Delta w = 0 & \text{in } \Omega, \\ w = g & \text{on } \partial D_0, \\ \partial_\nu w = f & \text{on } \partial\Omega. \end{cases}$$

The existence of such function is guaranteed by the usual elliptic regularity for both Dirichlet and Neumann BVP's for the Laplace operator. Similar argument will be discussed hereafter.

In light of the fact that $\tilde{u}^0_D - w$ is harmonic in $\Omega \setminus \overline{D}$, $\tilde{u}^0_D - w = -w \leq 0$ on ∂D and since $\partial_\nu(\tilde{u}^0_D - w) = 0$ on $\partial\Omega$, we find by applying the twice the maximum principle and Hopf's lemma that $(0 \leq)\tilde{u}^0_D \leq w$ in $\overline{\Omega} \setminus D$. Whence

$$\|\tilde{u}^0_D\|_{C(\overline{\Omega}\setminus D)} \leq \|w\|_{C(\overline{\Omega}\setminus D)}.$$

Let $\mu = \inf(\delta, \mathrm{dist}(\Omega, \Omega_0))/2$ and set

$$U = \{x \in \Omega \setminus \overline{D};\ \mathrm{dist}(x, \Omega_0) \leq \mu\}.$$

By [10, Lemma 3.11 in p. 118], we have

$$\|\tilde{u}^0_D\|_{C^{2,\alpha}(U)} \leq \mu^{-1}C(n)\|\tilde{u}^0_D\|_{C(\overline{\Omega}\setminus D)} \leq \mu^{-1}C(n)\|w\|_{C(\overline{\Omega}\setminus D)}. \tag{1.8}$$

Let $u_1 \in C^{2,\alpha}(\overline{\Omega_0} \setminus D)$ be the solution of the BVP

$$\begin{cases} \Delta u_1 = 0 \text{ in } \Omega_0 \setminus \overline{D}, \\ u_1 = \tilde{u}_D^0 \text{ on } \partial\Omega_0 \cup \partial D, \end{cases}$$

and $u_2 \in C^{2,\alpha}(\overline{\Omega} \setminus \Omega_0)$ be the solution of the BVP

$$\begin{cases} \Delta u_2 = 0 & \text{in } \Omega \setminus \overline{\Omega_0}, \\ \partial_\nu u_2 = \partial_\nu \tilde{u}_D^0 & \text{on } \partial\Omega \cup \partial\Omega_0. \end{cases}$$

A careful examination of the classical Schauder estimates in [18, Chap. 6], for both Dirichlet and Neumann problems, we see that the different constants only depend on $C^{2,\alpha}$ parameters of the domain. Therefore

$$\|u_1\|_{C^{2,\alpha}(\overline{\Omega_0}\setminus D)} \leq C(\mathfrak{d})\|u_1\|_{C^{2,\alpha}(\partial\Omega_0 \cup \partial D)}, \tag{1.9}$$

$$\|u_2\|_{C^{2,\alpha}(\overline{\Omega}\setminus \Omega_0)} \leq C(\mathfrak{d})\|\partial_\nu u_2\|_{C^{1,\alpha}(\partial\Omega \cup \partial\Omega_0)}. \tag{1.10}$$

We put (1.8) in (1.9) and (1.10) in order to get

$$\|\tilde{u}_D^0\|_{C^{2,\alpha}(\overline{\Omega}\setminus D)} \leq C.$$

The expected inequality follows then by noting that

$$\|\tilde{u}_D\|_{C^{2,\alpha}(\overline{\Omega}\setminus D)} \leq 2\|\tilde{u}_D^0\|_{C^{2,\alpha}(\overline{\Omega}\setminus D)}.$$

The proof is then complete. □

Let $w \in H^2(\Omega \setminus \overline{\Omega_0})$. We have from the usual interpolation inequalities and trace theorems, with $\mathfrak{d} = (\varrho, \aleph)$,

$$\|w\|_{H^1(\partial\Omega)} \leq C(\mathfrak{d})\|w\|_{L^2(\partial\Omega)}^{1/3}\|w\|_{H^{3/2}(\partial\Omega)}^{2/3} \leq C(\mathfrak{d})C_{\Omega\setminus\overline{\Omega_0}}\|w\|_{L^2(\partial\Omega)}^{1/3}\|w\|_{H^2(\Omega\setminus\overline{\Omega_0})}^{2/3},$$

the constant $C_{\Omega\setminus\overline{\Omega_0}}$ only depends on $\Omega \setminus \overline{\Omega_0}$.

In light of this inequality, Proposition 1.1 and inequalities (1.6) and (1.7), we can state the following result.

Theorem 1.2 *Set* $\mathfrak{d}_1 = (\varrho, \aleph, \varrho_0, \aleph_0, \delta, \mathfrak{b})$. *There exist three constants* $C = C(\mathfrak{d}_1) > 0$, $c = c(\mathfrak{d}_1) > 0$ *and* $\beta = \beta(\mathfrak{d}_1)$ *so that, for any* $0 < \epsilon < 1$ *and* $(D_1, D_2) \in \mathscr{C}_1(\mathfrak{d}_1)$ *with* $\partial D_1 \cap \partial D_2 \neq \emptyset$, *we have*

$$C\|\tilde{u}_{D_2}^0 - \tilde{u}_{D_1}^0\|_{L^\infty(\partial(D_1\cup D_2))} \leq \epsilon^\beta + e^{c/\epsilon}\|\tilde{u}_{D_2} - \tilde{u}_{D_1}\|_{L^2(\partial\Omega)}^{1/3}.$$

Theorem 1.3 *Let* $D \in \mathscr{D}_0(\varrho, \aleph)$. *For any sequence* (k_j) *in* $(0, \infty)$ *such that* $\lim_{j\to\infty} k_j = \infty$, *we have*

$$\lim_{j\to\infty} \|\mathfrak{t}u_D(k_j) - \mathfrak{t}\tilde{u}_D\|_{L^2(\partial\Omega)} = 0.$$

In particular, $(\mathfrak{t}u_D(k_j)) \in \ell^\infty(L^2(\partial\Omega))$.

Proof Let

$$W = \left\{ w \in H^1(\Omega \setminus \overline{D});\ \int_{\partial\Omega} w(x)d\sigma(x) = 0 \right\}.$$

W is a closed subspace of $H^1(\Omega \setminus \overline{D})$ and the norm $\|\nabla w\|_{L^2(\Omega\setminus\overline{D})}$ is equivalent on W to the norm $\|w\|_{H^1(\Omega\setminus\overline{D})}$. Moreover, for any $w \in V$, $w_{|\Omega\setminus\overline{D}} \in W$.

Clearly, $v_D(k) = u_D(k) - \tilde{u}_D$ is the variational solution of the BVP

$$\begin{cases} \operatorname{div}\left((k_0 + (k - k_0)\chi_D)\, \nabla v\right) = -\operatorname{div}\left((k_0 + (k - k_0)\chi_D)\, \nabla \tilde{u}_D\right) & \text{in } \Omega, \\ \partial_\nu v = 0 & \text{on } \partial\Omega. \end{cases}$$

As

$$\int_{\partial\Omega} u_D(k)d\sigma(x) = \int_{\partial\Omega} \tilde{u}_D d\sigma(x) = 0,$$

we have $v_D(k) \in V$.

By Green's formula, for any $w \in V$, we get

$$\begin{aligned} -k_0 \int_{\Omega\setminus\overline{D}} \nabla v_D(k) \cdot \nabla w dx - k \int_D \nabla v_D(k) \cdot \nabla w dx \\ = k_0 \int_{\Omega\setminus\overline{D}} \nabla \tilde{u}_D \cdot \nabla w dx - \int_{\partial\Omega} f w d\sigma(x). \end{aligned} \tag{1.11}$$

Take in this identity $w = v_D(k)$ and make use of Cauchy-Schwarz's identity in order to obtain

$$\begin{aligned} & k\|\nabla v_D(k)\|^2_{L^2(D)} + k_0\|\nabla v_D(k)\|^2_{L^2(\Omega\setminus\overline{D})} \\ & \quad \le k_0\|\nabla v_D(k)\|_{L^2(\Omega\setminus\overline{D})}\|\nabla\tilde{u}\|_{L^2(\Omega\setminus\overline{D})} + \|f\|_{L^2(\partial\Omega)}\|\mathfrak{t}v_D(k)\|_{L^2(\partial\Omega)}. \end{aligned} \tag{1.12}$$

But the trace operator $w \in H^1(\Omega \setminus \overline{D}) \mapsto w_{|\partial\Omega} \in L^2(\partial\Omega)$ is bounded. Hence there exists a constant $C_0 > 0$, depending on Ω and D, so that

$$\|w\|_{L^2(\partial\Omega)} \le C_0\|\nabla w\|_{L^2(\Omega\setminus\overline{D})}, \quad w \in W.$$

Therefore

$$\|\nabla v_D(k)\|_{L^2(\Omega\setminus\overline{D})} \le \|\nabla\tilde{u}\|_{L^2(\Omega\setminus\overline{D})} + C_0 k_0^{-1}\|f\|_{L^2(\partial\Omega)} := M. \tag{1.13}$$

This in (1.12) entails

$$\|\nabla v_D(k)\|_{L^2(D)} \le \frac{M}{\sqrt{k}}. \tag{1.14}$$

Pick (k_j) a sequence in $(0,\infty)\setminus\{k_0\}$ so that $k_j\to\infty$ as $j\to\infty$. Under the temporary notation $v_j = v_D(k_j)$, we have from (1.13) to (1.14) that v_j is bounded in $H^1(\Omega)$ and $\nabla v_j \to 0$ in $L^2(D)$ when $j\to\infty$. Subtracting if necessary a subsequence, we may assume that $v_j \to \overline{v}\in H^1(\Omega)$, strongly in $H^{3/4}(\Omega)$ and weakly in $H^1(\Omega)$. In consequence, $\nabla\overline{v} = 0$ in D.

Define

$$W_0 = \{w\in W;\ w = 0 \text{ on } \partial D\}.$$

An extension by 0 of a function in W_0 enables us to consider W_0 as a closed subspace of V.

It is not hard to see that (1.11) yields

$$-k_0\int_{\Omega\setminus\overline{D}} \nabla v_j\cdot\nabla w dx = k_0\int_{\Omega\setminus\overline{D}}\nabla\tilde{u}_D\cdot\nabla w dx - \int_{\partial\Omega} f w d\sigma(x),\quad w\in W_0.$$

Passing to the limit when $j\to\infty$, we obtain

$$-k_0\int_{\Omega\setminus\overline{D}} \nabla \overline{v}\cdot\nabla w dx = k_0\int_{\Omega\setminus\overline{D}}\nabla\tilde{u}_D\cdot\nabla w dx - \int_{\partial\Omega} f w d\sigma(x),\quad w\in W_0.$$

Taking in this identity an arbitrary $w\in C_0^\infty(\Omega\setminus\overline{D})$, we find that $\overline{v}$ is harmonic in $\Omega\setminus\overline{D}$. Applying then generalized Green's function to deduce that $\partial_\nu\overline{v}=0$. On the other hand we know that v_j converges strongly to $\overline{v}$ in $H^{1/4}(\partial\Omega)$ (thank to the continuity of the trace operator). Therefore $\overline{v}\in V$ and hence $\overline{v}$ is identically equal to zero, implying in particular that v_j converges strongly to 0 in $L^2(\partial\Omega)$. □

We get by combining Theorems 1.2 and 1.3 the following result.

Theorem 1.4 *If* $\mathfrak{d}_1 = (\varrho,\aleph,\varrho_0,\aleph_0,\delta,\mathfrak{b})$, *then there exist three constants* $C = C(\mathfrak{d}_1) > 0$, $c = c(\mathfrak{d}_1) > 0$ *and* $\beta = \beta(\mathfrak{d}_1)$ *so that, for any* $0<\epsilon<1$ *and* $(D_1, D_2)\in\mathscr{C}_1(\mathfrak{d}_1)$ *with* $\partial D_1\cap\partial D_2\neq\emptyset$, *and for any sequence* (k_j) *in* $(0,\infty)$ *such that* $\lim_{j\to\infty}k_j=\infty$, *we have*

$$C\|\tilde{u}^0_{D_2}-\tilde{u}^0_{D_1}\|_{L^\infty(\partial(D_1\cup D_2))}\le \epsilon^\beta + e^{c/\epsilon}\left[\sup_j\|u_{D_2}(k_j)-u_{D_1}(k_j)\|_{L^2(\partial\Omega)}\right]^{1/3}.$$

1.3 Proof of the Main Result

If $\mathcal{U}$ is a bounded domain of $\mathbb{R}^n$, we set

$$\mathcal{U}^\delta = \{x\in\mathcal{U};\ \mathrm{dist}(x,\partial\mathcal{U})>\delta\},\quad \delta>0.$$

Define then

$$\varkappa(\mathcal{U}) = \sup\{\delta > 0;\ \mathcal{U}^\delta \neq \emptyset\}.$$

We endow $C^{1,\alpha}(\overline{\mathcal{U}})$ with norm

$$|u|_{1,\alpha} := \|\nabla u\|_{L^2(\mathcal{U})^n} + [\nabla u]_\alpha,$$

where

$$[\nabla u]_\alpha = \sup\left\{\frac{|\nabla u(x) - \nabla u(y)|}{|x-y|^\alpha};\ x, y \in \overline{\mathcal{U}},\ x \neq y\right\}.$$

For $0 < \alpha < 1$, $\eta > 0$ and $M > 0$, define $\mathscr{S}(\mathcal{U}) = \mathscr{S}(\mathcal{U}, \alpha, \eta, M)$ by

$$\mathscr{S}(\mathcal{U}) = \{u \in C^{1,\alpha}(\overline{\mathcal{U}});\ |u|_{1,\alpha} \leq M,\ \|\nabla u\|_{L^\infty(\Gamma)} \geq \eta \text{ and } \Delta u = 0\}.$$

Denote by $\mathscr{U} = \mathscr{U}(\varrho, \aleph, \mathfrak{h}, \mathfrak{b})$ the set of bounded domains $\mathcal{U}$ of $\mathbb{R}^n$ that are of class $C^{0,1}$, with parameters $\varrho > 0$, $\aleph > 0$, and satisfy $\varkappa(\mathcal{U}) \geq \mathfrak{h} > 0$ and $d_g^{\mathcal{U}} \leq \mathfrak{b}$.

Theorem 1.5 *Let $\tilde{\mathfrak{d}} = (\varrho, \aleph, \mathfrak{h}, \mathfrak{b}, \alpha, \eta, M)$. There exist two constants $c = c(\tilde{\mathfrak{d}}) > 0$ so that, for any $\mathcal{U} \in \mathscr{U}$, $0 < \delta < \mathfrak{h}$, $x_0 \in \mathcal{U}^\delta$ and $u \in \mathscr{S}(\mathcal{U})$, we have*

$$e^{-e^{c/\delta}} \leq \|u\|_{L^2(B(x_0,\delta))}.$$

Proof We mimic the proof of [13, Theorem 2.1] in which we substitute the three-ball inequality of u by a three-ball inequality for ∇u. If one examines carefully the proof of [13, Theorem 2.1], he can see that the different constants do not depend on $\mathcal{U} \in \mathscr{U}$ but only on $(\varrho, \aleph, \mathfrak{h}, \mathfrak{b})$. We obtain

$$e^{-e^{c/\delta}} \leq \|\nabla u\|_{L^2(B(x_0,\delta))}.$$

This and Caccioppoli's inequality yield the expected inequality. □

Bearing in mind that (1.3) holds, the following corollary is a consequence of Theorem 1.5 and Proposition 1.1.

Corollary 1.2 *Set $\mathfrak{d} = (\varrho, \aleph, \delta)$. Then there exist two constants $c = c(\mathfrak{d}) > 0$ so that, for any $D \in \mathscr{D}_1(\mathfrak{d})$, $x_0 \in \Omega_0 \setminus \overline{D}$, we have*

$$e^{-e^{c/d_0}} \leq \|\tilde{u}_D^0\|_{L^2(B(x_0,d_0/4))},$$

where $d_0 = d(x_0, \overline{D})$.

Proof (*of Theorem* 1.1) Set $\tilde{u}_j = \tilde{u}_{D_j}^0$, $j = 1, 2$. According to the maximum principle, we have

$$\max_{\overline{D}_2 \setminus D_1} |\tilde{u}_1| = \max_{\overline{D}_2 \setminus D_1} |\tilde{u}_1 - \tilde{u}_2| = \max_{\partial(D_2 \setminus \overline{D}_1)} |\tilde{u}_1 - \tilde{u}_2| \leq \max_{\partial(D_1 \cup D_2)} |\tilde{u}_1 - \tilde{u}_2|. \quad (1.15)$$

Pick $x_0 \in \partial D_2$ so that

$$\max_{x\in\partial D_2} d(x, \overline{D}_1) = d(x_0, \overline{D}_1) := \overline{d}_1.$$

Noting that $\tilde{u}_1 \geq 0$, we apply Harnack's inequality (see [18, Proof of Theorem 2.5 in p. 16]) in order to get

$$\max_{B(x_0,\overline{d}/4)} \tilde{u}_1 \leq 3^n \min_{B(x_0,\overline{d}/4)} \tilde{u}_1 \leq 3^n \max_{\overline{D}_2\setminus D_1} |\tilde{u}_1|.$$

Whence

$$\int_{B(x_0,\overline{d}/4)} \tilde{u}_1^2(x)dx \leq \left|\mathbb{S}^{n-1}\right| \left(3\overline{d}_1/4\right)^n \left(\max_{\overline{D}_2\setminus D_1} |\tilde{u}_1|\right)^2.$$

This and the estimate in Corollary 1.2 yield, by changing if necessary the constant c,

$$e^{-e^{c/\overline{d}_1}} \leq \max_{\overline{D}_2\setminus D_1} |\tilde{u}_1|. \tag{1.16}$$

We have similarly

$$e^{-e^{c/\overline{d}_2}} \leq \max_{\overline{D}_1\setminus D_2} |\tilde{u}_2| \leq \max_{\partial(D_1\cup D_2)} |\tilde{u}_1 - \tilde{u}_2|, \tag{1.17}$$

where

$$\overline{d}_2 := \max_{x\in\partial D_1} d(x, \overline{D}_2).$$

In light of (1.15), (1.16) and (1.17) entail

$$e^{-e^{c/\overline{d}}} \leq \max_{\partial(D_1\cup D_2)} |\tilde{u}_1 - \tilde{u}_2|. \tag{1.18}$$

Here $\overline{d} = \max(\overline{d}_1, \overline{d}_2) = \mathbf{d}_H(\overline{D_1}, \overline{D_2})$.

Set

$$\Lambda := \sup_j \|u_{D_2}(k_j) - u_{D_1}(k_j)\|_{L^2(\partial\Omega)}.$$

Then the estimates in Theorem 1.4 in (1.18) give

$$e^{-e^{c/\overline{d}}} \leq \epsilon^\beta + e^{c/\epsilon}\Lambda^{1/3}, \quad 0 < \epsilon < 1. \tag{1.19}$$

Since the function $\epsilon \in (0, 1) \to \epsilon^\beta e^{-c\beta}$ is increasing, if $\Lambda < e^{-3c} := \Lambda_0$ then we can take ϵ in (1.19) so that $\epsilon^\beta e^{-c\beta} = \Lambda^{1/3}$. A straightforward computation shows that $\epsilon \leq 3(c + \beta)|\ln \Lambda|^{-1}$. Modifying if necessary c in (1.19), we obtain

$$e^{-e^{c/\overline{d}}} \leq |\ln \Lambda|^{-1}. \tag{1.20}$$

Therefore, if $\Lambda < \min(\Lambda_0, e^{-e})$ then (1.20) implies

$$\overline{d} \leq c\,(\ln \ln |\ln \Lambda|)^{-1}.$$

The proof is then complete. □

1.4 Uniqueness

We first observe that the following uniqueness result is an immediate consequence of Theorem 1.1.

Corollary 1.3 *Assume that Ω is of class $C^{2,\alpha}$. Let $D_0 \Subset D_j \Subset \Omega$ of class $C^{2,\alpha}$, $j = 1, 2$, so that $\partial D_1 \cap \partial D_2 \neq \emptyset$ and $\mathbf{d}_m(D_1, D_2) = \mathbf{d}_H(D_1, D_2)$. If $\mathfrak{t}u_{D_1}(k_\ell) = \mathfrak{t}u_{D_2}(k_\ell)$ for some sequence (k_ℓ) in $(0, \infty) \setminus \{k_0\}$ with $k_\ell \to \infty$ when $\ell \to \infty$, then $D_1 = D_2$.*

It is worth mentioning that this uniqueness result together with the analyticity of the mapping $k \mapsto \mathfrak{t}u_D(k)$ enable us to establish an uniqueness result when k varies in a subset of $(0, \infty) \setminus \{k_0\}$ possessing an accumulation point.

Lemma 1.2 *Assume that Ω and D are two Lipschitz domains of $\mathbb{R}^n$ so that $D \Subset \Omega$. Then the mapping*

$$k \in (0, \infty) \setminus \{k_0\} \mapsto \mathfrak{t}u_D(k) \in L^2(\partial\Omega)$$

is real analytic.

Proof Let $k \in (0, \infty) \setminus \{k_0\}$ and $|\ell| \leq k/2$ so that $k + \ell \in (0, \infty) \setminus \{k_0\}$. Since $u_D(k + \ell)$ is the solution of the variational problem

$$\int_\Omega \mathbf{a}_D(k+\ell)\nabla u_D(k+\ell) \cdot \nabla v dx = \int_{\partial\Omega} f v d\sigma(x), \quad v \in V,$$

we have

$$\min(k/2, k_0)\|u_D(k+\ell)\|_V \leq \|\mathfrak{t}\|\|f\|_{L^2(\partial\Omega)}. \tag{1.21}$$

Here $\|\mathfrak{t}\|$ denotes the norm of $\mathfrak{t}$ as a bounded operator acting on V with values in $L^2(\partial\Omega)$.

Next, let $w_D(k) \in V$ be the solution of the variational problem

$$\int_\Omega \mathbf{a}_D(k)\nabla w_D(k) \cdot \nabla v = -\int_\Omega \chi_D \nabla u_D(k) \cdot \nabla v, \quad v \in V.$$

Then elementary computations show that

$$\int_\Omega \mathbf{a}_D(k)\nabla\left[u_D(k+\ell)-u_D(k)-\ell w_D(k)\right]\cdot\nabla v$$
$$=\ell\int_\Omega \chi_D\nabla\left[u_D(k)-u_D(k+\ell)\right]\cdot\nabla v,\quad v\in V.$$

Hence

$$\min(k,k_0)\|u_D(k+\ell)-u_D(k)-\ell w_D(k)\|_V\le|\ell|\|u_D(k)-u_D(k+\ell)\|_V. \tag{1.22}$$

But

$$\int_\Omega \mathbf{a}_D(k)\nabla\left[u_D(k+\ell)-u_D(k)\right]\cdot\nabla v=-\ell\int_\Omega \chi_D\nabla u_D(k+\ell)\cdot\nabla v,\quad v\in V.$$

Whence

$$\min(k,k_0)\|u_D(k+\ell)-u_D(k)\|_V\le|\ell|\|u_D(k+\ell)\|_V,$$

which, combined with (6.1), yields

$$\|u_D(k+\ell)-u_D(k)\|_V\le C_0(k)|\ell|, \tag{1.23}$$

with $C_0(k)=\left[\min(k/2,k_0)\right]^{-1}\|\mathfrak{t}\|\|f\|_{L^2(\partial\Omega)}$.

Putting (6.71) into (6.70), we get

$$\|u_D(k+\ell)-u_D(k)-\ell w_D(k)\|_V\le C_1(k)\ell^2,$$

where $C_1(k)=\left[\min(k,k_0)\right]^{-1}C_0(k)$.

In other words, we proved that the mapping

$$k\in(0,\infty)\setminus\{k_0\}\mapsto u_D(k)\in V$$

is differentiable and its derivative $u_D'(k)$ is the solution of the variational problem

$$\int_\Omega \mathbf{a}_D(k)\nabla u_D'(k)\cdot\nabla v=-\int_\Omega \chi_D\nabla u_D(k)\cdot\nabla v,\quad v\in V.$$

Therefore, we have the a priori estimate

$$\|u_D'(k)\|_V\le C\phi(k)^2, \tag{1.24}$$

where we set $C=\|\mathfrak{t}\|\|f\|_{L^2(\partial\Omega)}$ and $\phi(k)=\left[\min(k,k_0)\right]^{-1}$.

Now an induction argument in j shows that $k\in(0,\infty)\setminus\{k_0\}\mapsto u_D(k)\in V$ is j-times differentiable and $u_D^{(j)}(k)$ is the solution of the variational problem

$$\int_\Omega \mathbf{a}_D(k)\nabla u_D^{(j)}(k)\cdot\nabla v = -j\int_\Omega \chi_D\nabla u_D^{(j-1)}(k)\cdot\nabla v, \quad v\in V.$$

In light of this identity and (6.72) we show, again by using an induction argument in j, that

$$\|u_D^{(j)}(k)\|_V \le Cj!\phi(k)^{j+1}.$$

Consequently, if $|k-\ell| < \phi(k)^{-1} = \min(k, k_0)$, the series

$$\sum_{j\ge 0}\frac{1}{j!}\|u_D^{(j)}(k)\|_V(k-\ell)^j$$

converges and hence, think to the completeness of V, the series

$$\sum_{j\ge 0}\frac{1}{j!}u_D^{(j)}(k)(k-\ell)^j$$

converges in V. That is we proved that $k\in(0,\infty)\setminus\{k_0\}\mapsto u_D(k)\in V$ is real analytic and, since $\mathfrak{t}\in\mathscr{B}(V, L^2(\partial\Omega))$, we conclude that $k\in(0,\infty)\setminus\{k_0\}\mapsto \mathfrak{t}u_D(k)\in L^2(\partial\Omega)$ is also real analytic. □

In light of Corollary 1.3 and the fact that a real analytic function $F:(0,\infty)\setminus\{k_0\}\to V$ cannot vanish in a subset of $(0,\infty)\setminus\{k_0\}$ possessing an accumulation point in $(0,\infty)\setminus\{k_0\}$ without being identically equal to zero, we get the following uniqueness result.

Corollary 1.4 *Assume that* Ω *is of class* $C^{2,\alpha}$. *Let* $D_0 \Subset D_j \Subset \Omega$ *of class* $C^{2,\alpha}$, $j=1,2$ *so that* $\partial D_1\cap\partial D_2\neq\emptyset$ *and* $\mathbf{d}_m(D_1,D_2)=\mathbf{d}_H(D_1,D_2)$. *If* $\mathfrak{t}u_{D_1}=\mathfrak{t}u_{D_2}$ *in some subset of* $(0,\infty)\setminus\{k_0\}$ *having an accumulation point in* $(0,\infty)\setminus\{k_0\}$, *then* $D_1=D_2$.

The uniqueness results in Corollaries 1.3 and 1.4 are different from those existing in the literature in the case of single measurement (compare with [19, Theorems 4.3.2 and 4.3.5]).

1.5 Expansion in the Eigenfunction Basis of the NP Operator

We introduce some definitions and results that we prove in Appendix 1.6.

Let N be the Neumann–Green function on Ω. That is N satisfies the following properties, where $y\in\Omega$ is arbitrary,

$$\Delta N(\cdot, y)=\delta_y, \quad \partial_\nu N(\cdot, y)_{|\partial\Omega}=0.$$

We normalize N so that

$$\int_{\partial\Omega} N(x, y)d\sigma(x) = 0, \quad y \in \Omega.$$

Denote by E the usual fundamental solution of the Laplacian in the whole space. That is

$$E(x) = \begin{cases} -\ln|x|/(2\pi) & \text{if } n = 2. \\ |x|^{2-n}/((n-2)\omega_n) & \text{if } n \geq 3. \end{cases}$$

Here $\omega_n = |\mathbb{S}^{n-1}|$.

We establish in Appendix 1.6 that the Neumann–Green function is symmetric and has the form

$$N(x, y) = E(x - y) + R(x, y), \quad x, y \in \Omega, \ x \neq y.$$

With $R \in C^\infty(\Omega \times \Omega)$ and $R(\cdot, y) \in H^{3/2}(\Omega)$, $y \in \Omega$.

Consider the integral operator acting on $L^2(\partial D)$ as follows

$$\mathscr{S}_D f(x) = \int_{\partial D} N(x, y) f(y) d\sigma(y), \quad f \in L^2(\partial D), \ x \in \Omega.$$

We will see later that $\mathscr{S}_D$ is extended to a bounded operator from $H^{-1/2}(\partial D)$ into the space

$$\mathscr{H}_D = \{u \in H^1(\Omega); \ \Delta u = 0 \text{ in } \Omega \setminus \partial D \text{ and } \partial_\nu u = 0 \text{ on } \partial\Omega\}$$

endowed with the norm $\|\nabla u\|_{L^2(\Omega)}$. Note that according to the usual trace theorem $\partial_\nu u$ is an element of $H^{-1/2}(\partial\Omega)$.

We define the NP operator $\mathscr{K}_D$ as the integral operator acting on $L^2(\partial D)$ with weakly singular kernel

$$L(x, y) = \frac{(x - y) \cdot \nu(y)}{|x - y|^n} + \partial_{\nu(y)} R(x, y), \quad x, y \in \partial D, \ x \neq y.$$

Finally, we establish in Appendix 1.6 that $S_D = \mathscr{S}_{D|\partial D}$ defines an isomorphism from $H^{-1/2}(\partial D)$ onto $H^{1/2}(\partial D)$.

Theorem 1.6 *Define successively, as long as the maximum is positive, the energy quotients*

$$\lambda_j^+(D) = \max_{g \perp \{g_{D,0}^+, \ldots, g_{D,j-1}^+\}} \frac{\|\nabla \mathscr{S}_D g\|^2_{L^2(\Omega\setminus\overline{D})} - \|\nabla \mathscr{S}_D g\|^2_{L^2(D)}}{\|\nabla \mathscr{S}_D g\|^2_{L^2(\Omega)}}.$$

Here the orthogonality is with respect to the scalar product $(\nabla \mathscr{S}_D \cdot |\nabla \mathscr{S}_D \cdot)_{L^2(\Omega)}$. *The maximum is attained at* $g_{D,j}^+ \in H^{1/2}(\partial D)$.

Define similarly

$$\lambda_j^-(D) = \min_{g \perp \{g_{D,0}^-, \ldots, g_{D,j-1}^-\}} \frac{\|\nabla \mathscr{S}_D g\|^2_{L^2(\Omega \setminus \overline{D})} - \|\nabla \mathscr{S}_D g\|^2_{L^2(D)}}{\|\nabla \mathscr{S}_D g\|^2_{L^2(\Omega)}}.$$

The minimum is attained at $g_{D,j}^- \in H^{1/2}(\partial D)$.

The potentials $\mathscr{S}_D g_{D,j}^\pm$ *together with all* $\mathscr{S}_D h$, $h \in \ker(\mathscr{K}_D S_D) \subset H^{-1/2}(\partial D)$, *are mutually orthogonal and complete in* $\mathscr{H}_D$.

This eigenvalue variational problem is correlated to the eigenvalue problem of the NP operator $\mathscr{K}_D$. Precisely, we have

Corollary 1.5 *The spectrum of* $\mathscr{K}_D$ *consists in the eigenvalues* $\mu_j^\mp(D) = -\lambda_j^\pm(D)/2$, $j \geq 1$, *multiplicities included, together with possibly the point zero. The extremal functions* $g_{D,j}^\pm$ *are exactly the eigenfunctions of* $\mathscr{K}_D$.

Set $\varphi_{D,j}^\pm = \mathscr{S}_D g_{D,j}^\pm / \|\nabla \mathscr{S}_D g_{D,j}^\pm\|_{L^2(\Omega)}$, $j \geq 1$ and

$$\mathscr{H}_D^\pm = \text{span}\{\varphi_{D,j}^\pm;\ j \geq 1\}.$$

Let $\varphi_{D,j}^0$, $1 \leq j \leq \aleph$ if $0 < \aleph < \infty$ and $j \geq 1$ if $\aleph = \infty$, be an orthonormal basis of $\mathscr{H}_D^0 = \{\psi = \mathscr{S}_D h;\ h \in \ker(\mathscr{K}_D S_D)\}$. Here $\aleph \in [0, \infty]$ is the dimension of $\mathscr{H}_D^0$.

For simplicity convenience we only treat the case $\aleph = \infty$. The results in the case $\aleph < \infty$ are quite similar.

The preceding theorem says that $\mathscr{H}_D = \mathscr{H}_D^+ \oplus \mathscr{H}_D^- \oplus \mathscr{H}_D^0$.

We are now ready to give the expansion of the solution of the BVP (1.1) in the basis $\{\varphi_{D,j}^\epsilon,\ j \in I,\ \epsilon \in \{+, -, 0\}\}$, where we set $I = \{j \geq 1\}$.

Proposition 1.2 *Let* $u_D(k)$ *be the solution of the BVP (1.1). Then*
(i) u_D *admits the following expansion, where* $\tilde{u}_D$ *is as in the beginning of Sect. 1.2,*

$$u_D(k) = \tilde{u}_D + \sum_{\substack{j \in I \\ \epsilon \in \{+,-,0\}}} A_{D,j}^\epsilon(k) \varphi_{D,j}^\epsilon,$$

the coefficients $A_{D,j}^\epsilon(k)$ *satisfies, for* $j \in I$ *and* $\epsilon \in \{+, -, 0\}$,

$$(k - k_0) \sum_{\substack{i \in I \\ \eta \in \{+,-,0\}}} A_{D,i}^\eta(k) (\nabla \varphi_{D,i}^\eta | \nabla \varphi_{D,j}^\epsilon)_{L^2(D)} + k_0 A_{D,j}^\epsilon(k) = k_0 B_{D,j}^\epsilon,$$

where

$$B_{D,j}^\epsilon = (f | \varphi_{D,j}^\epsilon)_{L^2(\partial\Omega)} - (\nabla \tilde{u}_D | \nabla \varphi_{D,j}^\epsilon)_{L^2(\Omega \setminus \overline{D})}.$$

(ii) $k \in (0, \infty) \setminus \{k_0\} \to A_{D,j}^\epsilon(k)$, $j \in I$ *and* $\epsilon \in \{+, -, 0\}$, *is real analytic. Moreover, for any* $\tilde{k} \in (0, \infty) \setminus \{k_0\}$, *there exists* $\delta > 0$ *so that, for any* $j \in I, \epsilon \in \{+, -, 0\}$

and $|k-\tilde{k}|<\delta$, *the series*

$$\sum_{\ell\in\mathbb{N}}\frac{1}{\ell!}\frac{d^\ell}{dk^\ell}A^\epsilon_{D,j}(\tilde{k})(k-\tilde{k})^\ell$$

converges.

Proof (i) Recall that u_D has the following decomposition $u_D(k)=\tilde{u}_D+v_D(k)$. Observing that $v_D(k)$ satisfies $\partial_\nu v_D(k)=0$ (as an element of $H^{-1/2}(\partial\Omega)$) and

$$\operatorname{div}(\mathbf{a}_D(k)\nabla v_D(k))=-\operatorname{div}(\mathbf{a}_D(k)\tilde{u}_D)\quad\text{in } \mathscr{D}'(\Omega),$$

we find by using the generalized Green's formula

$$\begin{aligned} k(\nabla v_D(k)|\nabla\varphi)_{L^2(D)}&+k_0(\nabla v_D(k)|\nabla\varphi)_{L^2(\Omega\setminus\overline{D})} \\ &=k_0(f|\varphi)_{L^2(\partial\Omega)}-k_0(\nabla\tilde{u}_D|\nabla\varphi)_{L^2(\Omega\setminus\overline{D})},\quad \varphi\in H^1(\Omega). \end{aligned}\tag{1.25}$$

We expand $v_D(k)$ in the basis $\{\varphi^\epsilon_{D,j},\ j\in I,\ \epsilon\in\{+,-,0\}\}$:

$$v_D(k)=\sum_{\substack{j\in I\\ \epsilon\in\{+,-,0\}}}A^\epsilon_{D,j}(k)\varphi^\epsilon_{D,j}.$$

Taking $\varphi=\varphi^\epsilon_{D,j}$ in (1.25), we obtain

$$(k-k_0)(\nabla v_D(k)|\nabla\varphi^\epsilon_{D,j})_{L^2(D)}+k_0A^\epsilon_{D,j}(k)=k_0B^\epsilon_{D,j},\tag{1.26}$$

for $j\in I$ and $\epsilon\in\{+,-,0\}$.

Whence

$$(k-k_0)\sum_{\substack{i\in I\\ \eta\in\{+,-,0\}}}A^\eta_{D,i}(k)(\nabla\varphi^\eta_{D,i}|\nabla\varphi^\epsilon_{D,j})_{L^2(D)}+k_0A^\epsilon_{D,j}(k)=k_0B^\epsilon_{D,j},\tag{1.27}$$

for $j\in I$ and $\epsilon\in\{+,-,0\}$.

(ii) We know from the preceding section that $k\in(0,\infty)\setminus\{k_0\}\to v_D(k)\in V$ is real analytic. Then so is $k\in(0,\infty)\setminus\{k_0\}\to A^\epsilon_{D,j}(k)\in\mathbb{C}$ and $k\in(0,\infty)\setminus\{k_0\}\to (\nabla v_D(k)|\nabla\varphi^\epsilon_{D,j})_{L^2(D)}\in\mathbb{C}$, $j\in I$, $\epsilon\in\{+,-,0\}$.

We get by taking successively the derivative in (1.26) with respect to k

$$\begin{aligned} (k-k_0)(\nabla v_D^{(\ell)}(k)|\nabla\varphi^\epsilon_{D,j})_{L^2(D)}&+k_0\frac{d^\ell}{dk^\ell}A^\epsilon_{D,j}(k) \\ &=-\ell(\nabla v_D^{(\ell-1)}(k)|\nabla\varphi^\epsilon_{D,j})_{L^2(D)}, \end{aligned}\tag{1.28}$$

for $j\in I$, $\epsilon\in\{+,-,0\}$ and $\ell\in\mathbb{N}\setminus\{0\}$.

The choice of $k = \tilde{k}$ in (1.28) entails

$$k_0 \frac{d^\ell}{dk^\ell} A^\epsilon_{D,j}(\tilde{k}) = -(\tilde{k} - k_0)(\nabla v_D^{(\ell)}(\tilde{k}) | \nabla \varphi^\epsilon_{D,j})_{L^2(D)} + \\ - \ell(\nabla v_D^{(\ell-1)}(\tilde{k}) | \nabla \varphi^\epsilon_{D,j})_{L^2(D)}, \tag{1.29}$$

for $j \in I$, $\epsilon \in \{+, -, 0\}$ and $\ell \in \mathbb{N} \setminus \{0\}$.

We have

$$\left| (\nabla v_D^{(\ell)}(\tilde{k}) | \nabla \varphi^\epsilon_{D,j})_{L^2(D)} \right| \leq \| \nabla v_D^{(\ell)}(\tilde{k}) \|_{L^2(\Omega)}, \quad \ell \in \mathbb{N}.$$

On the other hand, the series

$$\sum_{\ell \in \mathbb{N}} \frac{1}{\ell!} v_D^{(\ell)}(\tilde{k})(k - \tilde{k})^\ell$$

converges in V provided that $|k - \tilde{k}| \leq \delta$, for some δ. Therefore, in light of (1.29), we can assert that the series

$$\sum_{\ell \in \mathbb{N}} \frac{1}{\ell!} \frac{d^\ell}{dk^\ell} A^\epsilon_{D,j}(\tilde{k})(k - \tilde{k})^\ell$$

also converges whenever $|k - \tilde{k}| \leq \delta$. The proof is then complete. □

Remark 1.1 Unfortunately, computing all the terms $(\nabla \varphi^\eta_{D,i} | \nabla \varphi^\epsilon_{D,j})_{L^2(D)}$ seems to be not possible, especially for $\eta \neq \epsilon$. Let us compute those equal to zero. As $\varphi^\pm_{D,j}$ is a solution of a minimisation problem, we obtain in a standard way

$$\lambda_k^\pm(D) \int_\Omega \nabla \varphi^\pm_{D,j} \cdot \nabla \varphi dx = \int_{\Omega \setminus \overline{D}} \nabla \varphi^\pm_{D,j} \cdot \nabla \varphi dx - \int_D \nabla \varphi^\pm_{D,j} \cdot \nabla \varphi dx,$$

for any $\varphi \in \text{span}\{\varphi^\pm_{D,1}, \dots \varphi^\pm_{D,j-1}\}^\perp$.

In particular,

$$0 = \int_{\Omega \setminus \overline{D}} \nabla \varphi^\pm_{D,j} \cdot \nabla \varphi^\pm_{D,i} dx - \int_D \nabla \varphi^\pm_{D,j} \cdot \nabla \varphi^\pm_{D,i} dx, \quad i > j. \tag{1.30}$$

But

$$0 = \int_{\Omega \setminus \overline{D}} \nabla \varphi^\pm_{D,j} \cdot \nabla \varphi^\pm_{D,i} dx + \int_D \nabla \varphi^\pm_{D,j} \cdot \nabla \varphi^\pm_{D,i} dx, \quad i > j.$$

This and (1.30) yield

$$(\nabla \varphi^\pm_{D,i} | \nabla \varphi^\pm_{D,j})_{L^2(D)} = (\nabla \varphi^\pm_{D,i} | \nabla \varphi^\pm_{D,j})_{L^2(\Omega \setminus \overline{D})} = 0, \quad i > j. \tag{1.31}$$

We have similarly

$$0 = \int_{\Omega\setminus\overline{D}} \nabla\varphi^0_{D,j} \cdot \nabla\varphi dx - \int_D \nabla\varphi^0_{D,j} \cdot \nabla\varphi dx,$$

for any $\varphi \in \mathscr{H}^0_D$. Hence

$$0 = \int_{\Omega\setminus\overline{D}} \nabla\varphi^0_{D,j} \cdot \nabla\varphi^0_{D,i} dx - \int_D \nabla\varphi^0_{D,j} \cdot \nabla\varphi^0_{D,i} dx, \quad i, j \in I. \tag{1.32}$$

As before, we deduce from (1.32)

$$(\nabla\varphi^0_{D,i}|\nabla\varphi^0_{D,j})_{L^2(D)} = (\nabla\varphi^0_{D,i}|\nabla\varphi^0_{D,j})_{L^2(\Omega\setminus\overline{D})} = 0, \quad i, j \in I, \ i \neq j. \tag{1.33}$$

Acknowledgements CJ is supported by NSFC (no. 11971121, key projects no.11331004, no.11421110002) and the Programme of Introducing Talents of Discipline to Universities (number B08018). MC is supported by the grant ANR-17-CE40-0029 of the French National Research Agency ANR (project MultiOnde). LS is supported by NSFC (No.11925104), Program of Shanghai Academic/Technology Research Leader (19XD1420500) and National Key Research and Development Program of China (No. 2017YFC1404103). This work started during the stay of MC at Fudan University on February 2017. He warmly thanks Fudan University for hospitality.

Appendix 1.6: Spectral Analysis of the NP Operator

Before we proceed to the spectral analysis, we define some integral operators with weakly singular kernels. Let $D \Subset \Omega$ be a bounded domain of $\mathbb{R}^n$, $n \geq 2$, of class $C^{1,\alpha}$, for some $0 < \alpha < 1$.

Denote by ν the unit normal outward vector field on $\partial\Omega$. Then a slight modification of the proof of [15, Lemma 3.15, p. 124] yields

$$|(x-y)\cdot\nu(x)| \leq C|x-y|^{1+\alpha}, \quad x, y \in \partial D, \ x \neq y.$$

Here the constant C only depends on D. Hence

$$\left|\frac{(x-y)\cdot\nu(x)}{|x-y|^n}\right| \leq \frac{C}{|x-y|^{n-1-\alpha}}, \quad x, y \in \partial D. \tag{1.34}$$

Define the integral operator $K^*_D : L^2(\partial D) \to L^2(\partial D)$ by

$$K^*_D f(x) = \int_{\partial D} \frac{(x-y)\cdot\nu(x)}{|x-y|^n} f(y) d\sigma(y), \quad f \in L^2(D).$$

Estimate (1.34) says that the kernel of K^*_D is weakly singular and therefore it is compact (see for instance [23, Sect. 2.5.5, p. 128]).

Note that K_D^* is nothing but the adjoint of the operator $K : L^2(\partial D) \to L^2(\partial D)$ given as follows

$$K_D f(x) = \int_{\partial D} \frac{(x-y)\cdot\nu(y)}{|x-y|^n} f(y)d\sigma(y), \quad f \in L^2(D).$$

As K_D^*, K_D is also an integral operator with weakly singular kernel and then it is also compact.

Denote by E the usual fundamental solution of the Laplacian in the whole space. That is

$$E(x) = \begin{cases} -\ln|x|/(2\pi) & \text{if } n = 2. \\ |x|^{2-n}/((n-2)\omega_n) & \text{if } n \geq 3. \end{cases}$$

Here $\omega_n = |\mathbb{S}^{n-1}|$.

Recall that the single layer potential S_D is the integral operator with kernel $E(x - y)$:

$$S_D f(x) = \int_{\partial D} E(x-y) f(y)d\sigma(y), \quad f \in L^2(\partial D),\ x \in \mathbb{R}^n \setminus \partial D.$$

Before stating a jump relation satisfied by $S_D f$, we introduce the following notations, where $w \in C^1(\mathbb{R}^n)$ and $x \in \partial D$,

$$w_{|\pm}(x) = \lim_{t\searrow 0} w(x \pm t\nu(x)),$$
$$\partial_\nu w(x)_{|\pm} = \lim_{t\searrow 0} \nabla w(x \pm t\nu(x)) \cdot \nu(x).$$

For any $f \in L^2(\partial D)$, $\partial_\nu S_D f(\cdot)_{|\pm}$ exists as an element of $L^2(\partial D)$ and the following jump relation holds

$$\partial_\nu S_D f(x)_{|\pm}(x) = \left(\pm\frac{1}{2} + K_D^*\right) f(x) \quad \text{a.e. } x \in \partial\Omega. \tag{1.35}$$

We refer for instance to [2, Theorem 2.4 in p. 16] and its comments.

For $y \in \Omega$, consider the following BVP

$$\begin{cases} \Delta R = 0 & \text{in } \Omega, \\ \partial_\nu R = -\partial_\nu E(\cdot - y) & \text{on } \partial\Omega. \end{cases} \tag{1.36}$$

As $\partial_\nu E(\cdot - y) \in L^2(\partial\Omega)$, in light of [20, Theorem 2, p. 204 and Remarks (b) p. 206], the BVP (1.36) has a unique solution $R(\cdot, y) \in H^{3/2}(\Omega)$ so that

$$\int_{\partial\Omega} R(x, y)dx = \varkappa(y),$$

where the constant $\varkappa(y)$ is to be determined hereafter.

Define then N by

$$N(x, y) = E(x - y) + R(x, y), \quad x, y \in \Omega, \ x \neq y.$$

The function N possesses the following properties, where $y \in \Omega$ is arbitrary,

$$\Delta N(\cdot, y) = \delta_y, \quad \partial_\nu N(\cdot, y)_{|\partial\Omega} = 0.$$

We fix in the rest of this text $\varkappa(y)$ in such a way that

$$\int_{\partial\Omega} N(x, y) d\sigma(x) = 0. \tag{1.37}$$

The function N is usually called the Neumann–Green function.

Mimicking the proof of [2, Lemma 2.14, p. 30], we get $N(x, y) = N(y, x), x, y \in \Omega, x \neq y$. Hence $R(x, y) = R(y, x), x, y \in \Omega$. By interior regularity for harmonic functions $R(\cdot, y), y \in \Omega$, belongs to $C^\infty(\Omega)$ and consequently $R(x, \cdot), x \in \Omega$ is also in $C^\infty(\Omega)$ implying that $R \in C^\infty(\Omega \times \Omega)$.

Consider the integral operator acting on $L^2(\partial D)$ as follows

$$\mathscr{S}_D f(x) = \int_{\partial D} N(x, y) f(y) d\sigma(y), \quad f \in L^2(\partial D), \ x \in \Omega.$$

Clearly $\mathscr{S}_D = S_D + S_D^0$, where S_D^0 is the integral with (smooth) kernel R, i.e.

$$S_D^0 f(x) = \int_{\partial D} R(x, y) f(y) d\sigma(y,) \quad f \in L^2(\partial D), \ x \in \Omega.$$

Using (1.35) we find that $\mathscr{S}_D$ satisfies the following jump condition:

$$\partial_\nu \mathscr{S}_D f(x)_{|\pm}(x) = \left(\pm\frac{1}{2} + \mathscr{K}_D^*\right) f(x) \quad \text{a.e. } x \in \partial\Omega. \tag{1.38}$$

Here $\mathscr{K}_D^* = K_D^* + K_{D,0}^*$, where $K_{D,0}^*$ is the integral operator with kernel $\partial_{\nu(x)} R$, which is the dual of the integral operator $K_{D,0}$ whose kernel is $\partial_{\nu(y)} R$ (think to the symmetry of R).

We get in particular that $\mathscr{K}_D^* : L^2(\partial D) \to L^2(\partial D)$ is compact. More specifically, $\mathscr{K}_D^* : L^2(\partial D) \to H^1(\partial D)$ is bounded (see for instance [2, Theorem 2.11, p. 28]).

We defined in Sect. 1.5 $S_D = \mathscr{S}_{D|\partial D}$ that we consider as a bounded operator on $L^2(\partial D)$ and set

$$\mathcal{H}_D = \{u \in H^1(\Omega); \ \Delta u = 0 \text{ in } \mathscr{D}'(\Omega \setminus \partial D)\}.$$

Define on $\mathcal{H}_D$ the positive hermitian form

$$(u|v)_{\mathcal{H}_D} = \int_\Omega \nabla u \cdot \overline{\nabla v} dx.$$

The corresponding semi-norm is denoted by

$$\|u\|_{\mathcal{H}_D} = (u|u)^{1/2}_{\mathcal{H}_D}.$$

Let $u_- = u_{|D}$ and $u_+ = u_{|\Omega\setminus\overline{D}}$. As $u_- \in H^1(D)$ and $\Delta u = 0$ in $\mathscr{D}'(D)$, we know from the usual trace theorem that $\partial_\nu u_- \in H^{-1/2}(D)$. Similarly, we have $\partial_\nu u_+ \in H^{-1/2}(\partial D) \cap H^{-1/2}(\partial\Omega)$. Therefore, according to generalized Green's formula, for any $u, v \in \mathcal{H}_D$, we have

$$\int_D \nabla u_- \cdot \nabla v_- = \langle \partial_\nu u_- | \overline{v_-}\rangle_{1/2,\partial D}, \tag{1.39}$$

$$\int_{\Omega\setminus\overline{D}} \nabla u_+ \cdot \nabla v_+ = -\langle \partial_\nu u_+ | \overline{v_+}\rangle_{1/2,\partial D} + \langle \partial_\nu u_+ | \overline{v_+}\rangle_{1/2,\partial\Omega}. \tag{1.40}$$

The symbol $\langle\cdot|\cdot\rangle_{1/2,\Gamma}$ denotes the duality pairing between $H^{1/2}(\Gamma)$ and its dual $H^{-1/2}(\Gamma)$, with $\Gamma = \partial D$ or $\Gamma = \partial\Omega$.

Taking the sum side by side in inequalities (1.39) and (1.40), we find

$$(u|v)_{\mathcal{H}_D} = \langle \partial_\nu u_- - \partial_\nu u_+ | \overline{v}\rangle_{1/2,\partial D} + \langle \partial_\nu u_+ | \overline{v_+}\rangle_{1/2|\partial\Omega}, \tag{1.41}$$

where we used that $v_- = v_+ = v$ in ∂D.

We apply (1.41) to $u = v = \mathscr{S}_D f$, with $f \in L^2(\partial D)$. Taking into account that $\partial_\nu u_+ = 0$ on $\partial\Omega$, we obtain

$$\|\mathscr{S}_D f\|_{\mathcal{H}_D} = \langle \partial_\nu \mathscr{S}_D f_{|-} - \partial_\nu \mathscr{S}_D f_{|+} | \overline{\mathcal{S}_D f}\rangle_{1/2,\partial D}.$$

This and the jump condition (1.38) entail

$$\|\mathscr{S}_D f\|^2_{\mathcal{H}_D} = \langle f | \overline{\mathcal{S}_D f}\rangle_{1/2,\partial D} = (f|\mathcal{S}_D f)_{L^2(\partial D)}. \tag{1.42}$$

Here $(\cdot|\cdot)_{L^2(\partial D)}$ is the usual scalar product on $L^2(\partial D)$.

In other words, we proved that $\mathcal{S}_D : L^2(\partial D) \to L^2(\partial D)$ is strictly positive operator. Define, as in [21], H_D^- to be the completion of $L^2(\partial D)$ with respect to the norm $\|\sqrt{\mathcal{S}_D} f\|_{L^2(\partial D)} = (f|\mathcal{S}_D f)_{L^2(\partial D)}$. Let $H_D^+ = \mathrm{R}(\sqrt{\mathcal{S}_D})$, the range of $\sqrt{\mathcal{S}_D}$, that can be regarded as the domain of the unbounded operator $\sqrt{\mathcal{S}_D}^{-1}$. We observe that H_D^+ is complete for the norm induced by the form $(\sqrt{\mathcal{S}_D}^{-1} f, \sqrt{\mathcal{S}_D}^{-1} f)_{L^2(\partial D)}$. Therefore $\mathcal{S}_D$ can be extended by continuity as an isomorphism, still denoted by $\mathcal{S}_D$, from H_D^- onto H_D^+ and the relation

$$(\sqrt{\mathcal{S}_D} f|g)_{L^2(\partial D)} = (f|\sqrt{\mathcal{S}_D} g)_{L^2(\partial D)}$$

defines a duality pairing between H_D^+ and H_D^-, with respect to the pivot space $L^2(\partial D)$.

Recall that the following closed subspace of $\mathcal{H}_D$ was introduced in Sect. 1.5

$$\mathscr{H}_D = \{u \in H^1(\Omega);\ \ \Delta u = 0 \text{ in } \mathscr{D}'(\Omega \setminus \partial D) \text{ and } \partial_\nu u_{|\partial\Omega} = 0\}.$$

Note that we have seen before that $\partial_\nu u_{|\partial\Omega}$ is an element of $H^{-1/2}(\partial\Omega)$. In that case (1.41) takes the form

$$(u|v)_{\mathcal{H}_D} = \langle \partial_\nu u_- - \partial_\nu u_+ | \overline{v}\rangle_{1/2,\partial D},\quad u \in \mathcal{H}_D,\ v \in \mathscr{H}_D. \tag{1.43}$$

Lemma 1.3 *For any $g \in H^{1/2}(\partial D)$, there exists a unique $v = \mathscr{E}_D g \in \mathscr{H}_D$ so that $v_{|\partial D} = g$ and*

$$\|\mathscr{E}_D g\|_{H^1(\Omega)} \le C\|g\|_{H^{1/2}(\partial D)}, \tag{1.44}$$

the constant C only depends on Ω and D.

Proof Take $v \in H^1(\Omega)$ so that $v_{|D} = v_-$ and $v_{\Omega\setminus\overline{D}} = v_+$, where v_- and v_+ are the respective variational solutions of the BVP's

$$\begin{cases} \Delta v_- = 0 \text{ in } D, \\ v_- = g \quad \text{on } \partial D, \end{cases}$$

and

$$\begin{cases} \Delta v_+ = 0 \text{ in } \Omega \setminus \overline{D}, \\ v_+ = g \quad \text{on } \partial D, \\ \partial_\nu v_+ = 0 \text{ on } \partial\Omega. \end{cases}$$

Clearly, $v \in \mathscr{H}_D$ and (1.44) holds. □

Lemma 1.4 *We have $H_D^+ = H^{1/2}(\partial D)$.*

Proof Let $g \in H^{1/2}(\partial D)$ and $f \in L^2(\partial D)$. Apply then (1.43), with $u = \mathscr{S}_D f$ and $v = \mathscr{E}_D g$, in order to obtain

$$(f|g)_{L^2(\partial D)} = (\mathscr{S}_D f | \mathscr{E}_D g)_{\mathcal{H}_D}.$$

Whence

$$\left|(f|g)_{L^2(\partial D)}\right| \le C\|\mathscr{E}_D g\|_{\mathcal{H}_D}\|\mathscr{S}_D f\|_{\mathcal{H}_D} \le C\|g\|_{H^{1/2}(\partial D)}\|\sqrt{S_D} f\|_{L^2(\partial D)}.$$

In other words, the linear form $f \in L^2(\partial D) \mapsto (f|g)_{L^2(\partial D)}$ is bounded for the norm $\|\sqrt{S_D} f\|_{L^2(\partial D)}$. Therefore, according to Riesz's representation theorem, there exists $k \in L^2(\partial D)$ so that

$$(f|g)_{L^2(\partial D)} = (f|\sqrt{S_D} k).$$

Since f is arbitrary, we deduce that $g = \sqrt{S_D} k$ and hence $g \in H_D^+$. That is $H^{1/2}(\partial D) \subset H_D^+$.

Conversely, as $f \in H_D^+$ has the form $f = \mathcal{S}_D g$ for some $g \in H_D^-$, the coupling $(\mathcal{S}_D g|g) := (\sqrt{\mathcal{S}_D} g|\sqrt{\mathcal{S}_D} g)_{L^2(\partial D)}$ is well defined. Let (g_k) be a sequence in $L^2(\partial D)$ converging to g in the topology of H_D^-. We have

$$\|\mathscr{S}_D g_k - \mathscr{S}_D g_\ell\|_{\mathcal{H}_D} = (\mathcal{S}_D g_k - \mathcal{S}_D g_\ell | g_k - g_\ell)_{L^2(\partial D)}.$$

Hence $(\mathscr{S}_D g_k)$ is a Cauchy sequence in $\mathscr{H}_D$ which is complete with respect to the norm $\|\cdot\|_{\mathcal{H}_D}$. The limit u of the sequence $(\mathscr{S}_D g_k)$ satisfies $u_{|\partial D} = \mathcal{S}_D g = f$ by the continuity of the trace map. Whence $f \in H^{1/2}(\partial D)$. □

As byproduct of the preceding proof we see that $\mathscr{S}_D$ is extended as a bounded operator from $H^{-1/2}(\partial D)$ onto $\mathscr{H}_D$ by setting

$$\mathscr{S}_D f = \lim_k \mathscr{S}_D f_k, \quad f \in H^{-1/2}(\partial D),$$

where (f_k) is an arbitrary sequence in $L^2(\partial D)$ converging to f in $H^{-1/2}(\partial D)$. Moreover, we have

$$u = \mathscr{S}_D \mathcal{S}_D^{-1}(u_{|\partial D}), \quad u \in \mathscr{H}_D.$$

Introduce the double layer type operator

$$\mathscr{D}_D f(x) = \int_{\partial D} \partial_{\nu(y)} N(x, y) f(y) d\sigma(y), \quad f \in L^2(\partial D),\ x \in \Omega \setminus \partial D.$$

From [2, Theorem 2.4, p. 16], we easily obtain, where $f \in L^2(D)$,

$$\mathscr{D}_D f_{|\pm} = (\mp 1/2 + \mathscr{K}_D) f \quad \text{a.e. on } \partial D. \tag{1.45}$$

As in [21, Lemma 2, p. 154], our spectral analysis is based on the Plemelj's symmetrization principle. We have

Lemma 1.5 *For $\mathcal{S}_D, \mathscr{K}_D : L^2(\partial D) \to L^2(\partial D)$, the following identity holds*

$$\mathscr{K}_D \mathcal{S}_D = \mathcal{S}_D \mathscr{K}_D^*.$$

Proof For $f \in \mathscr{D}(\partial D)$ and $x \in \Omega \setminus \partial D$, we have

$$\begin{aligned}\mathscr{D}_D \mathcal{S}_D f(x) &= \int_{\partial D} \partial_{\nu(y)} N(x, y) \left(\int_{\partial D} N(y, z) f(z) d\sigma(z) \right) d\sigma(y) \\ &= \int_{\partial D} \left(\int_{\partial D} \partial_{\nu(y)} N(x, y) N(y, z) d\sigma(y) \right) f(z) d\sigma(z).\end{aligned}$$

As $N(x, \cdot) = N(\cdot, x)$ and $N(\cdot, z)$ are harmonic in D, Green's formula yields

$$\begin{aligned}
\mathscr{D}_D \mathbf{S}_D f(x) &= \int_{\partial D}\left(\int_{\partial D} N(x,y)\partial_{\nu(y)}N(y,z)d\sigma(y)\right) f(z)d\sigma(z)\\
&= \int_{\partial D} N(x,y)\partial_{\nu(y)}\left(\int_{\partial D} N(y,z)f(z)d\sigma(z)\right) d\sigma(y)\\
&= \mathscr{S}_D[\partial_{\nu(y)}\mathscr{S}_D f_{|-}].
\end{aligned}$$

In light of the jump conditions in (1.38) and (1.45) we get

$$\mathscr{D}_D \mathbf{S}_D f_{|+} = (-1/2 + \mathscr{K}_D)\,\mathbf{S}_D f = \mathscr{S}_D[\partial_{\nu(y)}\mathscr{S}_D f_{|-}]_{|+} = \mathbf{S}_D(-1/2 + \mathscr{K}_D^*) f.$$

This and the fact that $\mathscr{D}(\partial D)$ is dense in $L^2(\partial D)$ yield the expected inequality. □

We recall the definition of the Shatten class, also called Shatten-von Neumann class. To this end, we consider a complex Hilbert space H. If $T : H \to H$ is a compact operator and if $T^* : H \to H$ denotes its adjoint, then $|T| := \sqrt{T^*T} : H \to H$ is positive and compact and therefore it is diagonalizable. The non-negative sequence $(s_k(T))_{k\geq 1}$ of the eigenvalues of $|T|$ is usually called the sequence of the singular values of T. For $1 \leq p < \infty$, if

$$\sum_{k=1}^{\infty} [s_k(T)]^p < \infty$$

we say that T belongs to the Shatten class $S_p(H)$. It is worth mentioning that $S_p(H)$ is an ideal of $\mathscr{B}(H)$, $S_1(H)$ is known as the trace class and $S_2(H)$ corresponds to the Hilbert-Schmidt class.

Lemma 1.6 *We have $\mathscr{K}_D \in S_p(L^2(\partial D))$, for any $p \geq (n-1)/\alpha$.*

Proof We first note that ℓ, the kernel of $\mathscr{K}_D$, satisfies

$$\ell(x,y) = O\left(|x-y|^{-(n-1)+\alpha}\right). \tag{1.46}$$

Let ω be an open subset of $\mathbb{R}^{n-1}$ and ρ continuous on $\omega \times \omega \setminus \{(x,x);\ x \in \omega\}$ and $\rho(x,y) = O\left(|x-y|^{-(n-1)+\alpha}\right)$, $x, y \in \omega$.

Let $1 < q \leq \alpha/(n-1-\alpha)$. Then elementary computations show that

$$\sup_{x\in\omega}\int_{\omega} |\rho(x,y)|^q dx < \infty.$$

From this we get, in light of (1.46) and using local cards and partition of unity, that $\ell \in L_x^\infty(\partial D; L_y^q(\partial D))$. Consequently, according to [22, Theorem 1], $\mathscr{K}_D$ belongs to $S_p(L^2(\partial D))$, where $p = q/(q-1)$ is the conjugate exponent of q. We complete the proof by noting that the condition on q yields $p \geq (n-1)/\alpha$. □

We remark that [21, Theorem 4.2] is obtained as an immediate consequence of the abstract theorem [21, Theorem 3.1]. Since all the assumptions of [21, Theorem 3.1] hold in our case, we get, similarly to [21, Theorem 4.2], Theorem 1.6.

In Theorem 1.6 we anticipated the regularity of the extremal functions $g^{\pm}_{D,j}$. This result can be proved by following the same arguments as in [21, p. 164].

Following [21], $\lambda^{\pm}_j(D)$, $j \geq 1$, are called the eigenvalues of the spectral variational Poincaré problem: find those λ's for which there exists $u \in \mathscr{H}_D$, $u \neq 0$, so that

$$\int_{\Omega\setminus\overline{D}} \nabla u \cdot \overline{\nabla v} dx - \int_D \nabla u \cdot \overline{\nabla v} dx = \lambda \int_\Omega \nabla u \cdot \overline{\nabla v} dx, \ v \in \mathscr{H}_D.$$

As we already mentioned in Sect. 1.5, this variational eigenvalue problem is correlated to eigenvalue problem for the NP operator $\mathscr{K}_D$. To see this, we observe that as straightforward consequence of (1.39), (1.40) together with the jump condition (1.35), we have

$$\int_{\Omega\setminus\overline{D}} \nabla \mathscr{S}_D g \cdot \overline{\nabla \mathscr{S}_D h} dx - \int_D \nabla \mathscr{S}_D g \cdot \overline{\nabla \mathscr{S}_D h} dx$$
$$= -2(\mathscr{K}_D^* S_D g | S_D h)_{L^2(\partial D)} = -2(S_D g | \mathscr{K}_D S_D h)_{L^2(\partial D)}.$$

These identities are first established for $g, h \in L^2(\partial D)$ and then extended by density to $g, h \in H^{-1/2}(\partial D)$. Thus, we have in particular

$$\frac{\|\nabla \mathscr{S}_D g\|^2_{L^2(\Omega\setminus\overline{D})} - \|\nabla \mathscr{S}_D g\|^2_{L^2(D)}}{\|\nabla \mathscr{S}_D g\|^2_{L^2(\Omega)}} = \frac{-2(S_D g | \mathscr{K}_D S_D g)_{L^2(\partial D)}}{(S_D g | g)_{L^2(\partial D)}},$$

with $g \in H^{-1/2}(\partial D)$.

These comments enable us to deduce Corollary 1.5 from Theorem 1.6.

References

1. G. Alessandrini, E. Beretta, E. Rosset, S. Vessella, Optimal stability for inverse elliptic boundary value problems with unknown boundaries. Ann. Scuola Norm. Sup. Pisa Cl. Sci. (4) **29**(4), 755–806 (2000)
2. H. Ammari, H. Kang, *Reconstruction of Small Inhomogeneities from Boundary Measurements*, vol. 1846. Lecture Notes in Mathematics (Springer, Berlin, 2004)
3. H. Ammari, F. Triki, Identification of an inclusion in multifrequency electric impedance tomography. Comm. Part. Differ. Equ. **42**(1), 159–177 (2017)
4. G. Bao, P. Li, Inverse medium scattering problems for electromagnetic waves. SIAM J. Appl. Math. **65**, 2049–2066 (2005)
5. G. Bao, P. Li, J. Lin, F. Triki, Inverse scattering problems with multi-frequencies. Inverse Probl. **31**(9), 093001 (2015), 21 pp
6. G. Bao, F. Triki, Error estimates for the recursive linearization of inverse medium problems. J. Comput. Math. **28**, 725–744 (2010)

7. M. Bellassoued, M. Choulli, Global logarithmic stability of the Cauchy problem for anisotropic wave equations, arXiv:1902.05878
8. A.L. Bukhgeim, J. Cheng, M. Yamamoto, Conditional stability in an inverse problem of determining a non-smooth boundary. J. Math. Anal. Appl. **242**(1), 57–74 (2000)
9. M. Choulli, Local stability estimate for an inverse conductivity problem. Inverse Probl. **19**(4), 895–907 (2003)
10. M. Choulli, *Analyse fonctionnelle: équations aux dérivées partielles* (Vuibert, Paris, 2013)
11. M. Choulli, *Applications of Elliptic Carleman Inequalities to Cauchy and Inverse Problems*. BCAM SpringerBriefs (Springer, Berlin, 2016)
12. M. Choulli, An introduction to the analysis of elliptic partial differential equations, book under review
13. M. Choulli, F. Triki, Hölder stability for an inverse medium problem with internal data. Res. Math. Sci. **6**(1), 9 (2019), 15 pp
14. M. Choulli, M. Yamamoto, Logarithmic global stability of parabolic Cauchy problems, arXiv:1702.06299
15. G.B. Folland, *Introduction to Partial Differential Equations* (Princeton University Press, Princeton, 1976)
16. J. Elschner, G. Hu, M. Yamamoto, Single logarithmic conditional stability in determining unknown boundaries. Appl. Anal. (2018). In press
17. L.C. Evans, *Partial Differential Equations*. Graduate Studies in Mathematics, vol. 19 (American Mathematical Society, Providence, 1998), xviii+662 pp
18. D. Gilbarg, N.S. Trudinger, *Elliptic Partial Differential Equations of Second Order* (Springer, Berlin, 1998)
19. V. Isakov, *Inverse Problems for Partial Differential Equations* (Springer, New York, 1998)
20. D.S. Jerison, C. Kenig, The Neumann problem in Lipschitz domains. Bull. Amer. Math. Soc. **4**, 203–207 (1981)
21. D. Khavinson, M. Putinar, H.S. Shapiro, Poincaré's variational problem in potential theory. Arch. Rational Mech. Anal. **185**, 143–184 (2007)
22. B. Russo, On the Hausdorff-Young theorem for integral operators. Pac. J. Math. **68**(1), 241–253 (1977)
23. H. Triebel, *Higher Analysis* (Johann Ambrosius Barth Verlag GmbH, Leipzig, 1992)

Chapter 2
Superexponential Stabilizability of Degenerate Parabolic Equations via Bilinear Control

Piermarco Cannarsa and Cristina Urbani

Abstract The aim of this paper is to prove the superexponential stabilizability to the ground state solution of a degenerate parabolic equation of the form

$$u_t(t,x) + (x^\alpha u_x(t,x))_x + p(t)x^{2-\alpha}u(t,x) = 0, \qquad t \geq 0, x \in (0,1)$$

via bilinear control $p \in L^2_{loc}(0,+\infty)$. More precisely, we provide a control function p that steers the solution of the equation, u, to the ground state solution in small time with doubly-exponential rate of convergence. The parameter α describes the degeneracy magnitude. In particular, for $\alpha \in [0,1)$ the problem is called weakly degenerate, while for $\alpha \in [1,2)$ strong degeneracy occurs. We are able to prove the aforementioned stabilization property for $\alpha \in [0,3/2)$. The proof relies on the application of an abstract result on rapid stabilizability of parabolic evolution equations by the action of bilinear control. A crucial role is also played by Bessel's functions.

Keywords Stabilization · Bilinear control · Degenerate equations · Parabolic equations · Bessel's functions

2.1 Introduction

The control of degenerate parabolic equations has received increasing attention by the mathematical community in recent years. In our opinion this fact is due to, at least, two reasons. First, degenerate parabolic operators occur in several applied contexts, such as population genetics [10, 16, 19, 20], fluids flows [27], and climate

P. Cannarsa (✉)
Univeristà di Roma Tor Vergata, 00133 Roma, Italy
e-mail: cannarsa@mat.uniroma2.it

C. Urbani
Gran Sasso Science Institute, 67100 L'Aquila, Italy
e-mail: cristina.urbani@gssi.it

Laboratoire Jacques-Louis Lions, Sorbonne Université, 75005 Paris, France

J. Cheng et al. (eds.), *Inverse Problems and Related Topics*,
Springer Proceedings in Mathematics & Statistics 310,
https://doi.org/10.1007/978-981-15-1592-7_2

models [17, 18, 23]. Second, compared to uniformly parabolic problems, degenerate equations exhibit different behaviors from the point of view of controllability. Indeed, it is known that under the action of an additive control—locally distributed or located at the boundary—exact null controllability may fail if degeneracy is too violent, or else be true in any time $T > 0$ (see [13]), or even be true after some critical time $T^* > 0$, related to the distance of the control support from the degeneracy set, as proved, for instance, in [5, 7, 8].

In this paper, however, we are not interested in an additive control problem but rather in a bilinear one. More precisely, we investigate the response of the degenerate parabolic equation

$$\begin{cases} u_t - \left(x^\alpha u_x\right)_x + p(t)\mu(x)u = 0, & (t,x) \in (0,+\infty) \times (0,1) \\ u(t,1) = 0, \quad \begin{cases} u(t,0) = 0, & \text{if } \alpha \in [0,1), \\ \left(x^\alpha u_x\right)(t,0) = 0, & \text{if } \alpha \in [1,2), \end{cases} \\ u(0,x) = u_0(x). \end{cases} \tag{2.1}$$

to the action of a scalar control $p \in L^2_{loc}(0,\infty)$. We observe that the importance of bilinear control problems is due to the fact that they refer to materials that are able to react to control inputs by changing their principal parameters. This process is called *catalysis* and it is described in some examples in [24].

A stronger kind of control, which is intermediate—in some sense—between additive and bilinear control, is multiplicative control, where one uses a zero order coefficient, $p(t,x)$, to act upon the equation. In this direction, we recall the approximate controllability results by Khapalov et al. [11, 12] for uniformly parabolic equations, and [21] for degenerate parabolic models.

To understand the difference between bilinear and additive control it suffices to recall the celebrated negative result by Ball, Marsden and Slemrod [3] for abstract evolution equations of the form

$$\begin{cases} u'(t) + Au(t) + p(t)Bu(t) = 0, \ t > 0 \\ u(0) = u_0, \end{cases} \tag{2.2}$$

where A is the infinitesimal generator of a C^0-semigroup of bounded linear operators on a Banach space X, $B : X \to X$ is a bounded operator, and $p \in L^r_{loc}(0,\infty)$ for some $r > 1$. Denoting the unique solution of (2.2) by $u(\cdot; u_0, p)$, it was proved in [3] that the attainable set from u_0, defined by

$$S(u_0) = \{u(t; u_0, p) \ : \ t \geq 0, \ \ p \in L^r_{loc}(0,\infty)\},$$

has a dense complement. Therefore, (2.2) fails to be controllable.

For hyperbolic and dispersive models, however, some positive results were later obtained. We would like to mention, in this respect, the results concerning attainable sets for the Schrödinger and wave equations near the ground state solution, obtained in [4, 6], respectively.

So, returning to the abstract problem (2.2) for a densely defined linear operator $A : D(A) \subset X \to X$, a natural question to investigate is the possibility of stabilizing the system near some specific solution. We recall below a possible solution to such a problem in case X is a Hilbert space, which consists of the *superexponential stabilizability* property obtained in [2] under the following assumptions:

$$\begin{aligned} &(a)\ A \text{ is self-adjoint,} \\ &(b)\ A \text{ is accretive: } \langle Au, u\rangle \geq 0,\ \forall u \in D(A), \\ &(c)\ \exists\, \lambda > 0, \text{ such that } (\lambda I + A)^{-1} : X \to X \text{ is compact.} \end{aligned} \tag{2.3}$$

We denote by $\{\lambda_k\}$ $(0 \leq \lambda_k \leq \lambda_{k+1})$ the eigenvalues of A and by $\{\varphi_k\}$ the associated eigenvectors. Recalling that φ_1 is usually called the *ground state* of A, we will refer to $\psi_1(t) := e^{-\lambda_1 t}\varphi_1$ as the *ground state solution* of (2.2) (with $p \equiv 0$). Finally, we denote by $B_R(u)$ the open ball of radius $R > 0$, centered at $u \in X$.

Theorem 2.1 *Let $A : D(A) \subset X \to X$ be a linear operator on the Hilbert space X satisfying hypothesis* (2.3). *Suppose that, for some $\alpha > 0$,*

$$\sqrt{\lambda_{k+1}} - \sqrt{\lambda_k} \geq \alpha, \quad \forall k \in \mathbb{N}^*. \tag{2.4}$$

Let $B : X \to X$ be a bounded linear operator with the following properties:

$$\begin{aligned} &\langle B\varphi_1, \varphi_k\rangle \neq 0, \qquad \forall k \in \mathbb{N}^*, \\ &\exists\, \tau > 0 \textit{ such that } \sum_{k\in\mathbb{N}^*} \frac{e^{-2\lambda_k \tau}}{|\langle B\varphi_1, \varphi_k\rangle|^2} < \infty. \end{aligned} \tag{2.5}$$

Then, for every $\rho > 0$ there exists $R > 0$ such that any $u_0 \in B_R(\varphi_1)$ admits a control $p \in L^2_{loc}(0, \infty)$ such that the corresponding solution $u(\cdot; u_0, p)$ of (2.2) *satisfies*

$$||u(t) - \psi_1(t)|| \leq M e^{-\rho e^{\omega t} - \lambda_1 t} \qquad \forall t \geq 0, \tag{2.6}$$

where M and ω are positive constants depending only on A and B.

The purpose of this paper is to apply Theorem 2.1 to the degenerate control system (2.1), deducing local superexponential stabilizability for such a system.

From the technical point of view, we will have to check that operator A, given by the realization of the elliptic part of the equation in (2.1), and the multiplication operator B, associated to the coefficient $\mu(x) = x^{\alpha-1}$, satisfy the assumptions (2.3), (2.4) and (2.5). For this purpose, the properties of Bessel's functions of the first kind will play a crucial role. Indeed, the eigenvalues and eigenfunctions of A are related to such special functions and their zeros, as observed in [14, 15, 22].

This paper is organized as follows. In Sect. 2.2, we assemble preliminary material on degenerate parabolic equations and Bessel's functions. In Sect. 2.3, we state and prove our main result.

2.2 Preliminaries

Let $I = (0, 1)$, $X = L^2(I)$ and consider the following degenerate parabolic equation

$$\begin{cases} u_t - (a(x)u_x)_x + p(t)\mu(x)u = 0, & x \in I, \ t > 0 \\ u(0) = u_0, & x \in I \end{cases} \tag{2.7}$$

where p is the bilinear control function and $a(x)$ is the degenerate coefficient. Depending on the type of degeneracy, it is customary to assign different boundary conditions to the problem.

Let us recall the definition of two different kinds of degenerate problems. Let

$$a \in C^0([0, 1]) \cap C^1((0, 1]), \ a > 0 \text{ on } (0, 1] \text{ and } a(0) = 0. \tag{2.8}$$

Consider $u_0 \in X$ and $p \in L^2_{loc}([0, \infty))$.

Definition 2.1 If (2.8) holds and moreover

$$\frac{1}{a} \in L^1(I) \tag{2.9}$$

we say that the controlled equation

$$\begin{cases} u_t - (a(x)u_x)_x + p(t)\mu(x)u = 0, & x \in I, \ t > 0 \\ u(t, 0) = 0, \ u(t, 1) = 0, & t > 0 \\ u(0) = u_0, & x \in I. \end{cases} \tag{2.10}$$

is *weakly degenerate*.

Definition 2.2 If (2.8) holds and moreover

$$a \in C^1([0, 1]) \text{ and } \frac{1}{\sqrt{a}} \in L^1(I) \tag{2.11}$$

we say that the controlled equation

$$\begin{cases} u_t - (a(x)u_x)_x + p(t)\mu(x)u = 0, & x \in I, \ t > 0 \\ (au_x)(t, 0) = 0, \ u(t, 1) = 0, & t > 0 \\ u(0) = u_0, & x \in I. \end{cases} \tag{2.12}$$

is *strongly degenerate*.

In particular, we will be interested in treating the degenerate coefficient $a(x) = x^\alpha$. Following the above definitions, we have a weakly degenerate problem for $\alpha \in [0, 1)$ and a strongly degenerate one for $\alpha \in [1, 2)$.

We will treat separately the cases of weak and strong degeneracy.

2.2.1 Weak Degeneracy

Let $\alpha \in [0, 1)$ and consider the degenerate bilinear control problem

$$\begin{cases} u_t - (x^\alpha u_x)_x + p(t)\mu(x)u = 0, \ x \in I, \ t > 0 \\ u(t,0) = 0, \ u(t,1) = 0, \qquad t > 0 \\ u(0) = u_0, \qquad x \in I \end{cases} \tag{2.13}$$

with Dirichlet boundary conditions. The natural spaces for the well-posedness of degenerate problems are weighted Sobolev spaces. Let $X = L^2(I)$, we define the spaces

$$\begin{array}{l} H^1_\alpha(I) = \left\{u \in X : u \text{ is absolutely continuous on } [0,1], x^{\alpha/2}u_x \in X\right\} \\ H^1_{\alpha,0}(I) = \left\{u \in H^1_\alpha(I) : \ u(0) = 0, \ u(1) = 0\right\} \\ H^2_\alpha(I) = \left\{u \in H^1_\alpha(I) : x^\alpha u_x \in H^1(I)\right\}, \end{array} \tag{2.14}$$

and the linear operator $A : D(A) \subset X \to X$ by

$$\begin{cases} \forall u \in D(A), \quad Au := -(x^\alpha u_x)_x, \\ D(A) := \{u \in H^1_{\alpha,0}(I), \ x^\alpha u_x \in H^1(I)\}. \end{cases} \tag{2.15}$$

It is possible to prove that $D(A)$ is dense in X and $A : D(A) \subset X \to X$ is a self-adjoint accretive operator (see, for instance, [9]). Therefore $-A$ is the infinitesimal generator of an analytic C^0-semigroup of contraction e^{-tA} on X.

To determine the spectrum of A, we need to solve the eigenvalue problem

$$\begin{cases} -(x^\alpha \varphi_x(x))_x = \lambda\varphi(x), \ x \in I \\ \varphi(0) = 0, \quad \varphi(1) = 0, \end{cases} \tag{2.16}$$

and it turns out that Bessel functions play a fundamental role in this circumstance.

Indeed, for $\alpha \in [0, 1)$ let

$$\nu_\alpha := \frac{1-\alpha}{2-\alpha}, \qquad k_\alpha := \frac{2-\alpha}{2}. \tag{2.17}$$

Given $\nu \geq 0$, we denote by J_ν the Bessel function of the first kind and order ν and by $j_{\nu,1} < j_{\nu,2} < \cdots < j_{\nu,k} < \ldots$ the sequence of all positive zeros of J_ν. It is possible to prove that the pairs eigenvalue/eigenfunction $(\lambda_{\alpha,k}, \varphi_{\alpha,k})$ that satisfy (2.16) are given by

$$\lambda_{\alpha,k} = k_\alpha^2 j_{\alpha,k}^2, \tag{2.18}$$

$$\varphi_{\alpha,k}(x) = \frac{\sqrt{2k_\alpha}}{|J'_{\nu_\alpha}(j_{\nu_\alpha,k})|} x^{(1-\alpha)/2} J_{\nu_\alpha}\left(j_{\nu_\alpha,k} x^{k_\alpha}\right) \tag{2.19}$$

for every $k \in \mathbb{N}^*$. Moreover, the family $\left(\varphi_{\alpha,k}\right)_{k\in\mathbb{N}^*}$ is an orthonormal basis of X, see [22].

2.2.2 Strong Degeneracy

In the case of strong degeneracy, that is, when $\alpha \in [1, 2)$, we consider the following degenerate bilinear control problem

$$\begin{cases} u_t - (x^\alpha u_x)_x + p(t)\mu(x)u = 0, \ x \in I, \ t > 0 \\ (x^\alpha u_x)(t, 0) = 0, \ \ u(t, 1) = 0, \quad t > 0 \\ u(0) = u_0, \quad\quad\quad\quad\quad\quad\quad\quad\quad x \in I \end{cases} \tag{2.20}$$

with a Neumann condition at the extremum where degeneracy occurs, $x = 0$, and a Dirichlet condition at $x = 1$.

We define the Sobolev spaces

$$\begin{aligned} &H^1_\alpha(I) = \left\{u \in X : u \text{ is absolutely continuous on } (0, 1], \ \ x^{\alpha/2}u_x \in X\right\} \\ &H^1_{\alpha,0}(I) := \left\{u \in H^1_\alpha(I) : \ u(1) = 0\right\}, \\ &H^2_\alpha(I) = \left\{u \in H^1_\alpha(I) : \ x^\alpha u_x \in H^1(I)\right\} \end{aligned} \tag{2.21}$$

and the linear operator $A : D(A) \subset X \to X$ by

$$\begin{cases} \forall u \in D(A), \quad Au := -(x^\alpha u_x)_x, \\ D(A) := \left\{u \in H^1_{\alpha,0}(I) : \ x^\alpha u_x \in H^1(I)\right\} \\ \quad\quad = \left\{u \in X : \ u \text{ is absolutely continuous in (0,1]}, \ x^\alpha u \in H^1_0(I), \right. \\ \quad\quad\quad \left. x^\alpha u_x \in H^1(I) \text{ and } (x^\alpha u_x)(0) = 0\right\}. \end{cases} \tag{2.22}$$

It can be proved that $D(A)$ is dense in X and that A is self-adjoint and accretive (see, for instance, [13]) and thus $-A$ is the infinitesimal generator of an analytic semigroup of contractions e^{tA} on X.

To compute the eigenvalues and eigenfunctions of A, we should solve the eigenvalue problem

$$\begin{cases} -(x^\alpha \varphi_x(x))_x = \lambda\varphi(x), \ x \in I \\ (x^\alpha \varphi_x)(0) = 0, \\ \varphi(1) = 0. \end{cases} \tag{2.23}$$

For $\alpha \in [1, 2)$, if we define the quantities

$$\nu_\alpha := \frac{\alpha - 1}{2 - \alpha}, \qquad k_\alpha := \frac{2 - \alpha}{2}, \tag{2.24}$$

the eigenvalues and eigenfunctions that solve (2.23) have the same structure as in the case of the weakly degenerate problem (2.18) and (2.19). Therefore, the family $(\varphi_{\alpha,k})_{k\in\mathbb{N}^*}$ still forms an orthonormal basis of X.

Proposition 2.1 *Let $\alpha \in [1, 2)$. The following properties holds true:*

1. $|v(x)| \leq \frac{2\|v\|_{D(A)}}{\alpha-1} x^{1-\alpha}, \ \forall v \in D(A)$,
2. $|x^\alpha v(x)| \leq C\sqrt{x}, \ \forall v \in D(A)$,
3. *for $\alpha \in [1, 3/2)$ it holds that*

$$\lim_{x\to 0} x^2 v(x) w_x(x) = 0, \quad \forall v, w \in D(A),$$

4. *for* $\alpha \in [1, 3/2)$ *it holds that*

$$\lim_{x\to 0} x v(x) w(x) = 0, \quad \forall v, w \in D(A),$$

5. *let* $\{\varphi_{\alpha,k}\}_{k\in\mathbb{N}^*}$ *be the family of eigenfunctions of* A. *For* $\alpha \in [1, 3/2)$ *and for every* $k, j \in \mathbb{N}^*$, *it holds that*

$$\lim_{x\to 0} \left(x (\varphi_{\alpha,j})_x(x)\right)_x x^\alpha \varphi_{\alpha,k}(x) = 0.$$

Proof 1. For all $v \in D(A)$ and $y \in I$, we have

$$\begin{aligned}|v(1) - v(y)| &= \left|\int_y^1 v_x(x)dx\right| = \left|\int_y^1 (x^\alpha v_x(x))\frac{1}{x^\alpha}dx\right| \\ &\leq \sup_{0<x<1} |x^\alpha v_x(x)| \frac{\left|1 - y^{1-\alpha}\right|}{\alpha - 1} \\ &\leq \frac{2||v||_{D(A)}}{\alpha - 1} y^{1-\alpha}\end{aligned}$$

where in the last inequality we have used that, for all $v \in D(A)$, it holds that

$$|a(y)v_x(y)| = \left|\int_0^y (a(x)v_x(x))_x\, dx\right| \leq ||(av_x)_x||_X \sqrt{y} \tag{2.25}$$

with $a(y) = y^\alpha$. Finally, recalling that $v(1) = 0$, we obtain the desired formula.

2. For every $v \in D(A)$ and $y \in I$, we have

$$|y^\alpha v(y)| \leq \left|\int_0^y (x^\alpha v(x))_x dx\right| \leq ||(av)_x||_X \sqrt{y}.$$

3. Let $v, w \in D(A)$. We can rewrite $x^2 v(x) w_x(x)$ as

$$x^{2-\alpha} v(x) x^\alpha w_x(x). \tag{2.26}$$

Thanks to (2.25), there exists a constant $C > 0$ such that

$$|x^\alpha w_x(x)| \leq C x^{1/2}. \tag{2.27}$$

Thus, using the first item and (2.27) we obtain that

$$|x^{2-\alpha} v(x) x^\alpha w_x(x)| \leq C x^{2-\alpha} x^{1-\alpha} x^{1/2} \tag{2.28}$$

and therefore the right-hand side tends to 0 as x goes to 0 for $\alpha < 3/2$.

4. Let $v \in D(A)$. It is sufficient to prove that $\lim_{x\to 0} x^{1/2}v(x) = 0$.
For this purpose, we observe that the function $x^{1/2}v(x)$ is integrable in I: indeed, using again the first point of the Proposition, we get

$$|x^{1/2}v(x)| \le Cx^{1/2+1-\alpha}$$

that is integrable in I. Moreover, the derivative of $x^{1/2}v(x)$ is integrable in I:

$$\left(x^{1/2}v(x)\right)_x = x^{1/2}v_x(x) + \frac{1}{2}x^{-1/2}v(x) \tag{2.29}$$

and we can bound the two terms on the right by

$$|x^{1/2}v_x(x)| \le |x^{\alpha}v_x(x)x^{1/2-\alpha}| \le Cx^{1-\alpha}$$

that is integrable for any $\alpha \in [1, 2)$ and by

$$|x^{-1/2}v(x)| \le Cx^{1/2-\alpha}$$

that is integrable for $\alpha \in [1, 3/2)$.
Thus, we can deduce that the function $x^{1/2}v(x)$ is absolutely continuous in I for $\alpha \in [1, 3/2)$. So, the limit

$$\lim_{x\to 0^+} x^{1/2}v(x) = L \tag{2.30}$$

does exist. If $L \ne 0$, then $v(x)$ would be of the same order as $\frac{1}{x^{1/2}}$ near 0. This contradicts the fact that $v \in X$. Thus, $L = 0$.
5. Recalling that $(x^{\alpha}(\varphi_{\alpha,k})_x(x))_x = -\lambda_k\varphi_{\alpha,k}(x)$, we have

$$\begin{aligned}(x(\varphi_{\alpha,j})_x(x))_x x^{\alpha}\varphi_{\alpha,k}(x) &= (x^{\alpha}(\varphi_{\alpha,j})_x(x)x^{1-\alpha})_x x^{\alpha}\varphi_{\alpha,k}(x)\\ &= (x^{\alpha}(\varphi_{\alpha,j})_x(x))_x x^{1-\alpha}x^{\alpha}\varphi_{\alpha,k}(x)\\ &\quad + (1-\alpha)x^{\alpha}(\varphi_{\alpha,j})_x(x)x^{-\alpha}x^{\alpha}\varphi_{\alpha,k}(x)\\ &= -\lambda_j x\varphi_{\alpha,j}(x)\varphi_{\alpha,k}(x) + (1-\alpha)x^{\alpha}(\varphi_{\alpha,j})_x(x)\varphi_{\alpha,k}(x).\end{aligned}$$

The first of the two terms in the last equation on the right-hand side of the above formula goes to 0 as $x \to 0$, for $\alpha < 3/2$, by the previous item. Moreover, we have

$$|x^{\alpha}(\varphi_{\alpha,j})_x(x)\varphi_{\alpha,k}(x)| \le Cx^{1/2}x^{1-\alpha}.$$

Therefore,

$$\lim_{x\to 0}\left(x(\varphi_{\alpha,j})_x(x)\right)_x x^{\alpha}\varphi_{\alpha,k}(x) = 0$$

for $\alpha \in [1, 3/2)$, as it was claimed.

□

2.3 Main Result

Theorem 2.2 *Let* $\alpha \in [0, 3/2)$. *Then, for every* $\rho > 0$ *there exists* $R > 0$ *such that any* $u_0 \in B_R(\varphi_1)$ *admits a control* $p \in L^2_{loc}(0, \infty)$ *such that the corresponding mild solution* $u \in C([0, 1]; X)$ *of*

$$\begin{cases} u_t - \left(x^\alpha u_x\right)_x + p(t)x^{2-\alpha}u = 0, & (t, x) \in (0, \infty) \times (0, 1) \\ u(t, 1) = 0, \quad \begin{cases} u(t, 0) = 0, & \text{if } \alpha \in [0, 1), \\ \left(x^\alpha u_x\right)(t, 0) = 0, & \text{if } \alpha \in [1, 3/2), \end{cases} \\ u(0, x) = u_0(x). \end{cases} \tag{2.31}$$

satisfies

$$||u(t) - \psi_1(t)|| \leq Me^{-\rho e^{\omega t} - \lambda_1 t}, \qquad \forall t \geq 0 \tag{2.32}$$

where M *and* ω *are positive constants depending only on* α.

Proof The proof of the Theorem consists in checking the validity of the hypotheses of Theorem 2.1. We have already observed that $D(A)$ is dense in X and that $A : D(A) \subseteq X \to X$ is self-adjoint and accretive, in both weakly and strongly degenerate cases. Moreover, it can be proved that A has a compact resolvent (see, for instance, [1, Appendix]).

Concerning the gap conditions for the eigenvalues (2.4), it has been proved (see [25], page 135) that

- if $\alpha \in [0, 1)$, $\nu_\alpha = \frac{1-\alpha}{2-\alpha} \in \left(0, \frac{1}{2}\right]$, the sequence $\left(j_{\nu_\alpha,k+1} - j_{\nu_\alpha,k}\right)_{k\in\mathbb{N}^*}$ is nondecreasing and converges to π. Therefore,

$$\sqrt{\lambda_{k+1}} - \sqrt{\lambda_k} = k_\alpha \left(j_{\nu_\alpha,k+1} - j_{\nu_\alpha,k}\right) \geq k_\alpha \left(j_{\nu_\alpha,2} - j_{\nu_\alpha,1}\right) \geq \frac{7}{16}\pi,$$

- if $\nu_\alpha \geq \frac{1}{2}$, the sequence $\left(j_{\nu_\alpha,k+1} - j_{\nu_\alpha,k}\right)_{k\in\mathbb{N}^*}$ is nonincreasing and converges to π. Thus,

$$\sqrt{\lambda_{k+1}} - \sqrt{\lambda_k} = k_\alpha \left(j_{\nu_\alpha,k+1} - j_{\nu_\alpha,k}\right) \geq k_\alpha \pi \geq \frac{\pi}{2}.$$

Therefore, the gap condition is satisfied in both weak and strong degenerate problems with different constants.

The operator $B : X \to X$ is the multiplication operator by the function $\mu(x) = x^{2-\alpha}$ and it is linear and bounded in I. What remains to prove in order to apply Theorem 2.1, is that there exists $\tau > 0$ such that

$$\begin{aligned} &\langle \mu\varphi_{\alpha,1}, \varphi_{\alpha,k} \rangle \neq 0, \qquad \forall k \in \mathbb{N}^*, \\ &\sum_{k\in\mathbb{N}^*} \frac{e^{-2\lambda_k \tau}}{|\langle \mu\varphi_{\alpha,1}, \varphi_{\alpha,k} \rangle|^2} < +\infty. \end{aligned} \tag{2.33}$$

We compute the scalar product $\langle \mu\varphi_{\alpha,1}, \varphi_{\alpha,k} \rangle$ for $k \neq 1$ and, from now on, we write φ_k instead of $\varphi_{\alpha,k}$ to lighten the notation:

$$
\begin{aligned}
\langle \mu\varphi_1, \varphi_k \rangle &= \int_0^1 \mu(x)\varphi_1(x)\varphi_k(x) = -\frac{1}{\lambda_k}\int_0^1 \mu(x)\varphi_1(x)\left(x^\alpha(\varphi_k)_x(x)\right)_x dx \\
&= -\frac{1}{\lambda_k}\left(\mu(x)\varphi_1(x)x^\alpha(\varphi_k)_x(x)\Big|_0^1 - \int_0^1 (\mu(x)\varphi_1(x))_x\, x^\alpha(\varphi_k)_x(x)dx\right) \\
&= \frac{1}{\lambda_k}\left(\int_0^1 \mu_x(x)\varphi_1(x)x^\alpha(\varphi_k)_x(x)dx + \int_0^1 \mu(x)(\varphi_1)_x(x)x^\alpha(\varphi_k)_x(x)dx\right) \\
&= \frac{1}{\lambda_k}\left(\int_0^1 \mu_x(x)\varphi_1(x)x^\alpha(\varphi_k)_x(x)dx + \mu(x)(\varphi_1)_x(x)x^\alpha\varphi_k(x)\Big|_0^1\right. \\
&\qquad\qquad \left. - \int_0^1 \left(\mu(x)(\varphi_1)_x(x)x^\alpha\right)_x \varphi_k(x)dx\right) \qquad (2.34)\\
&= \frac{1}{\lambda_k}\left(\int_0^1 \mu_x(x)\varphi_1(x)x^\alpha(\varphi_k)_x(x)dx - \int_0^1 \mu_x(x)(\varphi_1)_x(x)x^\alpha\varphi_k(x)dx\right. \\
&\qquad\qquad \left. - \int_0^1 \mu(x)(x^\alpha(\varphi_1)_x(x))_x\varphi_k(x)dx\right) \\
&= \frac{1}{\lambda_k}\left(\int_0^1 \mu_x(x)x^\alpha\left[\varphi_1(x)(\varphi_k)_x(x) - (\varphi_1)_x(x)\varphi_k(x)\right]dx\right. \\
&\qquad\qquad \left. + \lambda_1\int_0^1 \mu(x)\varphi_1(x)\varphi_k(x)dx\right).
\end{aligned}
$$

We observe that in the weakly degenerate case, thanks to the Dirichlet conditions in both extrema, the boundary terms vanish. We can deduce the same vanishing property at $x = 0$ for the strong degenerate case thanks to the first item of Proposition 2.1 and to (2.25).

Moving the last term of (2.34) to the left-hand side, we get

$$
\left(1 - \frac{\lambda_1}{\lambda_k}\right)\langle \mu\varphi_1, \varphi_k \rangle = \frac{1}{\lambda_k}\int_0^1 \mu_x(x)x^\alpha\varphi_1^2(x)\left(\frac{\varphi_k(x)}{\varphi_1(x)}\right)_x dx \qquad (2.35)
$$

and therefore, integrating by parts we obtain

$$
\begin{aligned}
\langle \mu\varphi_1, \varphi_k \rangle &= \frac{1}{\lambda_k - \lambda_1}\left(\mu_x(x)x^\alpha\varphi_1^2(x)\frac{\varphi_k(x)}{\varphi_1(x)}\Big|_0^1 - \int_0^1 \left(\mu_x(x)x^\alpha\varphi_1^2(x)\right)_x \frac{\varphi_k(x)}{\varphi_1(x)}dx\right) \\
&= \frac{-1}{\lambda_k - \lambda_1}\left(\int_0^1 (\mu_x(x)x^\alpha)_x\varphi_1^2(x)\frac{\varphi_k(x)}{\varphi_1(x)}dx\right. \\
&\qquad\qquad \left. + 2\int_0^1 \mu_x(x)x^\alpha\varphi_1(x)(\varphi_1)_x(x)\frac{\varphi_k(x)}{\varphi_1(x)}dx\right) \qquad (2.36)\\
&= -\frac{1}{\lambda_k - \lambda_1}\left(\int_0^1 (\mu_x(x)x^\alpha)_x\varphi_1(x)\varphi_k(x)dx\right. \\
&\qquad\qquad \left. + 2\int_0^1 \mu_x(x)x^\alpha(\varphi_1)_x(x)\varphi_k(x)dx\right).
\end{aligned}
$$

The boundary terms vanish for the Dirichet conditions if $\alpha \in [0, 1)$ and thanks to the second item in Proposition 2.1 for $\alpha \in [1, 3/2)$.

Recalling that $\mu(x) = x^{2-\alpha}$, we have that

$$
\begin{aligned}
\langle \mu\varphi_1, \varphi_k\rangle &= -\frac{2(2-\alpha)}{\lambda_k-\lambda_1}\int_0^1 x(\varphi_1)_x(x)\varphi_k(x)dx \\
&= \frac{2(2-\alpha)}{\lambda_k(\lambda_k-\lambda_1)}\int_0^1 x(\varphi_1)_x(x)\left(x^\alpha(\varphi_k)_x(x)\right)_x dx \\
&= \frac{2(2-\alpha)}{\lambda_k(\lambda_k-\lambda_1)}\left(x(\varphi_1)_x(x)x^\alpha(\varphi_k)_x(x)\big|_0^1 - \int_0^1 (x(\varphi_1)_x(x))_x x^\alpha(\varphi_k)_x(x)dx\right) \\
&= \frac{2(2-\alpha)}{\lambda_k(\lambda_k-\lambda_1)}\left(x^{1+\alpha}(\varphi_1)_x(x)(\varphi_k)_x(x)\Big|_0^1 - (x(\varphi_1)_x(x))_x x^\alpha\varphi_k(x)\big|_0^1\right. \\
&\qquad\left. + \int_0^1 \left((x(\varphi_1)_x(x))_x x^\alpha\right)_x \varphi_k(x)dx\right) \\
&= \frac{2(2-\alpha)}{\lambda_k(\lambda_k-\lambda_1)}\left(x^{1+\alpha}(\varphi_1)_x(x)(\varphi_k)_x(x)\Big|_0^1\right. \\
&\qquad\left. + \int_0^1 \left(((\varphi_1)_x(x) + x(\varphi_1)_{xx}(x))x^\alpha\right)_x \varphi_k(x)dx\right) \\
&= \frac{2(2-\alpha)}{\lambda_k(\lambda_k-\lambda_1)}\left(x^{1+\alpha}(\varphi_1)_x(x)(\varphi_k)_x(x)\Big|_0^1 - \lambda_1\int_0^1 \varphi_1(x)\varphi_k(x)dx\right. \\
&\qquad\left. + \int_0^1 \left(x^{1+\alpha}(\varphi_1)_{xx}(x)\right)_x \varphi_k(x)dx\right) \\
&= \frac{2(2-\alpha)}{\lambda_k(\lambda_k-\lambda_1)}\left(x^{1+\alpha}(\varphi_1)_x(x)(\varphi_k)_x(x)\Big|_0^1 + \int_0^1 \left(x^{1+\alpha}(\varphi_1)_{xx}(x)\right)_x \varphi_k(x)dx\right)
\end{aligned}
\tag{2.37}
$$

where we have used the fact that, for $\alpha \in [1, 3/2)$, $(x(\varphi_1)_x(x))_x x^\alpha\varphi_k(x)|_0^1$ vanishes in view of Proposition 2.1.

Since φ_k is an eigenfunction of A for all $k \in \mathbb{N}^*$, it satisfies the equation

$$
-\left(\alpha x^{\alpha-1}(\varphi_k)_x(x) + x^\alpha(\varphi_k)_{xx}(x)\right) = \lambda_k\varphi_k(x), \tag{2.38}
$$

then we can rewrite the expression of $(\varphi_k)_{xx}(x)$ in (2.37) using (2.38):

$$
\begin{aligned}
\langle \mu\varphi_1, \varphi_k\rangle &= \frac{2(2-\alpha)}{\lambda_k(\lambda_k-\lambda_1)}\left(x^{1+\alpha}(\varphi_1)_x(x)(\varphi_k)_x(x)\Big|_0^1 + \int_0^1 \left(x^{1+\alpha}(\varphi_1)_{xx}(x)\right)_x \varphi_k(x)dx\right) \\
&= \frac{2(2-\alpha)}{\lambda_k(\lambda_k-\lambda_1)}\left(x^{1+\alpha}(\varphi_1)_x(x)(\varphi_k)_x(x)\Big|_0^1\right. \\
&\qquad\left. - \int_0^1 \left(\lambda_1 x\varphi_1(x) + \alpha x^\alpha(\varphi_1)_x(x)\right)_x \varphi_k(x)dx\right) \\
&= \frac{2(2-\alpha)}{\lambda_k(\lambda_k-\lambda_1)}\left(x^{1+\alpha}(\varphi_1)_x(x)(\varphi_k)_x(x)\Big|_0^1 - \lambda_1\int_0^1 x(\varphi_1)_x(x)\varphi_k(x)dx\right. \\
&\qquad\left. - \lambda_1\int_0^1 \varphi_1(x)\varphi_k(x)dx - \alpha\int_0^1 \underbrace{(x^\alpha(\varphi_1)_x(x))_x}_{-\lambda_1\varphi_1(x)}\varphi_k(x)dx\right).
\end{aligned}
\tag{2.39}
$$

Recalling that $\{\varphi_k\}_{k\in\mathbb{N}^*}$ is an orthonormal base of $L^2(0,1)$, the last two terms on the right-hand side of the above equality are zero.

Thus, from the first equality of (2.37) and the last one of (2.39), we obtain that

$$-\frac{2(2-\alpha)}{\lambda_k-\lambda_1}\left(1-\frac{\lambda_1}{\lambda_k}\right)\int_0^1 x(\varphi_1)_x(x)\varphi_k(x)dx=\frac{2(2-\alpha)}{\lambda_k(\lambda_k-\lambda_1)}\,x^{1+\alpha}(\varphi_1)_x(x)(\varphi_k)_x(x)\Big|_0^1 \tag{2.40}$$

which implies

$$\langle\mu\varphi_1,\varphi_k\rangle=-\frac{2(2-\alpha)}{\lambda_k-\lambda_1}\int_0^1 x(\varphi_1)_x(x)\varphi_k(x)dx=\frac{2(2-\alpha)}{(\lambda_k-\lambda_1)^2}\,x^{1+\alpha}(\varphi_1)_x(x)(\varphi_k)_x(x)\Big|_0^1 \tag{2.41}$$

Recalling that the eigenvalues $\{\lambda_k\}_{k\in\mathbb{N}^*}$ of A are defined by (2.18) where $\nu_\alpha=|1-\alpha|/(2-\alpha)$, and the eigenfunctions, $\{\varphi_k\}_{k\in\mathbb{N}^*}$, by (2.10), we compute the right-hand side of (2.41):

$$\begin{aligned}x^{1+\alpha}&(\varphi_1)_x(x)(\varphi_k)_x(x)=\\&=\frac{2(2-\alpha)k_\alpha x^{1+\alpha}}{|J'_{\nu_\alpha}(j_{\nu_\alpha,1})||J'_{\nu_\alpha}(j_{\nu_\alpha,k})|}\left(\frac{1-\alpha}{2}x^{-(1+\alpha)/2}J_{\nu_\alpha}(j_{\nu_\alpha,1}x^{k_\alpha})\right.\\&\qquad\qquad\left.+\,j_{\nu_\alpha,1}k_\alpha x^{(1-2\alpha)/2}J'_{\nu_\alpha}(j_{\nu_\alpha,1}x^{k_\alpha})\right)\\&\quad\cdot\left(\frac{1-\alpha}{2}x^{-(1+\alpha)/2}J_{\nu_\alpha}(j_{\nu_\alpha,k}x^{k_\alpha})+j_{\nu_\alpha,k}k_\alpha x^{(1-2\alpha)/2}J'_{\nu_\alpha}(j_{\nu_\alpha,k}x^{k_\alpha})\right).\end{aligned} \tag{2.42}$$

Therefore

$$x^{1+\alpha}(\varphi_1)_x(x)(\varphi_k)_x(x)\mid_0^1=(\varphi_1)_x(1)(\varphi_k)_x(1)=\frac{2k_\alpha^3 j_{\nu_\alpha,1}j_{\nu_\alpha,k}}{|J'_{\nu_\alpha}(j_{\nu_\alpha,1})||J'_{\nu_\alpha}(j_{\nu_\alpha,k})|}J'_{\nu_\alpha}(j_{\nu_\alpha,1})J'_{\nu_\alpha}(j_{\nu_\alpha,k}). \tag{2.43}$$

Now, recall that the zeros of J'_{ν_α}, $j'_{\nu_\alpha,k}$, satisfy $\nu_\alpha<j'_{\nu_\alpha,1}<j_{\nu_\alpha,1}<j'_{\nu_\alpha,2}<j_{\nu_\alpha,2}\dots$, to conclude that the right-hand side of (2.43) does not vanish.

From (2.41) and (2.43) we deduce that there exists a constant C such that

$$|\langle\mu\varphi_1,\varphi_k\rangle|\geq\frac{C}{\lambda_k^{3/2}},\quad\forall k\in\mathbb{N}^*,\ k\neq 1. \tag{2.44}$$

For $k=1$, we have

$$\begin{aligned}\langle\mu\varphi_1,\varphi_1\rangle&=\frac{2k_\alpha}{|J'_{\nu_\alpha}(j_{\nu_\alpha,1})|^2}\int_0^1 x^{2-\alpha}x^{1-\alpha}J^2_{\nu_\alpha}(j_{\nu_\alpha,1}x^{k_\alpha})dx\\&=\frac{4k_\alpha j^4_{\nu_\alpha,1}}{(2-\alpha)|J'_{\nu_\alpha}(j_{\nu_\alpha,1})|^2}\int_0^{j_{\nu_\alpha,1}}z^3J^2_{\nu_\alpha}(z)dz.\end{aligned} \tag{2.45}$$

We now appeal to the identity

$$(\sigma+2)\int^{z} t^{\sigma+2}J_{\nu}(t)dt = (\sigma+1)\left\{\nu^2-\frac{1}{4}(\sigma+1)^2\right\}\int^{z} t^{\sigma}J_{\nu}^2(t)dt + \frac{1}{2}z^{\sigma+1}\left[\left\{zJ_{\nu}'(z)-\frac{1}{2}(\sigma+1)J_{\nu}(z)\right\}^2+\left\{z^2-\nu^2+\frac{1}{4}(\sigma+1)^2\right\}J_{\nu}^2(z)\right] \tag{2.46}$$

with $\sigma = 1$ (see [26], Eq. (17) page 256) to turn (2.46) into

$$\langle\mu\varphi_1,\varphi_1\rangle = \frac{4k_\alpha j_{\nu_\alpha,1}^4}{(2-\alpha)|J_{\nu_\alpha}'(j_{\nu_\alpha,1})|^2}\frac{2}{3}\left\{\nu_\alpha^2-1\right\}\int_0^{j_{\nu_\alpha,1}} zJ_{\nu_\alpha}^2(z)dz + \frac{1}{6}j_{\nu_\alpha,1}^3\left[\left\{j_{\nu_\alpha,1}J_{\nu_\alpha}'(j_{\nu_\alpha,1})-J_{\nu_\alpha}(j_{\nu_\alpha,1})\right\}^2+\left\{j_{\nu_\alpha,1}^2-\nu_\alpha^2+1\right\}J_{\nu_\alpha}^2(j_{\nu_\alpha,1})\right]. \tag{2.47}$$

Using Lommel's integral

$$\int_0^c zJ_\nu(az)^2dz = \frac{c^2}{2}\left[J_\nu^2(ac)-J_{\nu-1}(ac)J_{\nu+1}(ac)\right] \tag{2.48}$$

in (2.47), we obtain

$$\begin{aligned}
\langle\mu\varphi_1,\varphi_1\rangle &= \\
&= \frac{4k_\alpha j_{\nu_\alpha,1}^4}{(2-\alpha)|J_{\nu_\alpha}'(j_{\nu_\alpha,1})|^2}\left[\frac{2}{3}\left\{\nu_\alpha^2-1\right\}\left(\frac{j_{\nu_\alpha,1}^2}{2}\left(J_{\nu_\alpha}^2(j_{\nu_\alpha,1})-J_{\nu_\alpha-1}(j_{\nu_\alpha,1})J_{\nu_\alpha+1}(j_{\nu_\alpha,1})\right)\right)\right.\\
&\qquad\left.+\frac{1}{6}j_{\nu_\alpha,1}^3\left(j_{\nu_\alpha,1}J_{\nu_\alpha}'(j_{\nu_\alpha,1})\right)^2\right]\\
&= \frac{4k_\alpha j_{\nu_\alpha,1}^4}{(2-\alpha)|J_{\nu_\alpha}'(j_{\nu_\alpha,1})|^2}\left(-\frac{1}{3}j_{\nu_\alpha,1}^2\left\{\nu_\alpha^2-1\right\}J_{\nu_\alpha-1}(j_{\nu_\alpha,1})J_{\nu_\alpha+1}(j_{\nu_\alpha,1})\right.\\
&\qquad\left.+\frac{1}{24}j_{\nu_\alpha,1}^5\left(J_{\nu_\alpha-1}(j_{\nu_\alpha,1})-J_{\nu_\alpha+1}(j_{\nu_\alpha,1})\right)^2\right)\\
&= \frac{4k_\alpha j_{\nu_\alpha,1}^4}{(2-\alpha)|J_{\nu_\alpha}'(j_{\nu_\alpha,1})|^2}\left(\frac{1}{24}j_{\nu_\alpha,1}^5\left(J_{\nu_\alpha-1}^2(j_{\nu_\alpha,1})+J_{\nu_\alpha+1}^2(j_{\nu_\alpha,1})\right)\right.\\
&\qquad\left.-\left(\frac{1}{3}j_{\nu_\alpha,1}^2\left\{\nu_\alpha^2-1\right\}+\frac{j_{\nu_\alpha,1}^5}{12}\right)J_{\nu_\alpha-1}(j_{\nu_\alpha,1})J_{\nu_\alpha+1}(j_{\nu_\alpha,1})\right)\\
&\geq \frac{4k_\alpha j_{\nu_\alpha,1}^4}{(2-\alpha)|J_{\nu_\alpha}'(j_{\nu_\alpha,1})|^2}\left(\frac{1}{24}j_{\nu_\alpha,1}^5\left(J_{\nu_\alpha-1}^2(j_{\nu_\alpha,1})+J_{\nu_\alpha+1}^2(j_{\nu_\alpha,1})\right)\right.\\
&\qquad\left.-\left(\frac{1}{3}j_{\nu_\alpha,1}^2\left\{\nu_\alpha^2-1\right\}+\frac{j_{\nu_\alpha,1}^5}{12}\right)\frac{1}{2}\left(J_{\nu_\alpha-1}^2(j_{\nu_\alpha,1})+J_{\nu_\alpha+1}^2(j_{\nu_\alpha,1})\right)\right).
\end{aligned}$$

Thus, $\langle \mu\varphi_1, \varphi_1 \rangle > 0$ if

$$\frac{1}{24} j_{\nu_\alpha,1}^5 > \frac{1}{2}\left(\frac{1}{3} j_{\nu_\alpha,1}^2 \left\{\nu_\alpha^2 - 1\right\} + \frac{j_{\nu_\alpha,1}^5}{12}\right). \tag{2.49}$$

Since

$$\alpha \in [0,1) \;\Rightarrow\; \nu_\alpha \in (0, 1/2],$$
$$\alpha \in [1, 3/2) \;\Rightarrow \nu_\alpha \in [0, 1),$$

Equation (2.49) holds true for both weak and strong degeneracy.

Thus, since $\langle \mu\varphi_1, \varphi_k \rangle \neq 0$ for every $k \in \mathbb{N}^*$ and (2.44) is valid, the series (2.33) converges for every $\tau > 0$.

We have checked that every hypothesis of Theorem 2.1 holds for problem (2.31) if $\alpha \in [0, 3/2)$. Therefore, we conclude that, for any $\rho > 0$, if the initial condition u_0 is close enough to φ_1, the system is superexponentially stabilizable to the ground state solution ψ_1. Moreover, the following estimate holds true

$$||u(t) - \psi_1(t)|| \leq Me^{-(\rho e^{\omega t} + \lambda_1 t)}, \qquad \forall t \geq 0. \tag{2.50}$$

where $M, \omega > 0$ are suitable constants. □

Acknowledgements This paper was partly supported by the INdAM National Group for Mathematical Analysis, Probability and their Applications. The first author acknowledges support from the MIUR Excellence Department Project awarded to the Department of Mathematics, University of Rome Tor Vergata, CUP E83C18000100006. The second author is grateful to University Italo Francese (Vinci Project 2018) for partial support.

References

1. F. Alabau-Boussouira, P. Cannarsa, G. Fragnelli, Carleman estimates for degenerate parabolic operators with applications to null controllability. J. Evol. Equ. **6**(2), 161–204 (2006). Springer
2. F. Alabau-Boussouira, P. Cannarsa, C. Urbani, Superexponential stabilizability of evolution equations of parabolic type via bilinear control. Available on arXiv:1910.06802
3. J.M. Ball, J.E. Marsden, M. Slemrod, Controllability for distributed bilinear systems. SIAM J. Control Optim. **20**(4), 575–597 (1982). SIAM
4. K. Beauchard, Local controllability and non-controllability for a 1D wave equation with bilinear control. J. Differ. Equ. **250**(4), 2064–2098 (2011). Academic
5. K. Beauchard, P. Cannarsa, R. Guglielmi, Null controllability of Grushin-type operators in dimension two. J. Eur. Math. Soc. **16**(1), 67–101 (2014). Europeran Mathematical Society
6. K. Beauchard, C. Laurent, Local controllability of 1D linear and nonlinear Schrödinger equations with bilinear control. J. Math. Pures Appl. **94**(5), 520–554 (2010)
7. K. Beauchard, L. Miller, M. Morancey, 2D Grushin-type equations: minimal time and null controllable data. J. Differ. Equ. **259**(11), 5813–5845 (2015). Elsevier
8. K. Beauchard, J. Dardé, S. Ervedoza, Minimal time issues for the observability of Grushin-type equations. hal-01677037 (2018)
9. M. Campiti, G. Metafune, D. Pallara, Degenerate self-adjoint evolution equations on the unit interval. Semigroup Forum **57**(1), 1–36 (1998). Springer

10. M. Campiti, I. Rasa, Qualitative properties of a class of Fleming-Viot operators. Acta Math. Hung. **103**(1–2), 55–69 (2004). Akadémiai Kiadó, co-published with Springer Science+ Business Media BV
11. P. Cannarsa, G. Floridia, A.Y. Khapalov, Multiplicative controllability for semilinear reaction-diffusion equations with finitely many changes of sign. Journal de Mathématiques Pures et Appliquées **108**(4), 425–458 (2017). Elsevier
12. P. Cannarsa, A.Y. Khapalov, Multiplicative controllability for reaction-diffusion equations with target states admitting finitely many changes of sign. Discrete Contin. Dyn. Syst. Ser. B. **14**, 1293–1311 (2010). Citiseer
13. P. Cannarsa, P. Martinez, J. Vancostenoble, Carleman estimate for a class of degenerate parabolic operators. SIAM J. Control Optim. **47**(1), 1–19 (2008). SIAM
14. P. Cannarsa, P. Martinez, J. Vancostenoble, The cost of controlling strongly degenerate parabolic equations. J. Eur. Math. Soc. **16**(1), 67–101 (2013)
15. P. Cannarsa, P. Martinez, J. Vancostenoble, The cost of controlling weakly degenerate parabolic equations by boundary controls. Math. Control Relat. Fields **7**(2), 71–211 (2017)
16. S. Cerrai, P. Clément, On a class of degenerate elliptic operators arising from Fleming-Viot processes. J. Evol. Equ. **1**(3), 243–276 (2001). Springer
17. J.I. Díaz, On the mathematical treatment of energy balance climate models, *The Mathematics of Models for Climatology and Environment* (Springer, Berlin, 1997), pp. 217–251
18. J.I. Diaz, G. Hetzer, L. Tello, An energy balance climate model with hysteresis. Nonlinear Anal. Theory Methods Appl. **64**(9), 2053–2074 (2006). Elsevier
19. S.N. Ethier, A class of degenerate diffusion processes occurring in population genetics. Commun. Pure Appl. Math. **29**(5), 483–493 (1976). Wiley Online Library
20. S.N. Ethier, T.G. Kurtz, Fleming–Viot processes in population genetics. SIAM J. Control Optim. **31**(2), 345–386 (1993). SIAM
21. G. Floridia, Approximate controllability for nonlinear degenerate parabolic problems with bilinear control. J. Differ. Equ. **257**(9), 3382–3422 (2014). Elsevier
22. M. Gueye, Exact boundary controllability of 1-D parabolic and hyperbolic degenerate equations. SIAM J. Control Optim. **52**(4), 2037–2054 (2014). SIAM
23. G. Hetzer, The number of stationary solutions for a one-dimensional Budyko-type climate model. Nonlinear Anal. R. World Appl. **2**(2), 259–272 (2001)
24. A.Y. Khapalov, *Controllability of Partial Differential Equations Governed by Multiplicative Controls* (Springer, Berlin, 2010)
25. V. Komornik, P. Loreti, *Fourier Series in Control Theory* (Springer Science & Business Media, Berlin, 2005)
26. Y.L. Luke, *Integrals of Bessel Functions* (Courier Corporation, Mineola, 2014)
27. O.A. Oleinik, V.N. Samokhin, *Mathematical Models in Boundary Layer Theory*, vol. 15 (CRC Press, Boca Raton, 1999)

Chapter 3
Simultaneous Determination of Two Coefficients in Itô Diffusion Processes: Theoretical and Numerical Approaches

Michel Cristofol and Lionel Roques

Abstract In this paper, we consider a one-dimensional Itô diffusion process X_t with possibly nonlinear drift and diffusion coefficients. In a first part, we show that both coefficients are simultaneously uniquely determined by the observation of the expectation and variance of the process, during a small time interval, and starting from any values X_0 in a given subset of $\mathbb{R}$. Then in a second part, we present some numerical simulations which illustrate that this type of observation can be used in practice to estimate the coefficients of a diffusion process.

Keywords Inverse problems · Stochastic differential equations

3.1 Introduction

The following results are partially based on a recent work by the same authors [12]. We are interested with a one-dimensional Itô diffusion process $X_t \in \mathbb{R}$ satisfying stochastic differential equations of the form:

$$dX_t = b(X_t)\,dt + \sigma(X_t)dW_t,\ t \in [0, T];\ X_0 = x, \tag{3.1}$$

where $T > 0$, W_t is the one-dimensional Wiener process and $b : \mathbb{R} \to \mathbb{R}$, $\sigma : \mathbb{R} \to \mathbb{R}$, $\sigma > 0$, are Lipschitz-continuous functions. Under these assumptions, the solution of the Eq. (3.1) is unique in the sense of Theorem 5.2.1 in [18]. The term $b(X_t)\,dt$ can be interpreted as the deterministic part of the equation, while $\sigma(X_t)dW_t$ is the stochastic part of the equation. In the sequel, the functions b and σ are called the *drift term* and *diffusion term*, respectively.

M. Cristofol (✉)
Centrale Marseille, CNRS, I2M, Aix Marseille University, Marseille, France
e-mail: michel.cristofol@univ-amu.fr

L. Roques
BioSP, INRA, 84914 Avignon, France
e-mail: lionel.roques@inra.fr

J. Cheng et al. (eds.), *Inverse Problems and Related Topics*,
Springer Proceedings in Mathematics & Statistics 310,
https://doi.org/10.1007/978-981-15-1592-7_3

These equations arise in several domains of applications, such as biology, physics or financial mathematics with non-constant coefficients α, β, to model stock prices in the Black–Scholes model. The reconstruction of unknown parameters in stochastic differential equations has been largely addressed recently. We can refer among others to [1, 2, 16, 17] with different analytic strategies and to [22] for a survey on the topic, this list being far from to be exhaustive.

The aim of our study is to determine the drift term b and the diffusion term σ for general equations of the form (3.1), based on observations of the stochastic process X_t. Equivalently, this means showing the uniqueness of the coefficients b and σ that correspond to a given observation. The main type of observation that we consider is the expectation $E^x[f(X_t)] = E[f(X_t)|X_0 = x]$, of some function of the stochastic process X_t, for instance a momentum if $f(s) = s^k$ for some $k \geq 0$. The observation is carried out during a small time interval and for initial conditions X_0 in a small neighborhood of a given $x_0 \in \mathbb{R}$. In that respect we use parabolic partial differential equation (PDE) technics inspired from the theory of inverse problems.

The Itô diffusion processes are related to PDEs by the Kolmogorov's backward theorem (see e.g. Theorem 8.1.1 in [18]). Consider a function

$$f \in C^2(\mathbb{R}) \text{ such that } |f(x)| \leq C\, e^{\delta x^2}, \tag{3.2}$$

for $\delta > 0$ small enough and some $C > 0$. Define

$$u(t,x) = E^x\left[f(X_t)\right] = E\left[f(X_t)|X_0 = x\right], \tag{3.3}$$

where X_t is the solution of (3.1) with $X_0 = x$. The Kolmogorov's backward theorem implies that u is the unique solution in $C_1^2(\mathbb{R}_+ \times \mathbb{R})$ of:

$$\partial_t u = \frac{1}{2}\sigma^2(x)\partial_{xx}u + b(x)\partial_x u, \ t \geq 0; \ u(0,x) = f(x). \tag{3.4}$$

For parabolic equations of the form (3.4), several inverse problems have already been investigated. In all cases, the main question is to show the uniqueness of some coefficients in the equation, based on exact observations of the solution $u(t,x)$, for (t,x) in a given observation region $\mathcal{O} \subset [0,+\infty) \times \mathbb{R}$. Furthermore, one of the most challenging goal is to obtain such uniqueness results using the smallest possible observation region.

Most uniqueness results in inverse problems for parabolic PDEs have been obtained using the method of Carleman estimates [4] on bounded domains. This method requires, among other measurements, knowledge of the solution $u(\tau,x)$ at some time $\tau > 0$ and for all x in the domain [3, 10, 11, 14, 23, 24]. Other approaches are based on a semi-group formulation of the solutions, but use the same type of observations of the solution on the whole domain, at a given time [5]. More recent approaches [7, 19–21] lead to uniqueness results for one or several coefficients, under the assumption that u and its first spatial derivative are known at a *single point* x_0 of a bounded domain, and for all t in a small interval $(0,\varepsilon)$, and that the initial data

$u(0, x)$ is known over the entire domain. On the other hand, the case of unbounded domains is less addressed (see [8]).

Here, contrarily to most existing approaches:

- the domain is unbounded. All of the results in [7, 19–21] require a bounded domain assumption. Most of the studies based on other methods (e.g., Carleman estimates) also assume that the domain is bounded; see [8] or [9] for some results on unbounded domains, in the case of parabolic and hyperbolic operators;
- we determine simultaneously two coefficients in front of a second and a first order term in the PDE. The reconstruction of several coefficients, including a coefficient in front of a second order term, is very challenging not only from the theoretical point of view, but also from the numerical one [6, 13]. Up to our knowledge, theoretical studies on the determination of the diffusion and the drift coefficients from localized observations have not been proposed before;
- our results are interpreted in terms of nonlinear stochastic diffusion processes;
- as in the above-mentioned studies [7, 19–21], we assume that the observation set is reduced to a neighborhood of single point x_0, during a small time interval.

Our manuscript is organized as follows. In Sect. 3.2 we detail our assumptions on the unknown coefficients, on the observations and we state our main results. In Sect. 3.3 we present some numerical simulations which illustrate that the type of observation that we use in our theoretical results can be used in practice to estimate the coefficients of a diffusion process. Lastly, the results are proved in Sect. 3.4.

3.2 Assumptions and Main Results

3.2.1 Observations

We consider one main type of observations. Let $\varepsilon \in (0, T)$ and ω an open and nonempty subset in $\mathbb{R}$. The observation sets are of the form

$$\mathcal{O}_k[X_t] = \{E^x[(X_t)^k],\ \text{for } t \in (0, \varepsilon) \text{ and } x \in \omega\}, \tag{3.5}$$

for $k = 1, 2$. We assume $\varepsilon > 0$ and ω can be chosen as small as we want. For the sake of simplicity, and with a slight abuse of notation, for two processes X and $\tilde{X}$, we say that $\mathcal{O}_k[X_t] = \mathcal{O}_k[\tilde{X}_t]$ if and only if $E^x[(X_t)^k] = E^x[(\tilde{X}_t)^k]$ for $k = 1, 2$, for all $t \in (0, \varepsilon)$ and $x \in \omega$.

3.2.2 Unknown Functions

We assume that the unknown functions belong to the function space:

$$\mathcal{M} := \{\psi \text{ is Lipschitz-continuous and piecewise analytic in } \mathbb{R}\}. \tag{3.6}$$

A continuous function ψ is called piecewise analytic if there exist $n \geq 1$ and an increasing sequence $(\kappa_j)_{j\in\mathbb{Z}}$ such that $\lim_{j\to-\infty} \kappa_j = -\infty$, $\lim_{j\to+\infty} \kappa_j = +\infty$, $\kappa_{j+1} - \kappa_j > \delta$ for some $\delta > 0$, and $\psi(x) = \sum_{j\in\mathbb{Z}} \chi_{[\kappa_j,\kappa_{j+1})}(x)\varphi_j(x)$, for all $x \in \mathbb{R}$; here φ_j are some analytic functions defined on the intervals $[\kappa_j, \kappa_{j+1}]$, and $\chi_{[\kappa_j,\kappa_{j+1})}$ are the characteristic functions of the intervals $[\kappa_j, \kappa_{j+1})$ for $j \in \mathbb{Z}$.

In practice, the assumption $\psi \in \mathcal{M}$ is not very restrictive. For instance, the set of piecewise linear functions in $\mathbb{R}$ is a subset of $\mathcal{M}$.

3.2.3 Main Results

Determining several coefficients of parabolic PDEs is generally far more involved than determining a single coefficient. It requires more and well-chosen observations. For instance, four coefficients of a Lotka–Volterra system of parabolic equations have been determined in [21], based on the observation of the solution, starting with three different initial conditions. See also [5, 11] with different methods. Here, our result shows that, if the first momentum (expected value) and the second momentum of X_t are observed during a small time interval and for $X_0 = x$ in a small set $\omega \subset \mathbb{R}$, then both coefficients b and σ in (3.1) are uniquely determined.

Theorem 3.1 *Let $b, \tilde{b}, \sigma, \tilde{\sigma} \in \mathcal{M}$. Consider X_t, $\tilde{X}_t$ the solutions of* (3.1) *and of $d\tilde{X}_t = \tilde{b}(\tilde{X}_t)\, dt + \tilde{\sigma}(\tilde{X}_t) dW_t$, $t \in [0, T]$; $\tilde{X}_0 = x$, respectively. Assume that $\mathcal{O}_k[X_t] = \mathcal{O}_k[\tilde{X}_t]$ for $k = 1, 2$. Then, $b \equiv \tilde{b}$ and $\sigma \equiv \tilde{\sigma}$ in $\mathbb{R}$.*

An immediate corollary of Theorem 3.1 is that b and σ are uniquely determined by the observation of the mean and variance $V^x[X_t] = E^x[X_t^2] - (E^x[X_t])^2$ of the process X_t during a small time interval and for $X_0 = x$ in a small set $\omega \subset \mathbb{R}$. More precisely, define the set

$$\mathcal{O}_v[X_t] = \{V^x[X_t], \text{ for } t \in (0, \varepsilon) \text{ and } x \in \omega\}, \tag{3.7}$$

we have the following result.

Corollary 3.1 *Let $b, \tilde{b}, \sigma, \tilde{\sigma} \in \mathcal{M}$. Consider X_t, $\tilde{X}_t$ the solutions of* (3.1) *and $d\tilde{X}_t = \tilde{b}(\tilde{X}_t)\, dt + \tilde{\sigma}(\tilde{X}_t) dW_t$, $t \in [0, T]$; $\tilde{X}_0 = x$, respectively. Assume that $\mathcal{O}_1[X_t] = \mathcal{O}_1[\tilde{X}_t]$ and $\mathcal{O}_v[X_t] = \mathcal{O}_v[\tilde{X}_t]$. Then, $b \equiv \tilde{b}$ and $\sigma \equiv \tilde{\sigma}$ in $\mathbb{R}$.*

Remark 3.1 Our result remains true if the observation (3.5) is replaced by punctual observations at a given point $x_0 \in \mathbb{R}$ instead of observations in a subdomain ω. More precisely, if (3.5) is replaced by

$$\mathcal{O}_k'[X_t] = \{E^{x_0}[(X_t)^k],\ \partial_x E^x[(X_t)^k]_{|x=x_0}, \text{ for } t \in (0, \varepsilon) \text{ and } x \in \omega\}, \tag{3.8}$$

for $k = 1, 2$, the result of our theorem and corollary can still be obtained, by using the Hopf's Lemma in addition to the strong parabolic maximum principle.

Remark 3.2 All of our results are based on parabolic theory and cannot be extended to elliptic operators. However, it should be noted that our method remains valid even though the process X_t has converged to a stationary distribution. In such case, the Eq. (3.4) remains parabolic, as f is not a steady state. Thus, Theorem 3.1 and Corollary 3.1 can still be applied.

3.3 Numerical Simulations

We have shown that non-constant and possibly nonlinear drift and diffusion coefficients of a one-dimensional Itô diffusion process can be determined, based on observations of expectations of some functionals of the process, during a small time interval, and starting any from values $X_0 = x_0$ in a given small subset of $\mathbb{R}$. These results are based on ideal observations, in the sense that they assume that expectations are observed (and not only sample paths), and that these observations are not noisy.

The objective of this section is to illustrate that the type of observations in Corollary 3.1 can be used in practice to estimate the coefficients b and σ in

$$dX_t = b(X_t)\,dt + \sigma(X_t)dW_t,\ t \in [0, T];\ X_0 = x. \tag{3.9}$$

We recall that these observations correspond to the expectation

$$\mathcal{O}^1[X_t] = \{E^x[X_t],\ \text{for } t \in (0, \varepsilon) \text{ and } x \in \omega\}, \tag{3.10}$$

together with the variance

$$\mathcal{O}_v[X_t] = \{V^x[X_t],\ \text{for } t \in (0, \varepsilon) \text{ and } x \in \omega\}. \tag{3.11}$$

Generation of the observations. We fix two functions b and σ in some finite-dimensional subspaces $\mathcal{M}_1,\ \mathcal{M}_2 \subset \mathcal{M}$. For any fixed $x \in \mathbb{R}$ and $t \in (0, \varepsilon)$ we estimate $E^x[X_t]$ and $V^x[X_t]$ by simulating $N = 10^6$ trajectories X_t^i, for $t \in (0, \varepsilon)$. Namely, setting $\mathbf{X}_t^x = (X_t^1, \dots X_t^N)$, we compute the "empirical" mean and variance

$$\mathcal{E}(\mathbf{X}_t^x) = \frac{1}{N}\sum_{i=1}^{N} X_t^i,\ \mathcal{V}(\mathbf{X}_t^x) = \frac{1}{N}\sum_{i=1}^{N}(X_t^i)^2 - \mathcal{E}^2(\mathbf{X}_t^x),$$

at discrete times $t = t_1, \dots, t_K \in (0, \varepsilon)$.

In practice, we set

$$\mathcal{M}_1 = \left\{p(x) = \alpha_0 + \alpha_1\, x + \alpha_2\, x^2 + \alpha_3\, x^3,\ \alpha_0,\ \alpha_1,\ \alpha_2,\ \alpha_3 \in \mathbb{R}\right\},$$

and

$$\mathcal{M}_2 = \{q(x) = 0.1 + |\beta_0 + \beta_1 x|,\ \beta_0,\ \beta_1 \in \mathbb{R}\}.$$

The coefficients $b(x)$ and $\sigma(x)$ are defined by setting $\alpha_0 = 1$, $\alpha_1 = 1$, $\alpha_2 = 1$, $\alpha_3 = -1$ and $\beta_0 = 0.5$, $\beta_1 = 1$. We assume 100 observation times t_j, regularly spaced in [0, 1] ($K = 100$).

The simulation of the trajectories X_t^i is carried out with the *simulate* function in Matlab Financial Toolbox®. The observations are generated with two different initial conditions: $x = 0$ and $x = 0.1$ (see Remark 3.1 for some comment about punctual observations).

Functional to be minimized (objective function). For any $\tilde{b} \in \mathcal{M}_1$ and $\tilde{\sigma} \in \mathcal{M}_2$, we estimate $E^x[\tilde{X}_t^x]$ and $V^x[\tilde{X}_t^x]$ by simulating $\tilde{N}$ trajectories of

$$d\tilde{X}_t = \tilde{b}(\tilde{X}_t)\, dt + \tilde{\sigma}(\tilde{X}_t) dW_t,\ t \in [0, \varepsilon];\ \tilde{X}_0 = x, \tag{3.12}$$

and by computing the associated quantities $\mathcal{E}(\tilde{\mathbf{X}}_t^x)$ and $\mathcal{V}(\tilde{\mathbf{X}}_t^x)$, with $\tilde{\mathbf{X}}_t^x = (\tilde{X}_t^1, \ldots \tilde{X}_t^{\tilde{N}})$ (with initial condition x). For each initial condition $X_0 = \tilde{X}_0 = x$, the match between the trajectories generated with $(\tilde{b}, \tilde{\sigma})$ and the observations $\mathcal{E}(\mathbf{X}_t^x)$, $\mathcal{V}(\mathbf{X}_t^x)$ is measured by:

$$F(\tilde{b}, \tilde{\sigma}, x) = \sum_{j=1}^{k} (\mathcal{E}(\mathbf{X}_{t_j}^x) - \mathcal{E}(\tilde{\mathbf{X}}_{t_j}^x))^2 + \sum_{j=1}^{k} (\mathcal{V}(\mathbf{X}_{t_j}^x) - \mathcal{V}(\tilde{\mathbf{X}}_{t_j}^x))^2.$$

As the observations correspond to two initial conditions $x = 0$ and $x = 0.1$, the objective function is:

$$MS(\tilde{b}, \tilde{\sigma}) = F(\tilde{b}, \tilde{\sigma}, 0) + F(\tilde{b}, \tilde{\sigma}, 0.1).$$

Note that this objective function is stochastic, as it depends on the trajectories in $\mathbf{X}_t^x$. In our computations, $MS(\tilde{b}, \tilde{\sigma})$ is either computed with $\tilde{N} = 10^2$ or $\tilde{N} = 10^3$ trajectories.

Estimation algorithm. We check whether the coefficients b and σ can be numerically determined by minimizing $MS(\tilde{b}, \tilde{\sigma})$ over $\mathcal{M}_1 \times \mathcal{M}_2$. As the objective function MS is stochastic, the minimization of $MS(\tilde{b}, \tilde{\sigma})$ in $\mathcal{M}_1 \times \mathcal{M}_2$ is a stochastic optimization problem. To solve it, we use a the Matlab built-in simulated annealing algorithm [15] (*simulannealbnd* function in Matlab Global Optimization Toolbox®, with initial temperature 10^4, and at most 500 evaluations of the function MS).

We first check that the objective function MS tends to be smaller as the estimated coefficients tend to resemble the "true" coefficients (b, σ). In that respect, we depicted in Fig. 3.1, the squared L^2 distance (in $(-2, 2)$) between b and $\tilde{b}$ (Fig. 3.1a) and between σ and $\tilde{\sigma}$ (Fig. 3.1b) versus the corresponding value of $MS(\tilde{b}, \tilde{\sigma})$. The crosses corresponds to the (500) computations of MS carried out during the simulated annealing algorithm (with $\tilde{N} = 10^3$). We observe that, in spite of the fact that the objective function MS is stochastic, smaller values of MS are associated with coefficients $(\tilde{b}, \tilde{\sigma})$ which tend to be closer to (b, σ). Next, we depict in Fig. 3.2 the result of the minimization procedure, i.e., the functions $(\tilde{b}, \tilde{\sigma})$ associated with the

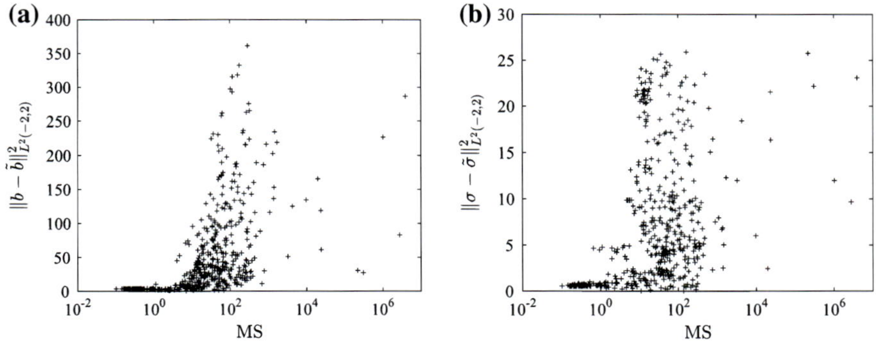

Fig. 3.1 **a** Squared L^2 distance (in $(-2, 2)$) between b and $\tilde{b}$ versus corresponding value of $MS(\tilde{b}, \tilde{\sigma})$; **b** $L^2(-2, 2)$ distance between σ and $\tilde{\sigma}$ versus corresponding value of $MS(\tilde{b}, \tilde{\sigma})$. Each cross correspond to one evaluation of the function MS carried out by the simulated annealing algorithm (with $\tilde{N} = 10^3$)

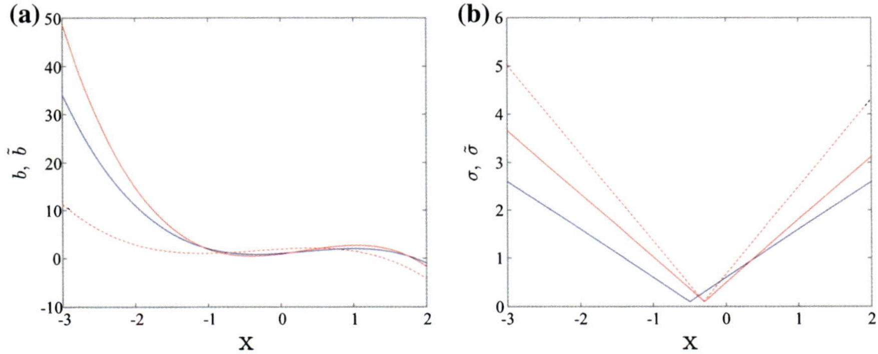

Fig. 3.2 **a** Blue line: unknown function b; dashed red line: function $\tilde{b}$ obtained by minimization of MS, with $\tilde{N} = 10^2$ ($\|b-\tilde{b}\|^2_{L^2(-2,2)} = 26.1$); continuous red line: function $\tilde{b}$ obtained by minimization of MS, with $\tilde{N} = 10^3$ ($\|b-\tilde{b}\|^2_{L^2(-2,2)} = 3.7$); **b** Blue line: unknown function σ; dashed red line: function $\tilde{\sigma}$ obtained by minimization of MS, with $\tilde{N} = 10^2$ ($\|\sigma-\tilde{\sigma}\|^2_{L^2(-2,2)} = 3.7$); continuous red line: function $\tilde{\sigma}$ obtained by minimization of MS, with $\tilde{N} = 10^3$ ($\|\sigma-\tilde{\sigma}\|^2_{L^2(-2,2)} = 0.6$)

smallest value of $MS(\tilde{b}, \tilde{\sigma})$ obtained with the simulated annealing algorithm, with either $\tilde{N} = 10^2$ and $\tilde{N} = 10^3$. In both cases, the global shape of the coefficients (b, σ) is well-estimated. Higher values of $\tilde{N}$ lead to a better accuracy of the estimation.

3.4 Proof of Theorem 3.1

In this case, we reconstruct simultaneously two coefficients from the principal part and the first order term in Eq. (3.4) and this implies to repeat the observations and to consider adapted weight functions in the form (3.5). We define, for all $t \in [0, T)$

and $x \in \mathbb{R}$, and for $f(s) = s^k$,

$$\begin{aligned} u(t,x) &= E^x[f(X_t)] = E[f(X_t)|X_0 = x], \\ \tilde{u}(t,x) &= E^x[f(\tilde{X}_t)] = E[f(\tilde{X}_t)|X_0 = x], \end{aligned} \tag{3.13}$$

and u and $\tilde{u}$ are respectively the unique solutions of:

$$\partial_t u = \frac{1}{2}\sigma^2(x)\partial_{xx}u + b(x)\partial_x u, \ t \in [0,T), \ x \in \mathbb{R}; \ u(0,x) = f(x), \tag{3.14}$$

and

$$\partial_t \tilde{u} = \frac{1}{2}\tilde{\sigma}^2(x)\partial_{xx}\tilde{u} + \tilde{b}(x)\partial_x \tilde{u}, \ t \in [0,T), \ x \in \mathbb{R} \ \tilde{u}(0,x) = f(x). \tag{3.15}$$

Define

$$B(x) = b(x) - \tilde{b}(x), \ \Sigma(x) = \frac{1}{2}(\sigma^2(x) - \tilde{\sigma}^2(x)), \text{ and } U(t,x) = u(t,x) - \tilde{u}(t,x).$$

Then $U(t,x)$ satisfies

$$\partial_t U = \frac{1}{2}\sigma^2(x)\partial_{xx}U + b(x)\partial_x U + B(x)\,\partial_x\tilde{u} + \Sigma(x)\,\partial_{xx}\tilde{u}, \ t \in [0,T), \ x \in \mathbb{R}, \tag{3.16}$$

and $U(0,x) = 0$ for all $x \in \mathbb{R}$.

Let $x_0 \in \omega$. We define:

$$\begin{aligned} x_B^* &= \sup\{x > x_0 \text{ such that } B \equiv 0 \text{ on } [x_0, x]\}, \\ x_\Sigma^* &= \sup\{x > x_0 \text{ such that } \Sigma \equiv 0 \text{ on } [x_0, x]\}. \end{aligned} \tag{3.17}$$

Then, four cases may occur.

Case 1: we assume that $x_B^* < x_\Sigma^*$. Using the piecewise analyticity of B, and from the definitions of x_B^*, we obtain the existence of some $x_2 \in (x_B^*, x_\Sigma^*)$ such that $B(x) \neq 0$ for all $x \in (x_B^*, x_2]$, i.e., B has a constant strict sign in $(x_B^*, x_2]$. Moreover, $\Sigma(x) = 0$ for all $x \in (x_B^*, x_2]$, thus U satisfies:

$$\partial_t U - \mathcal{L}U = B(x)\,\partial_x\tilde{u}, \ t \in [0,T), \ x \in (x_0, x_2), \tag{3.18}$$

where $\mathcal{L}U := \frac{1}{2}\sigma^2(x)\partial_{xx}U + b(x)\partial_x U$. Take $k = 1$ in the definition of $f(s) = s^k$. We have $\partial_x\tilde{u}(0,x) = f'(x) = 1$ for all $x \in \mathbb{R}$, which implies that there exists $\varepsilon' \in (0,\varepsilon)$ such that $\partial_x\tilde{u}(t,x)$ is positive on $[0,\varepsilon') \times [x_0, x_2]$. Finally, the term $B(x)\,\partial_x\tilde{u}$ in the right hand side of (3.18) has a constant sign in $(0,\varepsilon') \times [x_0, x_2]$. Without loss of generality, we can assume that:

$$B(x)\,\partial_x\tilde{u} \geq 0 \text{ for } (t,x) \text{ in } [0,\varepsilon') \times [x_0, x_2]. \tag{3.19}$$

We then observe that $\partial_t U(0, x_2) = B(x_2) \geq 0$ and, from the definition of x_2, the inequality is strict: $\partial_t U(0, x_2) > 0$. Thus, (even if it means reducing $\varepsilon' > 0$),

$$U(t, x_2) > 0 \text{ for } t \in (0, \varepsilon'). \tag{3.20}$$

Finally, U satisfies

$$\begin{cases} \partial_t U - \mathcal{L}U \geq 0, \ t \in (0, \varepsilon'), \ x \in (x_0, x_2), \\ U(t, x_0) = 0, \ U(t, x_2) > 0, \ t \in (0, \varepsilon'), \\ U(0, x) = 0, \ x \in (x_0, x_2). \end{cases} \tag{3.21}$$

From strong parabolic maximum principle $U(t, x) > 0$ in $(0, \varepsilon') \times (x_0, x_2)$. This contradicts the assumption $\mathcal{O}_1[X_t] = \mathcal{O}_1[\tilde{X}_t]$ of Theorem 3.1. Thus, Case 1 is ruled out.

Case 2: we assume that $x_B^* > x_\Sigma^*$. With the same type of arguments as in Case 1, we obtain the existence of some $x_2 \in (x_\Sigma^*, x_B^*)$ such that $\Sigma(x) \neq 0$ for all $x \in (x_\Sigma^*, x_2]$, i.e., Σ has a constant strict sign in $(x_\Sigma^*, x_2]$. Moreover, $B(x) = 0$ for all $x \in (x_\Sigma^*, x_2]$, thus U satisfies:

$$\partial_t U - \mathcal{L}U = \Sigma(x)\, \partial_{xx}\tilde{u}, \ t \in [0, T), \ x \in (x_0, x_2). \tag{3.22}$$

Take $k = 2$ in the definition of $f(s) = s^k$. We have $\partial_{xx}\tilde{u}(0, x) = f''(x) = 2$ for all $x \in \mathbb{R}$. Thus, with the same arguments as in Case 1, we get:

$$\Sigma(x)\, \partial_{xx}\tilde{u} \geq 0 \text{ for } (t, x) \text{ in } [0, \varepsilon') \times [x_0, x_2], \tag{3.23}$$

and $\partial_t U(0, x_2) > 0$. Thus, U again satisfies (3.21), and the strong parabolic maximum principle implies $U(t, x) > 0$ in $(0, \varepsilon') \times (x_0, x_2)$, leading to a contradiction with the assumption $\mathcal{O}_2[X_t] = \mathcal{O}_2[\tilde{X}_t]$ of Theorem 3.1. Thus, Case 2 is ruled out.

Case 3: we assume that $x_B^* = x_\Sigma^* < +\infty$. Let us set

$$G(t, x) = B(x)\, \partial_x \tilde{u} + \Sigma(x)\, \partial_{xx}\tilde{u},$$

corresponding to the right-hand side in (3.16). Then, set

$$l^* = \lim_{x \to x_B^*, x > x_B^*} \frac{\Sigma(x)}{B(x)}.$$

From the analyticity of Σ and B in a right neighborhood of x_B^*, l^* is well-defined and only two situations may occur: either $|l^*| < +\infty$ or $|l^*| = +\infty$.

Assume first that $|l^*| < +\infty$. Take $k = 1$ in the definition of $f(s) = s^k$. Thus,

$$\partial_x \tilde{u}(0, x) = 1 \text{ and } \partial_{xx}\tilde{u}(0, x) = 0. \tag{3.24}$$

Let $x_2 > x_B^*$ such that $B(x) \neq 0$ in $(x_B^*, x_2]$ and $\Sigma(x)/B(x)$ remains bounded in $(x_B^*, x_2]$. Without loss of generality, we can assume that $B > 0$ in $(x_B^*, x_2]$. Using (3.24), and since $|l^*| < +\infty$, we obtain the existence of $\varepsilon' \in (0, \varepsilon)$ such that

$$\frac{G(t,x)}{B(x)} = \partial_x \tilde{u} + \frac{\Sigma(x)}{B(x)} \partial_{xx} \tilde{u} > 0 \text{ for } (t,x) \text{ in } (0,\varepsilon') \times (x_0, x_2),$$

and $G(t,x)$ satisfies the same inequality. Thus, again, U satisfies (3.21), and the strong parabolic maximum principle implies that $U(t,x) > 0$ in $(0,\varepsilon') \times (x_0, x_2)$ and a contradiction with the assumption $\mathcal{O}_1[X_t] = \mathcal{O}_1[\tilde{X}_t]$ of Theorem 3.1. The assumption $|l^*| < +\infty$ is then ruled out.

Assume now that $|l^*| = +\infty$. Take $k = 2$ in the definition of $f(s) = s^k$. This time,

$$\partial_x \tilde{u}(0,x) = 2\,x \text{ and } \partial_{xx} \tilde{u}(0,x) = 2. \tag{3.25}$$

Let $x_2 > x_\Sigma^*$ such that

$$\Sigma(x) \neq 0 \text{ and } |2\,x\,(B(x)/\Sigma(x))| < 1 \text{ in } (x_\Sigma^*, x_2].$$

Without loss of generality, we assume that $\Sigma > 0$ in $(x_\Sigma^*, x_2]$. Using (3.25), and since $|l^*| = +\infty$, we can define $\varepsilon' \in (0, \varepsilon)$ such that

$$\frac{G(t,x)}{\Sigma(x)} = \frac{B(x)}{\Sigma(x)} \partial_x \tilde{u} + \partial_{xx} \tilde{u} > 0 \text{ for } (t,x) \text{ in } (0,\varepsilon') \times (x_0, x_2).$$

Again, using the strong parabolic maximum principle, we get a contradiction with the assumption $\mathcal{O}_2[X_t] = \mathcal{O}_2[\tilde{X}_t]$ of Theorem 3.1. Case 3 is then ruled out.

Finally, as Cases 1, 2, 3 are ruled out, we necessarily have $x_B^* = x_\Sigma^* = +\infty$, which show that $B \equiv \Sigma \equiv 0$ in $(x_0, +\infty)$. Using the same arguments with $(x_B^*)^- = \inf\{x < x_0 \text{ such that } B \equiv 0 \text{ on } [x, x_0]\}$ and $(x_\Sigma^*)^- = \inf\{x < x_0 \text{ such that } \Sigma \equiv 0 \text{ on } [x, x_0]\}$, instead of x_B^* and x_Σ^*, we also check that $B \equiv 0$ in $B \equiv \Sigma \equiv 0$ in $(-\infty, x_0)$ and consequently $B \equiv \Sigma \equiv 0$ in $\mathbb{R}$ which concludes the proof of Theorem 3.1. □

References

1. S. Albeverio, Ph. Blanchard, S. Kusuoka, L. Streit, An inverse problem for stochastic differential equations. J. Stat. Phys. **57**(1, 2) (1989)
2. G. Bao, C.C. Chen, P. Li, Inverse random source scattering problems in several dimensions. SIAM/ASA J. Uncertaint. Quantif **4**, 1263–1287 (2016)
3. M. Belassoued, M. Yamamoto, Inverse source problem for a transmission problem for a parabolic equation. J. Inverse Ill-Posed Probl. **14**, 47–56 (2006)
4. A.L. Bukhgeim, M.V. Klibanov, Uniqueness in the large of a class of multidimensional inverse problems. Soviet Mathematics - Doklady **24**, 244–247 (1981)

5. M. Choulli, M. Yamamoto, Uniqueness and stability in determining the heat radiative coefficient, the initial temperature and a boundary coefficient in a parabolic equation. Nonlinear Anal. Theory Methods Appl. **69**, 3983–3998 (2008). Elsevier
6. M. Cristofol, P. Gaitan, K. Niinimaki, O. Poisson, Inverse problem for a coupled parabolic system with discontinuous conductivities: one dimensional case. Inverse Probl. Imaging **7**, 159–182 (2013)
7. M. Cristofol, J. Garnier, F. Hamel, L. Roques, Uniqueness from pointwise observations in a multi-parameter inverse problem. Commun. Pur. Appl. Anal. **11**, 1–15 (2011)
8. M. Cristofol, I. Kaddouri, G. Nadin, L. Roques, Coefficient determination via asymptotic spreading speeds. Inverse Probl. **30**, 035005 (2014). IOP Publishing
9. M. Cristofol, S. Li, E. Soccorsi, Determining the waveguide conductivity in a hyperbolic equation from a single measurement on the lateral boundary. J. Math. Control. Relat. Fields **3**, 407–427 (2016)
10. M. Cristofol, L. Roques, Biological invasions: deriving the regions at risk from partial measurements. Math. Biosci. **215**, 158–166 (2008)
11. M. Cristofol, L. Roques, Stable estimation of two coefficients in a nonlinear Fisher-KPP equation. Inverse Probl. **29**, 095007 (2013)
12. M. Cristofol, L. Roques, Simultaneous determination of the drift and diffusion coefficients in stochastic differential equations. Inverse Probl. **33**, 095006 (2017)
13. M.S. Hussein, D. Leisnic, Simultaneous determination of time and space-dependent coefficients in a parabolic equation. Commun. Nonlinear Sci. Numer. Simul. **33**, 194–217 (2016)
14. O.Y. Immanuvilov, M. Yamamoto, Lipschitz stability in inverse parabolic problems by the Carleman estimate. Inverse Probl. **14**, 1229–1245 (1998)
15. S. Kirkpatrick, C.D. Gelatt, M.P. Vecchi, Optimization by simulated annealing. Science **220**, 671–680 (1983)
16. P.J. Li, An inverse random source scattering problem in inhomogeneous media. Inverse Probl. **27**, 035004 (2011)
17. P. Niu, S. Lu, J. Cheng, On periodic parameter identification in stochastic differential equation. Inverse Probl. Imaging **13**, 513–543 (2019)
18. B. Øksendal, *Stochastic Differential Equation* (Springer, Berlin, 2003)
19. L. Roques, M.D. Chekroun, M. Cristofol, S. Soubeyrand, M. Ghil, Parameter estimation for energy balance models with memory. Proc. R. Soc. A **470**, 20140349 (2014)
20. L. Roques, M. Cristofol, On the determination of the nonlinearity from localized measurements in a reaction-diffusion equation. Nonlinearity **23**, 675–686 (2010)
21. L. Roques, M. Cristofol, The inverse problem of determining several coefficients in a nonlinear Lotka-Volterra system. Inverse Probl. **28**, 075007 (2012)
22. H. Sorensen, Parametric inference for diffusion processes observed at discrete points in time: a survey. Int. Stat. Rev. **72**, 337–354 (2003)
23. M. Yamamoto, Carleman estimates for parabolic equations and applications. Inverse Probl. **25**, 123013 (2009). IOP Publishing
24. M. Yamamoto, J. Zou, Simultaneous reconstruction of the initial temperature and heat radiative coefficient. Inverse Probl. **17**, 1181–1202 (2001)

Chapter 4
On the Inverse Source Problem with Boundary Data at Many Wave Numbers

Victor Isakov and Shuai Lu

Abstract We review recent results on inverse source problems for the Helmholtz type equations from boundary measurements at multiple wave numbers combined with new results including uniqueness of obstacles. We consider general elliptic differential equations of the second order and arbitrary observation sites. We present some new results and outline basic ideas of their proofs. To show the uniqueness we use the analytic continuation, the Fourier transform with respect to the wave numbers and uniqueness in the lateral Cauchy problem for hyperbolic equations. To derive the increasing stability we utilize sharp bounds of the analytic continuation for higher wave numbers, the Huygens' principle, and boundary energy estimates in the initial boundary value problems for hyperbolic equations. Some numerical examples, based on a recursive Kaczmarz-Landweber iterative algorithm, shed light on theoretical results.

Keywords Helmholtz equation · Inverse source problems · Increasing stability

4.1 Introduction

We are interested in uniqueness and increasing stability of inverse source problems for elliptic equations when the (compactly supported) source function is to be determined from full/partial boundary measurements. More precisely, let $n = 2, 3$ and assume

---

Dedicated to Masahiro Yamamoto for his sixtieth birthday.

V. Isakov
Department of Mathematics, Statistics, and Physics, Wichita State University, Wichita, KS 67260-0033, USA
e-mail: victor.isakov@wichita.edu

S. Lu (✉)
Shanghai Key Laboratory for Contemporary Applied Mathematics, Key Laboratory of Mathematics for Nonlinear Sciences and School of Mathematical Science, Fudan University, Shanghai 200433, China
e-mail: slu@fudan.edu.cn

J. Cheng et al. (eds.), *Inverse Problems and Related Topics*,
Springer Proceedings in Mathematics & Statistics 310,
https://doi.org/10.1007/978-981-15-1592-7_4

that $u(x, k)$ is the solution of the following scattering problem,

$$(\Delta + k^2)u = -f \quad \text{in} \quad \mathbb{R}^n, \tag{4.1a}$$

$$\lim r^{\frac{n-1}{2}}(\partial_r u - \mathbf{i}ku) = 0 \quad \text{as} \quad r = |x| \to +\infty. \tag{4.1b}$$

We aim at the recovery of the source function f, compactly supported in a bounded Lipschitz domain Ω, from the near field data u, $\partial_\nu u$ on Γ_0 where ν is the outer unit normal vector and Γ_0 is an open non-void part of the boundary $\partial\Omega$. Here k is the wave number, which plays an essential role in current work. A straightforward and important example of such problems is the recovery of acoustic sources from boundary measurements of the pressure. This type of inverse source problems is also motivated by wide applications in antenna synthesis [2], biomedical imaging [1], and various kinds of tomography.

To recover the source function from boundary measurements at one wave number k in (4.1), it is not possible to determine it uniquely, i.e. [18, Chap. 4] and [6]. Nevertheless, in the case of a family of equations (like the Helmholtz equation (4.1) for various wave numbers $k \in (0, K)$) one can regain uniqueness in the inverse source problems [10]. The consequential crucial issue for real applications is then the stability of source recovery. In general, a feature of inverse problems for elliptic equations is a logarithmic type stability estimate which results in a robust recovery of only few parameters describing the source and hence yields very low resolution numerically. In [3, 4] the authors have considered the uniqueness and provided first increasing stability results for inverse source problems by spatial Fourier analysis methods. Recently, in [7] we used the Fourier transform in time to reduce our inverse source problem to the identification of the initial data in the hyperbolic initial value problem by lateral Cauchy data (observability in control theory) and, as a consequence, obtained most complete increasing (with growing K) stability results. At the same time in [21] we traced the dependence on the attenuation term where the following attenuated Helmholtz equation is considered

$$(\Delta + k^2 + \mathbf{i}kb)u = -f \quad \text{in} \quad \mathbb{R}^n, \tag{4.2}$$

with an attenuation positive constant b and the exponential decay of $u(x, k)$ at infinity. In all these results mentioned above, increasing stability bounds assume that the corresponding lateral Cauchy problem satisfies the so called non-trapping condition which is guaranteed by certain (pseudo) convexity of the domain and the surface with the Cauchy data with respect to the hyperbolic equation. In [15] we verify the uniqueness of the inverse source problems when the (pseudo) convexity or non-trapping conditions for the related hyperbolic problem are not satisfied.

Concerning the inverse problems at high wave numbers, more is known about uniqueness and increasing stability in the Cauchy problem for elliptic equations and on identification of the Schrödinger potential where boundary measurements at one fixed wave number are needed.

Carleman estimates imply some conditional logarithmic type global estimates and Hölder type stability estimates on compact sub domains for solutions of the elliptic Cauchy problem [18]. Classic results in [24] show that for the continuation of the Helmholtz equation from the unit disk onto any larger disk the stability estimate, which is uniform with respect to the wave numbers, is still of a logarithmic type. Various numerical examples in [20] are provided verifying the increasing stability for general domains and observation sites. Under (pseudo) convexity conditions on the geometry of the domain and on the coefficients of the elliptic equation, it was demonstrated that in a certain sense stability is always improving for larger k in [17], while in [14, 19, 20] it was shown that in some cases increasing stability holds without convexity/non-trapping conditions.

To study the identification of the Schrödinger potential, one needs measurements in form of the Dirichlet-to-Neumann map where increasing stability for the Schrödinger potential and conductivity coefficients was demonstrated in [22, 23]. In particular, in [22], the dependence of stability on the attenuation is well traced. A novel numerical reconstruction algorithm based on the linearised inverse Schrödinger potential problem is reported in [16] where increasing stability is numerically verified.

In this paper we review recent results on inverse source problems for the Helmholtz type equations from boundary (Cauchy) data with multiple wave numbers. We also present some new results and outline basic ideas of their proofs. In Sect. 4.2 we focus on the uniqueness of sources and obstacles without any convexity/non-trapping condition. In Sect. 4.3 a generalized increasing stability is provided by taking a standard Helmholtz equation and assuming certain regularity of the source functions. Section 4.4 collects several numerical examples showing the increasing stability with respect to different choices of dimensionality.

4.2 Uniqueness in Inverse Source Problems

We first consider the uniqueness in inverse source problems when the (pseudo) convexity or non-trapping conditions for the related hyperbolic problem are not satisfied, in addition to known results [15] we prove also simultaneous uniqueness of hard obstacles.

Let Ω be a bounded Lipschitz domain in $\mathbb{R}^n$, $n = 2, 3$. We introduce a sub domain Ω_0 such that $\bar{\Omega}_0 \subset \Omega$, whose boundary $\partial\Omega_0$ is of C^2 class, and $\mathbb{R}^n \setminus \bar{\Omega}_0$ is connected. Let D_0 be a sub domain of $\Omega \setminus \bar{\Omega}_0$ with connected $\Omega \setminus (\bar{\Omega}_0 \cup \bar{D}_0)$, $\partial D_0 \in C^2$ and D be a sub domain of Ω_0 with connected $\Omega_0 \setminus \bar{D}$, $\partial D \in C^2$. ν denotes the exterior unit normal to the boundary of a domain.

Let A be the second order elliptic operator

$$A := \sum_{j,m=1}^{n} \partial_j(a_{jm}\partial_m) + \mathbf{i}\sum_{j=1}^{n} b_j\partial_j + c, \quad a_{jm}, b_j \in C^1(\mathbb{R}^n),\ c \in L^\infty(\mathbb{R}^n),$$

and $A := \Delta$ in $\mathbb{R}^n \setminus \Omega$. We denote $b = (b_1, ...b_n)$. We assume the ellipticity condition $\epsilon_0|\xi|^2 \leq \sum_{j,m=1}^n a_{jm}(x)\xi_j\xi_m$ for some positive constant ϵ_0 and all $x, \xi \in \mathbb{R}^n$ as well as the additional conditions

$$\text{the imaginary part } \Im b_j = 0, \nabla \cdot b \leq 0 \text{ on } \Omega,\ \ b \cdot \nu \leq 0 \text{ on } (\partial D_0 \cup \partial D),$$

$$b_0 \in L_\infty(\mathbb{R}^n), 0 \leq b_0 \text{ on } \Omega,\ \text{ and } b_0 = 0 \text{ outside } \Omega, \tag{4.3}$$

and

$$0 \leq \Im c \text{ on } \Omega. \tag{4.4}$$

We consider the scattering problem

$$(A + b_0k\mathbf{i} + k^2)u = -f_1 - (\mathbf{i}k - b_0)f_0 \quad \text{in} \quad \mathbb{R}^n \setminus (\bar{D}_0 \cup \bar{D}), \tag{4.5}$$

where $f_1 \in L^2(\mathbb{R}^n)$, $f_0 \in H^1(\mathbb{R}^n)$ and $suppf_0 \cup suppf_1 \subset \Omega \setminus \Omega_0$ with the Neumann boundary condition

$$a\nabla u \cdot \nu = g(x, k) \quad \text{on} \quad \partial D_0,\ a\nabla u \cdot \nu = 0 \quad \text{on} \quad \partial D, \tag{4.6}$$

where a is the matrix (a_{jm}), so that $a\nabla u \cdot \nu = \sum_{j,m=1}^n a_{jm}\partial_j u\nu_m$, and the Sommerfeld radiation condition.

$$\lim r^{\frac{n-1}{2}}(\partial_r u - \mathbf{i}ku) = 0, \quad \text{as } r = |x| \to +\infty. \tag{4.7}$$

Here and in what follows, we denote $\|\cdot\|_{(\ell)}(\Omega)$, $\|\cdot\|_{(\ell)}(\partial\Omega)$ as the standard norms in the Sobolev spaces $H^\ell(\Omega)$, $H^\ell(\partial\Omega)$. We define

$$g(x, k) = \frac{1}{\sqrt{2\pi}}\int_0^\infty \mathcal{G}(x, t)e^{\mathbf{i}tk}dt \tag{4.8}$$

for some function $\mathcal{G} \in L^\infty((0,\infty); H^{\frac{1}{2}}(\partial D_0))$. We assume that $|\mathcal{G}(x, t)| + \|\mathcal{G}(, t)\|_{(\frac{1}{2})}(\partial D_0) \leq Ce^{-\delta t}$ for some positive numbers C, δ depending on $\mathcal{G}$. Then $g(x, k)$ is (complex) analytic with respect to $k = k_1 + \mathbf{i}k_2$ if $-\delta < k_2$.

The inverse problem of interest is to find f_0, f_1 on $\Omega \setminus (D_0 \cup \Omega_0)$, g, and the domain D entering (4.5)–(4.7) from the additional Cauchy data

$$u(, k) = u_0, \quad \partial_\nu u(, k) = u_1 \text{ on } \Gamma_0,\ K_0 < k < K \tag{4.9}$$

where Γ_0 is a non(empty) open subset of $\partial\Omega$ and K_0, K are some numbers, $0 \leq K_0 < K$.

Theorem 4.1 *Let the assumptions (4.3), (4.4) hold and g has the form* (4.8) *with the given assumptions on* $\mathcal{G}$. *Then a solution* (f_0, f_1, g, D) *to the inverse problem (4.5)–(4.7) with the additional Cauchy data (4.9) is unique in each of the cases*

(a) $f_0 = 0, b_0 = 0, D = \emptyset$;
(b) $g_1 = 0, \mathbb{R}^n \setminus \bar{\Omega}$ *is connected,* $D = \emptyset$;
(c) $f_0 = 0, b_0 = 0,\ supp f_1 \subset \Omega \setminus (\bar{\Omega}_0 \cup \bar{D}_0)$ *and is not empty.*

This result will be proven below by using analyticity of $u(x, k)$ with respect to k and an auxiliary hyperbolic initial boundary value problem obtained after the Fourier-Laplace transform of the scattering problem with respect to k. The source terms f_0, f_1 in the scattering problem will be the initial values for this hyperbolic problem and $\mathcal{G}$ will be the boundary value data on $\partial D_0 \times (0, +\infty)$. Using the John–Tataru's uniqueness of the continuation theorem one shows that the solution of the hyperbolic equation and hence the initial and boundary data are uniquely determined.

Solvability of the direct scattering problem and analyticity of its solution with respect to the wave number $k = k_1 + \mathbf{i}k_2$ in an open subset of the complex plane containing $\{k : 0 < k_1, 0 \leq k_2\}$ is shown in the cases (a), (b) in [15] by the Lax–Phillips method [18]. Similarly the solvability and analyticity can be demonstrated in the case (c). Theorem 4.1 in the cases (a), (b) is derived in [15]. Now we prove Theorem 4.1 in the case (c).

Proof Let $u(x, k; j), j = 1, 2$ solve the scattering problem (4.5), (4.6), (4.7) with $f_1 = f_1(; j), g_1 = g(; j), D = D(j)$ and have the same Cauchy data (4.9).

Let $U(x, t; j)$ be a solution to the hyperbolic mixed initial boundary value problem

$$-\partial_t^2 U(; j) + AU(; j) = 0 \quad \text{in } (\mathbb{R}^n \setminus (\bar{D}_0 \cup \bar{D}(j))) \times (0, \infty), \tag{4.10}$$

$$U(; j) = 0, \quad \partial_t U(; j) = \sqrt{2\pi} f_1(; j) \quad \text{on } (\mathbb{R}^n \setminus (\bar{D}_0 \cup \bar{D}(j))) \times \{0\}, \tag{4.11}$$

$$a\nabla U(; j) \cdot \nu = \mathcal{G}(; j) \quad \text{on } \partial D_0 \times (0, \infty), \ a\nabla U(; j) \cdot \nu = 0 \quad \text{on } \partial D(j) \times (0, \infty). \tag{4.12}$$

As known, there is a unique solution $U(; j) \in L^\infty((0, T); H^1(\mathbb{R}^n \setminus (\bar{D}_0 \cup \bar{D}(j))))$ of this problem for any $T > 0$ and moreover

$$\|\partial_t U(, t; j)\|_{(0)}(\mathbb{R}^n \setminus (\bar{D}_0 \cup \bar{D}(j))) + \|U(, t; j)\|_{(1)}(\mathbb{R}^n \setminus (\bar{D}_0 \cup \bar{D}(j))) \\ \leq C(U(; j))e^{\gamma_0 t} \tag{4.13}$$

for some positive $\gamma_0 = \gamma_0(U(; j))$. Then the following Fourier-Laplace transform is well defined

$$u^*(x, k; j) = \frac{1}{\sqrt{2\pi}} \int_0^\infty U(x, t; j) e^{\mathbf{i}kt} dt, \quad k = k_1 + \mathbf{i}\gamma, \quad \gamma_0 < \gamma. \tag{4.14}$$

Approximating $f_1(x; j), g(x; j)$ by smooth functions and integrating by parts we yield

$$0 = \int_0^{\infty} (-\partial_t^2 U(x, t; j) + AU(x, t; j))e^{\mathbf{i}kt} dt$$
$$= \partial_t U(x, 0; j) + k^2 \int_0^{\infty} U(x, t; j)e^{\mathbf{i}kt} dt + A \int_0^{\infty} U(x, t; j)e^{\mathbf{i}kt} dt.$$

Hence $u^*(x, k; j)$ solves (4.5)–(4.6) with $k = k_1 + \mathbf{i}\gamma, \gamma_0 < \gamma$. In addition, $u^*(x, k; j)$ decays exponentially for large $|x|$, in particular,

$$\left| \int_{\partial B(R)} \partial_\nu u^*(x, k; j)\bar{u}^*(x, k; j) d\Gamma(x) \right| \leq C(U(; j))e^{-2\varepsilon\theta R}. \tag{4.15}$$

Indeed, due to the finite speed of the propagation in the hyperbolic problems $U(x, t; j) = 0$ if $t < \theta|x| - R_0$ for some $R_0, \theta > 0$. Hence from (4.14)

$$\int_{B(R+4)\setminus B(R)} |u^*(x, k; j)|^2 dx$$
$$\leq \int_{B(R+4)\setminus B(R)} \left| \int_{\theta R - R_0}^{+\infty} U(x, t; j)e^{\mathbf{i}k_1 t - (\gamma-\varepsilon)t} e^{-\varepsilon t} dt \right|^2 dx$$
$$\leq \int_{B(R+4)\setminus B(R)} \left(\int_{\theta R - R_0}^{+\infty} |U(x, t; j)|^2 e^{-2(\gamma-\varepsilon)t} dt \int_{\theta R - R_0}^{+\infty} e^{-2\varepsilon t} dt \right) dx$$
$$\leq C(U(; j), \varepsilon)e^{-2\varepsilon\theta R} \int_{\theta R - R_0}^{+\infty} \|U(, t; j)\|_{(0)}^2 (\mathbb{R}^n \setminus (\bar{D}(j) \cup \bar{D}_0))e^{-2(\gamma-\varepsilon)t} dt$$
$$\leq C(U(; j), \varepsilon)e^{-2\varepsilon\theta R} \int_{\theta R - R_0}^{+\infty} e^{-2(\gamma-\gamma_0-\varepsilon)t} dt \leq C(U(; j), \gamma)e^{-2\varepsilon\theta R} \tag{4.16}$$

if we use (4.13) and choose $\varepsilon = \frac{\gamma-\gamma_0}{2}$. The same bound holds for $\partial_m u^*(x, k; j)$. Let us choose a cut off function $\chi \in C^2$, $|\chi|_2 \leq C$, $\chi(x) = 1$ when $|x| < R + 1$, $\chi(x) = 0$ when $R + 3 < |x|$. From the Green's formula,

$$\int_{\partial B(R)} \partial_\nu u^*(x, k; j)\bar{u}^*(x, k; j) dx$$
$$+ \int_{B(R+4)\setminus B(R)} (\nabla(\chi u^*(x, k; j)) \cdot \nabla\bar{u}^*(x, k; j) - k^2 \chi u^*(x, k; j)\bar{u}^*(x, k; j)) dx$$
$$= -\int_{B(R+4)\setminus B(R)} (\Delta(\chi u^*(x, k; j)) + k^2 \chi u^*(x, k; j))\bar{u}^*(x, k; j) dx$$
$$= -\int_{B(R+4)\setminus B(R)} (2\nabla(\chi) \cdot \nabla u^*(x, k; j)\bar{u}^*(x, k; j) + \Delta\chi u^*(x, k; j)\bar{u}^*(x, k; j)) dx$$

because $\Delta u^*(, k; j) + k^2 u^*(, k; j) = 0$ in $B(R + 4) \setminus B(R)$. From these equalities and from the above exponential decay (4.16) of $u^*(x, k; j)$, $\nabla u^*(x, k; j)$ for large $|x|$ it follows the bound (4.15).

As mentioned before the proof of Theorem 4.1, it follows from [15], ((15), (16)), that the function $u(x, k; j)$ has a complex analytic extension from $(0, \infty)$ into a neighbourhood S of the quarter plane $\{0 < \Re k, 0 \leq \Im k\}$. This extension satisfies (4.5)–(4.6) and exponentially decays as $|x| \to \infty$ when $0 < \Im k$ due to the known exponential decay of the Hankel function $H_0^{(1)}((k_1 + \mathbf{i}\gamma)r)$ for large r. By uniqueness in the direct scattering problem

$$u(x, k) = u^*(x, k) \text{ when } k = k_1 + \mathbf{i}\gamma, 0 < k_1, \gamma_0 < \gamma. \tag{4.17}$$

We first show uniqueness of f_1, g. Letting $U(x, -t; j) = -U(x, t; j)$, $0 < t$ (i.e. introducing the odd reflection with respect to t) we conclude that the extended $U(x, t; j)$ satisfies the hyperbolic equation (4.10) in $(\mathbb{R}^n \setminus (\bar{D}_0 \cup \bar{D}(j)) \times (-T, T)$. Using (4.14) and (4.9) we conclude that $u^*(x, k; 1) = u^*(x, k; 2)$, $\partial_\nu u^*(x, k; 1) = \partial_\nu u^*(x, k; 2)$, $x \in \Gamma_0$, $K_0 < k < K$ and hence by analyticity for all $k = k_1 + \mathbf{i}k_2$, $\gamma_0 < k_2$. By uniqueness for the Fourier-Laplace transform

$$U(; 1) = U(; 2), \partial_\nu U(; 1) = \partial_\nu U(; 2)$$

on $\Gamma_0 \times (-T, T)$ for any positive T. By the John–Tataru's Theorem [18, 27] we conclude that $U = 0$ in $(\mathbb{R}^n \setminus (\bar{\Omega}_0 \cup \bar{D}_0)) \times (-T, T)$ for any $T > 0$. Hence $f_1(; 1) = f_1(; 2)$ on $\Omega \setminus (\bar{\Omega}_0 \cup D_0)$ and $G(; 1) = G(; 2)$ on $\partial D_0 \times (0, +\infty)$. Again by the uniqueness for the Fourier-Laplace transform $g(; 1) = g(; 2)$ on ∂D_0.

Now we show uniqueness of D. Let $D(1)$, $D(2)$ be two different domains producing the same Cauchy data for the solutions $u(x, k; 1)$, $u(x, k; 2)$ to the scattering problems (4.5), (4.6), (4.7) with $D = D(1)$ and $D = D(2)$. Let $\Omega(0)$ be the connected component of $\Omega_0 \setminus (\bar{D}(1) \cup \bar{D}(2))$ with $\partial\Omega_0 \subset \bar{\Omega}(0)$. Let $\Omega_c(0)$ be $\Omega_0 \setminus \bar{\Omega}(0)$. We can assume that $D(0) = \Omega_c(0) \setminus \bar{D}(1)$ is not empty. As showed above, $u(, k; 1) = u(, k; 2)$ on $\Omega \setminus \Omega_0$, hence by the uniqueness in the Cauchy problem for the second order elliptic equation $u(x, k; 1) = u(x, k; 2)$ when $x \in \Omega(0)$, $K_0 < k < K$. Due to uniqueness of the analytic continuation with respect to k, $u(, k; 1) = u(, k; 2)$ in $\Omega(0)$ when $0 < k_1 < \infty$, $0 < k_2$. As follows from the standard elliptic theory, $u(, k; 1) \in H^2(\Omega_0 \setminus \bar{D}(1))$. Since $\partial D(0)$ consists only of points of $\partial D(1)$ or of $\partial D(2) \subset \bar{\Omega}(0)$ where $a\nabla u(, k; 1) \cdot \nu = a\nabla u(, k; 2) \cdot \nu = 0$, we have $a\nabla u(, k; 1) \cdot \nu = 0$ on $\partial D(0)$ when $0 < k_1 < \infty$, $0 < k_2$. Multiplying the Eq. (4.5) for $u = u(x, k; 1)$ by $\bar{u}(x, k; 1)$ and integrating by parts we yield

$$\int_{D(0)} \Big(- \sum_{j,m=1,\ldots,n} a_{jm} \partial_j u(x, k; 1) \partial_m \bar{u}(x, k; 1)$$

$$+ \mathbf{i} \sum_{j=1,\ldots,n} b_j \partial_j u(x, k; 1) \bar{u}(x, k; 1) + (c + k^2) u(x, k; 1) \bar{u}(x, k; 1) \Big) dx = 0.$$

Using the ellipticity and the inequality $|\partial_j u||u| \leq \delta |\nabla u|^2 + \delta^{-1} |u|^2$ we derive from the integral equality with $k = 1 + \mathbf{i}\gamma$ that

$$\int_{D(0)} (\epsilon_0|\nabla u(x,k;1)|^2 + (\gamma^2 - C)|u(x,k;1)|^2)dx$$
$$\leq C\int_{D(0)} (\delta|\nabla u(x,k;1)|^2 + \delta^{-1}|u(x,k;1)|^2)dx.$$

Then choosing $\delta < C^{-1}$ and $C(1+\delta^{-1}) < \gamma^2$ we conclude that $u(,k;1) = 0$ on $D(0)$. Using again analyticity with respect to k (due to (4.14)) we yield $u(,k;1) = 0$ on $D(0)$ when $k = k_1 + \mathbf{i}\gamma$, $k_1 \in \mathbb{R}$. By the uniqueness in the Fourier-Laplace transform $U(;1) = 0$ on $D(0) \times (-T,T)$ for any positive T. Using again the John–Tataru theorem we conclude that $U(;1) = 0$ on $(\Omega \setminus (\bar{\Omega}_0 \cup \bar{D}_0)) \times (-T,T)$ for any positive T and hence $f_1(;1) = 0$ on $\Omega \setminus (\bar{\Omega}_0 \cup \bar{D}_0)$ which contradicts our assumptions.

The proof is complete. □

Observe that the conditions $D \subset \Omega_0$, $supp f_1 \cap (\bar{\Omega}_0 \cup \bar{D}_0) = \emptyset$ were used in the proof of Theorem 4.1, but it is not clear whether these conditions are necessary for uniqueness of f_1, g, D. Using the results of [9] instead of [27] one can obtain generalization of Theorem 4.1 onto isotropic Maxwell and elasticity systems.

As suggested by known results (for example, [8], Theorem 2.2 and its application in Sect. 4.2), at fixed K one expects a logarithmic stability in this inverse source problem which is quite discouraging for applications. At present, there are no analytic results showing increasing stability for obstacles or sources for larger K in the general case.

4.3 Increasing Stability for Sources

In this section we further investigate the increasing stability for a standard 3D Helmholtz equation. Let

$$\varepsilon_0^2 = \int_0^K \|u(,\omega)\|^2_{(0)}(\partial\Omega)d\omega,$$
$$\varepsilon_1^2 = \int_0^K \left(\omega^2\|u(,\omega)\|^2_{(0)}(\partial\Omega) + \|u(,\omega)\|^2_{(1)}(\partial\Omega)\right)d\omega,$$

$E_j = -\ln \varepsilon_j$, $j = 0, 1$ and $\|f_0\|_{(1)}(\Omega) + \|f_1\|_{(0)}(\Omega) \leq M_1$, $\|f_0\|_{(2)}(\Omega) + \|f_1\|_{(1)}(\Omega) \leq M_2$, $1 \leq M_1, M_2$. In addition, we assume that $\partial\Omega \in C^2$. In this section we denote by C generic constants depending only on Ω.

Theorem 4.2 *Let $n = 3$, $A = \Delta$, $b_0 = 0$, D, D_0 be void, $1 < K$, and $0 < \alpha < \frac{\pi}{2}$. There exists C such that*

$$\|f_0\|^2_{(0)}(\Omega) + \|f_1\|^2_{(-1)}(\Omega) \leq C\left(\varepsilon_0^2 + \frac{M_1^2}{1+K^{2(1-\theta)}|E_0|^{2\theta}}\right), \tag{4.18}$$

$$\|f_0\|^2_{(1)}(\Omega) + \|f_1\|^2_{(0)}(\Omega) \leq C\left(\varepsilon_1^2 + \frac{M_2^2}{1+K^{2(1-\theta)}|E_0|^{2\theta}}\right), \ \theta = \frac{2\alpha}{\pi+2\alpha}, \tag{4.19}$$

for all $u \in H^2(\Omega)$ *solving* (4.5), (4.7).

The bounds (4.18), (4.19) show better (with larger K) stability in the inverse source problem (4.5), (4.7), (4.9) with $K_0 = 0$, $\Gamma_0 = \partial\Omega$ and minimal data (only u_0).

A first version of this result is given in [7], and a less precise form can be found in [11]. In [7, 11] $\alpha = \frac{\pi}{4}$, so $\theta = \frac{1}{3}$, but instead of $|E_0|^{\frac{2}{3}}$, in [7, 11] there is $|E_0|^{\frac{1}{2}}$. Changing α one can get various rates of dependence on K, ε_0. Observe, that in any event $\theta < \frac{1}{2}$. In the two-dimensional case there are additional difficulties due to the absence of the Huygens' principle and a more complicated behaviour of the fundamental solution to the Helmholtz equation. In this case Theorem 4.2 in a less precise form was obtained in [13].

Now we outline a proof of Theorem 4.2.

The well-known integral representation for (4.5), (4.7) yields

$$u(x,k) = \frac{1}{4\pi}\int_\Omega (f_1(y) + \mathbf{i}kf_0(y))\frac{e^{\mathbf{i}k|x-y|}}{|x-y|}dy. \tag{4.20}$$

Due to (4.20),

$$\int_{-\infty}^{\infty} \|u(x,\omega)\|^2_{(0)}(\partial\Omega)d\omega = I_0(k) + \int_{k<|\omega|} \|u(x,\omega)\|^2_{(0)}(\partial\Omega)d\omega, \tag{4.21}$$

where

$$I_0(k) = \frac{1}{8\pi^2}\int_0^k \int_{\partial\Omega}\left(\int (f_1(y) + \mathbf{i}\omega f_0(y))\frac{e^{\mathbf{i}\omega|x-y|}}{|x-y|}dy\right)$$
$$\left(\int (f_1(y) - \mathbf{i}\omega f_0(y))\frac{e^{-\mathbf{i}\omega|x-y|}}{|x-y|}dy\right)d\Gamma(x)d\omega, \tag{4.22}$$

and we used that $\overline{u(x,\omega)} = u(x,-\omega)$.

Due to the definitions of the norms of the boundary data,

$$\varepsilon_0^2 = I_0(K). \tag{4.23}$$

The truncation level k in (4.21) is important to keep balance between the known data and the unknown information when $k \in [K,\infty)$.

Since the integrands are entire analytic functions of ω, the integrals in (4.22) with respect to ω can be taken over any path joining points 0 and k of the complex plane. Thus $I_0(k)$ is an entire analytic function of $k = k_1 + \mathbf{i}k_2$.

Lemma 4.1 *Let* $suppf_0, suppf_1 \subset \Omega$ *and* $f_1 \in H^1(\Omega),\ f_0 \in H^1(\Omega)$. *Then*

$$|I_0(k)| \leq C\left(|k|\|f_1\|^2_{(0)}(\Omega) + |k|^3\|f_0\|^2_{(0)}(\Omega)\right)e^{2d|k_2|}, \tag{4.24}$$

where $d = \sup|x-y|$ *over* $x, y \in \Omega$.

A straightforward proof in [7] is based on (4.22).

Let $S(\alpha)$ be the sector $\{k : -\alpha < \arg k < \alpha\}, 0 < \alpha < \frac{\pi}{2}$ of the complex plane and $\mu(k)$ be the harmonic measure in $S(\alpha) \setminus [0, K]$.

Noticing that

$$|I_0(k)e^{-2(d+1)\tan\alpha k}| \leq 2\varepsilon_0^2 \text{ on } [0, K],$$

and

$$|I_0(k)e^{-2(d+1)\tan\alpha k}| \leq 2CM_0^2 \text{ on } S(\alpha),$$

we conclude that

$$|I_0(k)| \leq CM_0^2 e^{2(d+1)\tan\alpha k}\varepsilon^{2\mu(k)}, \tag{4.25}$$

when $K < k < +\infty$.

We need the following generalization of the lower bound of the harmonic measure $\mu(k)$ given in [7, Lemma 2.2].

Lemma 4.2 *If* $K < k < 2^{\frac{\alpha}{\pi}}K$, *then*

$$\frac{1}{2} \leq \mu(k). \tag{4.26}$$

If $2^{\frac{\alpha}{\pi}}K < k$, *then*

$$\frac{1}{\pi}\left(\left(\frac{k}{K}\right)^{\frac{\pi}{\alpha}} - 1\right)^{-\frac{1}{2}} \leq \mu(k). \tag{4.27}$$

Proof We will use the conformal mapping $w = \sqrt{k^{\frac{\pi}{\alpha}} - K^{\frac{\pi}{\alpha}}}$ of the domain $S(\alpha) \setminus [0, K]$ onto the right half-plane $\{w : \Re w > 0\}$. The interval $[0, K]$ corresponds to the interval $[-K^{\frac{\pi}{2\alpha}}\mathbf{i}, K^{\frac{\pi}{2\alpha}}\mathbf{i}]$ and the function $\mu(k(w))$ is the harmonic measure of $[-K^{\frac{\pi}{2\alpha}}\mathbf{i}, K^{\frac{\pi}{2\alpha}}\mathbf{i}]$ in $\{\Re w > 0\}$. As known,

$$\mu(k(w)) = \frac{2}{\pi}\tan^{-1}\frac{K^{\frac{\pi}{2\alpha}}}{w},$$

when $0 < w < +\infty$, and hence

$$\mu(k) = \frac{2}{\pi}\tan^{-1}\frac{1}{\sqrt{\left(\frac{k}{K}\right)^{\frac{\pi}{\alpha}} - 1}}, \tag{4.28}$$

when $K < k < +\infty$.

If $K < k < 2^{\frac{\alpha}{\pi}} K$, then $(\frac{k}{K})^{\frac{\alpha}{\pi}} - 1 \leq 1$ and (4.28) yields (4.26), since $\tan^{-1} 1 = \frac{\pi}{4}$. If $2^{\frac{\alpha}{\pi}} K < k$, then $1 < (\frac{k}{K})^{\frac{\alpha}{\pi}} - 1$. Using that

$$\tan^{-1} x = \int_0^x \frac{1}{1+s^2} ds$$

we conclude that $\frac{x}{2} < \tan^{-1} x$, provided $0 < x < 1$. From this inequality and from (4.28), we obtain (4.27). □

We consider the hyperbolic initial value problem

$$\partial_t^2 U - \Delta U = 0 \text{ on } \mathbb{R}^3 \times (0, +\infty),\ U(,0) = -\sqrt{2\pi} f_0,\ \partial_t U(,0) = \sqrt{2\pi} f_1 \text{ on } \mathbb{R}^3. \tag{4.29}$$

We define $U(x,t) = 0$ when $t < 0$. As shown in [7, Sect. 4], the solution of (4.29) coincides with the Fourier transform of U, namely

$$u(x,k) = \frac{1}{\sqrt{2\pi}} \int_{-\infty}^{+\infty} U(x,t) e^{\mathrm{i}kt} dt. \tag{4.30}$$

To proceed, an estimate for remainders in (4.21) is used and summarized in the next result similar to Lemma 4.1 in [7].

Lemma 4.3 *Let u be a solution to the forward problem (4.5), (4.7) with $f_1 \in H^1(\Omega)$, $f_0 \in H^2(\Omega)$, $suppf_0, suppf_1 \subset \Omega$, $A = \Delta$, $b_0 = 0$, void D, D_0, and $1 < K$. Then*

$$\int_{k<|\omega|} \|u(,\omega)\|_{(0)}^2(\partial\Omega) d\omega \leq Ck^{-2} \left(\|f_0\|_{(1)}^2(\Omega) + \|f_1\|_{(0)}^2(\Omega) \right), \tag{4.31}$$

$$\int_{k<|\omega|} \omega^2 \|u(,\omega)\|_{(0)}^2(\partial\Omega) d\omega + \int_{k<|\omega|} \|\nabla u(,\omega)\|_{(0)}^2(\partial\Omega)$$
$$\leq Ck^{-2} \left(\|f_0\|_{(2)}^2(\Omega) + \|f_1\|_{(1)}^2(\Omega) \right). \tag{4.32}$$

Proof We claim that

$$\|U\|_{(1)}^2(\partial\Omega \times (0,d)) \leq C \left(\|f_0\|_{(1)}^2(\Omega) + \|f_1\|_{(0)}^2(\Omega) \right). \tag{4.33}$$

Indeed, the initial value problem (4.29) can be viewed as the (strictly) hyperbolic transmission problem

$$\partial_t^2 U^- - \Delta U^- = 0 \text{ in } \Omega \times (0,+\infty),\ \partial_t^2 U^+ - \Delta U^+ = 0 \text{ in } (\mathbb{R}^3 \setminus \bar{\Omega}) \times (0,+\infty),$$

$$U^- = -\sqrt{2\pi} f_0,\ \partial_t U^- = \sqrt{2\pi} f_1 \text{ in } \Omega \times \{0\},\ U^+ = 0,\ \partial_t U^+ = 0 \text{ in } (\mathbb{R}^3 \setminus \bar{\Omega}) \times \{0\},$$

$$U^+ - U^- = g_0,\ \partial_\nu U^+ - \partial_\nu U^- = g_1 \text{ on } \partial\Omega \times (0,+\infty),$$

with $g_0 = 0$, $g_1 = 0$. Then (4.33) follows from a generalization of Sakamoto results onto transmission problems.

Obviously, the following inequalities hold

$$\int_{k<|\omega|} \|u(,\omega)\|^2_{(0)}(\partial\Omega)d\omega \leq k^{-2}\int_{k<|\omega|} \omega^2\|u(,\omega)\|^2_{(0)}(\partial\Omega)d\omega$$
$$\leq k^{-2}\int_{\mathbb{R}} \omega^2\|u(,\omega)\|^2_{(0)}(\partial\Omega)d\omega$$
$$= k^{-2}\int_{\mathbb{R}} \|\partial_t U(,t)\|^2_{(0)}(\partial\Omega)dt$$

by the Parseval's identity. Since according to the Huygens' principle $U(,t) = 0$ on Ω when $d < t$, the last inequality combined with (4.33) implies the bound (4.31).

Equation (4.32) is similarly derived as in [7, Lemma 4.1] and we skip the detail.

The next result follows almost immediately from the Huygens' principle for the initial value problem and the known bounds for initial boundary value hyperbolic problems.

Lemma 4.4 *Let U be a solution to (4.29) with $f_1 \in L^2(\Omega), f_0 \in H^1(\Omega)$, $suppf_0, suppf_1 \subset \Omega$. Then*

$$\|f_0\|^2_{(0)}(\Omega) + \|f_1\|^2_{(-1)}(\Omega) \leq C\|U\|^2_{(0)}(\partial\Omega \times (0,d)), \tag{4.34}$$

$$\|f_0\|^2_{(1)}(\Omega) + \|f_1\|^2_{(0)}(\Omega) \leq C\|U\|^2_{(1)}(\partial\Omega \times (0,d)). \tag{4.35}$$

Proof Since $suppf_0, suppf_1 \subset \Omega$ from the Huygens' principle it follows that $U = 0$ on $\Omega \times (d, +\infty)$. Now (4.34), (4.35) follow from the generalizations [25, Theorems 2.1, 2.3], of Sakamoto energy estimates for the initial boundary value problem applied to the initial boundary value problem

$$\partial_t^2 U - \Delta U = 0 \text{ on } \mathbb{R}^3 \times (0,d),\ U(,d) = 0,\ \partial_t U(,d) = 0 \text{ on } \Omega.$$

Finally, we are ready to prove the increasing stability estimate of Theorem 4.2.

Proof We can assume that $\varepsilon_0 < 1$, otherwise the bound (4.18) is straightforward.

In (4.21) we let

$$k = \delta K^{1-\theta}E_0^{\theta},\ \theta = \frac{2\alpha}{\pi + 2\alpha},\ \delta = (2\pi(d+1)\tan\alpha)^{-\theta}, \quad \text{when } \delta^{-\frac{1}{\theta}}K < E_0, \tag{4.36}$$

and

$$k = K, \quad \text{when } E_0 \leq \delta^{-\frac{1}{\theta}}K. \tag{4.37}$$

In the case (4.36) $K < k$, then from (4.25), (4.27), and (4.36) we obtain

$$|I_0(k)| \leq CM_0^2 e^{2(d+1)\tan\alpha k - \frac{2}{\pi}\left(\left(\frac{k}{K}\right)^{\frac{\pi}{\alpha}}-1\right)^{-\frac{1}{2}}E_0} \leq CM_0^2 e^{2(d+1)\tan\alpha k - \frac{2}{\pi}\left(\frac{K}{k}\right)^{\frac{\pi}{2\alpha}}E_0}$$
$$= CM_0^2 e^{2(d+1)\tan\alpha\delta K^{1-\theta}E_0^{\theta} - \frac{2}{\pi}\delta\delta^{-\frac{1}{\theta}}K^{1-\theta}E_0^{\theta}} = CM_0^2 e^{-2(d+1)\tan\alpha\delta K^{1-\theta}E_0^{\theta}}.$$

Using the assumption at the beginning of the proof and the elementary inequality $e^{-y} \leq \frac{2}{y^2}$ when $0 < y$, we conclude that

$$|I_0(k)| \leq CM_0^2 \frac{1}{K^{2(1-\theta)} E_0^{2\theta}}. \tag{4.38}$$

On the other hand, in the case (4.37), $k = K$ and we derive that

$$|I_0(k)| \leq 2\varepsilon_0^2. \tag{4.39}$$

Hence, using (4.21), (4.38), and (4.39) we yield

$$\int_{\partial\Omega} \int_{-\infty}^{+\infty} |u(x,\omega)|^2 d\omega d\Gamma(x) = I_0(k) + \int_{\partial\Omega} \int_{k<|\omega|} |u((x,\omega)|^2 d\omega d\Gamma(x)$$
$$\leq 2\varepsilon_0^2 + CM_0^2 \frac{1}{K^{2(1-\theta)} E_0^{2\theta}} + C \frac{\|f_0\|_{(1)}^2 + \|f_1\|_{(0)}^2}{1 + K^{2(1-\theta)} E_0^{2\theta}}, \tag{4.40}$$

where we used also (4.31).

By Lemma 4.4, we finally derive

$$\|f_0\|_{(0)}^2(\Omega) + \|f_1\|_{(-1)}^2(\Omega) \leq C\|U\|_{(0)}^2(\partial\Omega \times (0,d)) \leq C\|U\|_{(0)}^2(\partial\Omega \times \mathbb{R})$$
$$\leq C \int_{\partial\Omega} \int_{-\infty}^{+\infty} |u(x,\omega)|^2 d\omega d\Gamma(x) \leq C \left(\varepsilon_0^2 + \frac{\|f_0\|_{(1)}^2 + \|f_1\|_{(0)}^2}{1 + K^{2(1-\theta)} E_0^{2\theta}} \right),$$

due to the Parseval's identity and (4.40).

The proof of (4.18) is complete.

Equation. (4.19) can be proved similarly when we use (4.35) instead of (4.34). □

A generalization of the increasing stability onto the important case of attenuation, shown in [21], is presented below.

Theorem 4.3 *Let $n = 3$ and $\|f_0\|_{(3)}(\Omega) + \|f_1\|_{(3)}(\Omega) \leq M_3$, $1 \leq M_3$. There exists a constant $C = C(\Omega)$ such that*

$$\|f_0\|_{(1)}^2(\Omega) + \|f_1\|_{(0)}^2(\Omega) \leq Ce^{Cb_0^2} \left(\varepsilon^2 + \frac{M_3^2}{1 + K^{\frac{4}{3}} |E_1|^{\frac{1}{2}}} \right) \tag{4.41}$$

for all $u \in H^2(\Omega)$ solving (4.5), (4.7) with $A = \Delta$, a non-negative constant b_0, void D, D_0, $1 < K$, and

$$\varepsilon^2 = \int_0^K (\omega^2 \|u(,\omega)\|_{(0)}^2(\partial\Omega) + \|\nabla u(,\omega)\|_{(0)}^2(\partial\Omega)) d\omega.$$

The attenuation in the plane case was considered in [12].

4.4 Numerical Tests

In this section, we present some numerical examples verifying the theoretical results in Sects. 4.2–4.3. The basic 2D and 3D setting of these examples can be found in [7, 15] and we will briefly recall them below. In the first setting, we focus on the 2D problem and consider the situation that the closure of the observation site Γ_0 does not have intersection with the compactly supported domain. Whereas in the second setting, we choose the 3D problem and compare the performance between the full Dirichlet/Neumann data and Dirichlet data alone.

Recalling (4.1), in both setting, we set the radiating field $u(x, k)$ satisfying the Helmholtz equation and the Sommerfeld radiation condition

$$\Delta u(x; k) + k^2 u(x; k) = -f^\dagger(x), \quad x \in \mathbb{R}^n,$$
$$\lim_{r_x \to \infty} r_x^{\frac{n-1}{2}} \left(\frac{\partial u(x,k)}{\partial r_x} - \mathbf{i} k u(x,k) \right) = 0$$

with the wave number $k > 0$. We assume that the source function $f^\dagger(x) \in L^2(\mathbb{R}^n)$ and is compactly supported in a bounded open domain Ω. Denoting the observation site as Γ_0 we define the following operators

$$(\mathcal{L}_{1,k} f^\dagger)(x) = u(x; k) = g_1(x; k) = \int_\Omega f^\dagger(y)\Phi(x, y; k)\mathrm{d}y, \quad x \in \Gamma_0, \quad (4.42a)$$
$$(\mathcal{L}_{2,k} f^\dagger)(x) = \frac{\partial u(x; k)}{\partial n} = g_2(x; k) = \int_\Omega f^\dagger(y)\frac{\partial \Phi(x, y; k)}{\partial n}\mathrm{d}y, \quad x \in \Gamma_0 \quad (4.42b)$$

where the fundamental solution $\Phi(x, y; k)$ is

$$\Phi(x, y; k) = \begin{cases} H_0^{(1)}(k|x-y|), & n = 2, \\ \frac{1}{4\pi}\frac{e^{ik|x-y|}}{|x-y|}, & n = 3, \end{cases}$$

with the Hankel function $H_0^{(1)}(k|x-y|)$.

To stably reconstruct the source function by multi frequency boundary measurements, i.e. $g_1(x, k)$, $g_2(x, k)$ in (4.42) for $k \in [1, K]$, we need to recursively refine the reconstruction within the multi frequency regimes. One of the most efficient algorithms is the Landweber-Kaczmarz iteration developed in [4] where the convergence analysis can be found in [5]. Such an algorithm starts with a Kaczmarz iteration in low wave numbers and provides rough approximation of the unknown source function by an inner Landweber iteration. Notice that the residual between the exact solution and low frequency approximants remains in the next wave number, the algorithm switches to a larger wave number in the Kaczmarz iteration and refine the low frequency approximants with further details. The whole algorithm continues recursively up to a sufficiently large wave number and the dominant information of the unknown source is then recovered. To have extended discussion on this algorithm and some other numerical schemes we refer to [4, 5] and references therein.

4.4.1 2D Numerical Examples

In the first setting, we consider 2D numerical examples which aim to recover a source function out of the convex hull of the observation site Γ_0. The annular domain, which contains a compactly supported source function, is $\Omega \setminus \Omega_0 = B_{1.5}(0,0) \setminus B_{0.5}(0,0)$ and the observation site is the circle $\Gamma_0 = \partial B_{0.45}(0,0)$, c.f. Fig. 4.1. Hereafter we denote $B_r(x, y)$ and $\partial B_r(x, y)$ as a disk and its boundary centred at (x, y) and with a radius of r. The domain $\Omega \setminus \Omega_0$ is discretized both radially and spherically. More precisely, we discretize the radius equally in $(0.5, 1.5)$ with 30 points and the angle $[0, 2\pi)$ equally with 100 points. Then the mesh-points for the domain $\Omega \setminus \Omega_0$ is generated by the products of radical discretization and angle discretization. The observation site Γ_0 is discretized equally with 50 points. We use the collocation method to generate the discrete forward operators $\mathcal{L}_{1,k}$ and $\mathcal{L}_{2,k}$.

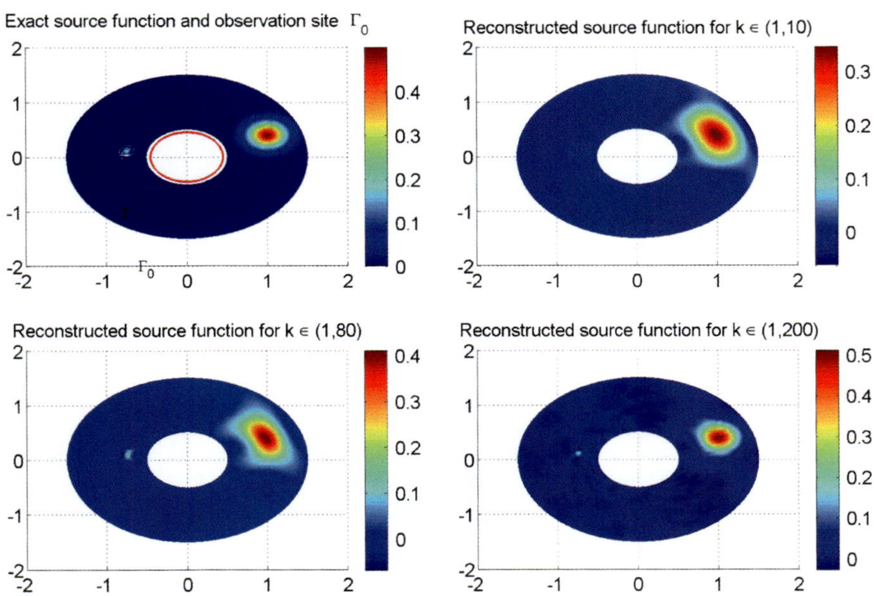

Fig. 4.1 Exact source function (4.43). Upper left: annular domain of the inverse source problem and the observation site Γ_0 (red solid line). Upper right: reconstructed source function by multifrequency data $k \in (1, 10)$. Bottom left: reconstructed source function by multifrequency data $k \in (1, 80)$. Bottom right: reconstructed source function by multifrequency data $k \in (1, 200)$

As shown in the upper left panel of Fig. 4.1 we consider a mixed-point source in the following form

$$f(x, y) = 0.5e^{-20((y-0.4)^2+(x-1)^2)} + 0.3e^{-800((y-0.1)^2+(x+0.75)^2)}. \tag{4.43}$$

One may realize that the above source function contains two point sources which have different scales. The classic uncertainty principle asserts that one could not identify small scale sources if the wave number is too small. Such a result is also confirmed in our numerical examples. More precisely, if the wave number interval is small, i.e. $k \in (1, K)$ with $K = 10$, the reconstruction algorithm does not detect the small scale point source in (4.43), see the upper right panel of Fig. 4.1. Nevertheless, if we increase the wave number interval to $k \in (1, K)$ with $K = 80, 200$, we visualize the small scale source shown in the bottom two panels of Fig. 4.1.

To further illustrate the efficiency of the proposed reconstruction algorithm, we choose a much more sophisticated source function (4.44) in [15] where point sources with different scales are included

$$\begin{aligned} f(x, y) = & 0.5e^{-50((y-0.4)^2+(x-1)^2)} + 0.3e^{-150((y-0.1)^2+(x-0.75)^2)} + 0.2e^{-80((y-0.2).^2+(x-1.1)^2)} \\ & + 0.6e^{-120((y-0.6)^2+(x-0.6)^2)} + 0.5e^{-50((y+0.8)^2+(x-0.5)^2)} + 0.3e^{-150((y+0.4)^2+(x-0.7)^2)} \\ & + 0.2e^{-80((y+0.2)^2+(x-0.8)^2)} + 0.6e^{-120((y+0.6)^2+(x-0.6)^2)} + 0.5e^{-50((y-0.7)^2+(x+0.7)^2)} \\ & + 0.3e^{-150((y-0.4)^2+(x+0.5)^2)} + 0.2e^{-80((y-0.2)^2+(x+0.8)^2)} + 0.6e^{-120((y-0.6)^2+(x+0.6)^2)} \\ & + 0.5e^{-50((y+0.4)^2+(x+1)^2)} + 0.3e^{-150((y+0.4)^2+(x+0.5)^2)} + 0.2e^{-80((y+0.8)^2+(x+0.2)^2)} \\ & + 0.6e^{-120((y+0.6)^2+(x+0.6)^2)}. \end{aligned} \tag{4.44}$$

We display the exact source function and its reconstruction with different wave number intervals in Fig. 4.2. As one can observe from the upper right panel there, by using the measurement $u(x, k)$, $\partial_n u(x, k)$ on Γ_0 for $k \in (1, 10)$, the reconstructed source function only provides a rough average information where one can not separate small scale point sources. When we increase the interval to $k \in (1, 100)$, the reconstructed source function is still blurred as shown in the bottom left panel of Fig. 4.2. Such phenomena can be essentially improved if we enlarge the wave number interval to $k \in (1, 400)$. As confirmed in the bottom right panel of Fig. 4.2, we obtain very fine approximation of the reconstructed source function with high resolution.

The above observation is also confirmed in a piecewise constant source function chosen below, which has been considered in [15] as well,

$$f(r, \theta) = \begin{cases} 1 \text{ if } 0.8 \le r \le 0.9 \text{ and } \theta \in [-\frac{\pi}{4}, \frac{\pi}{4}]; \\ 1 \text{ if } 0.9 \le r \le 1 \text{ and } \theta \in [-\frac{\pi}{4}, -\arctan\frac{1}{2}] \cup [\arctan\frac{1}{2}, \frac{\pi}{4}]; \\ 1 \text{ if } 1 \le r \le 1.1 \text{ and } \theta \in [-\frac{\pi}{4}, \frac{\pi}{4}]; \\ 0 \text{ elsewhere.} \end{cases} \tag{4.45}$$

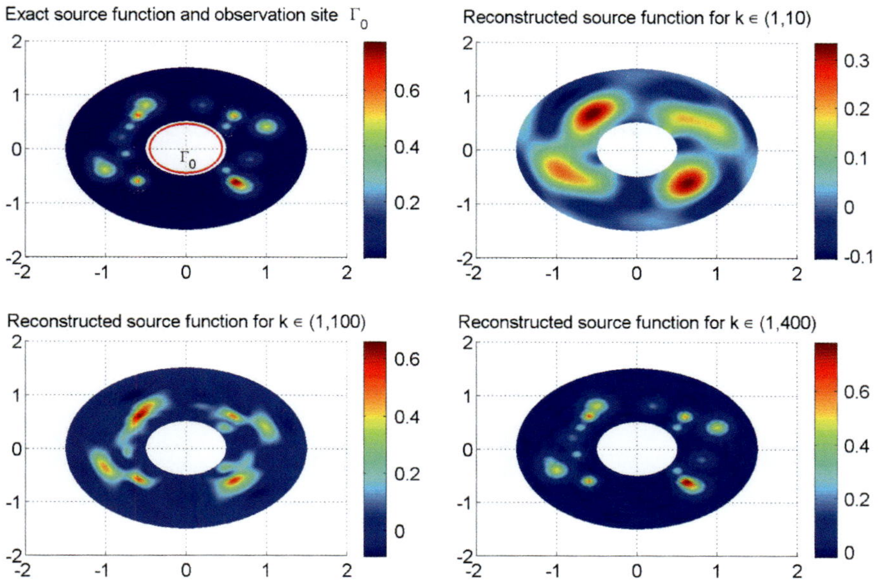

Fig. 4.2 Exact source function (4.44). Upper left: annular domain of the inverse source problem and the observation site Γ_0 (red solid line). Upper right: reconstructed source function by multifrequency data $k \in (1, 10)$. Bottom left: reconstructed source function by multifrequency data $k \in (1, 100)$. Bottom right: reconstructed source function by multifrequency data $k \in (1, 400)$

Here we use polar coordinates with $x = r\cos(\theta)$ and $y = r\sin(\theta)$ for simplicity's sake. As shown in Fig. 4.3, we can observe a hole inside the source function. Nevertheless, such property is neglected if we choose a small wave number interval, i.e. $k \in (1, 10)$. One can finely reconstruct the source function when the wave number interval is chosen sufficiently large. In current example, we have chosen $k \in (1, 400)$.

We also mention that the relative error slope for Figs. 4.2, 4.3 can be found in [15] and we skip these details.

4.4.2 3D Numerical Examples

To further incorporate the results in Sect. 4.3, we also consider a 3D cubic domain $\Omega = [-0.35, 0.35]^3$ with an empty set Ω_0 and the observation site $\Gamma_0 = \partial\Omega$. In the numerical calculation, we choose the cubic domain $[-0.3, 0.3]^3$ be the compact support domain of the source function to avoid the difficulties arising in the discretization. In the latter cubic domain we discretize along each axis with 11 points whereas the observation site Γ is discretized with 5 points along each axis.

As shown in Theorem 4.2 the source function can be determined only by the Dirichlet data on the whole boundary $\partial\Omega$. We then compare the performance of the

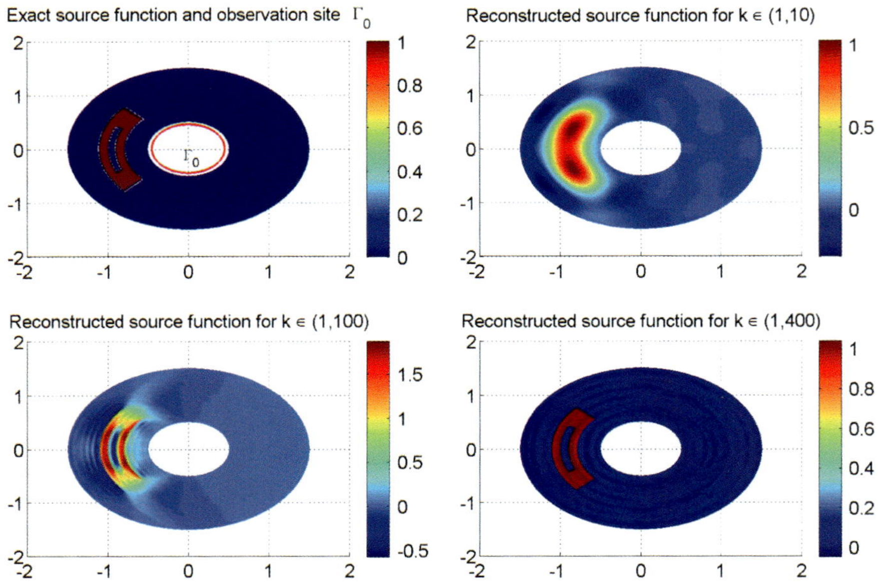

Fig. 4.3 Exact source function (4.45). Upper left: annular domain of the inverse source problem and the observation site Γ_0 (red solid line). Upper right: reconstructed source function by multifrequency data $k \in (1, 10)$. Bottom left: reconstructed source function by multifrequency data $k \in (1, 100)$. Bottom right: reconstructed source function by multifrequency data $k \in (1, 400)$

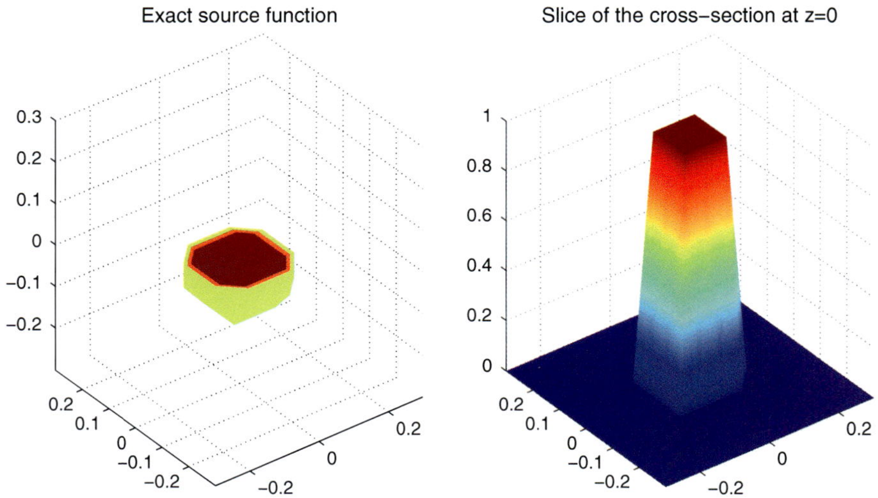

Fig. 4.4 3D domain. Left: the exact solution in lower half space. Right: slice of the cross-section at $z = 0$

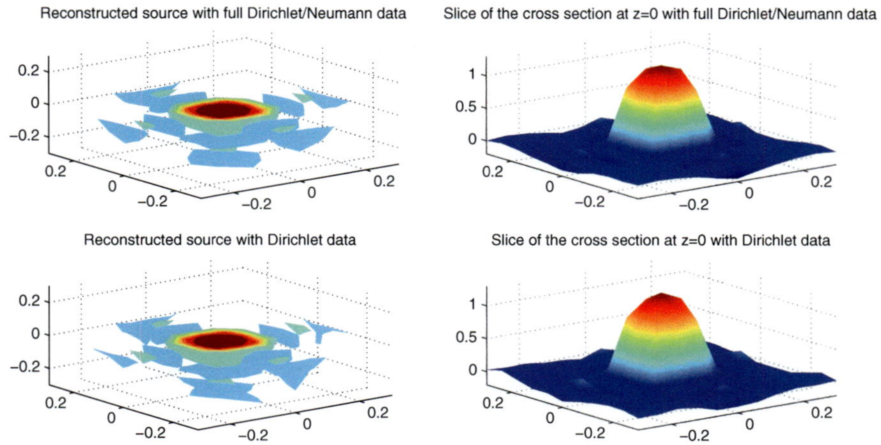

Fig. 4.5 3D domain. Upper left: the reconstructed solution in lower half space with full Dirichlet/Neumann data. Upper right: slice of the cross-section at $z = 0$ with full Dirichlet/Neumann data. Bottom left: the reconstructed solution in lower half space only with Dirichlet data. Bottom right: slice of the cross-section at $z = 0$ only with Dirichlet data

reconstruction algorithm by full Dirichlet/Neumann data and Dirichlet data alone. The source function is chosen below as a piecewise constant function in Fig. 4.4, such that

$$f(x, y, z) = \begin{cases} 1 \text{ if } |x| \le 0.1 \text{ and } |y| \le 0.1 \text{ and } |z| \le 0.1; \\ 0 \text{ elsewhere.} \end{cases}$$

To reconstruct the source function uniquely and stably, we have considered the full Dirichlet and Neumann data in the above Sect. 4.4.1. In current subsection, we focus on the performance of the reconstruction algorithm if we only use the Dirichlet data referring to Theorem 4.2. These results are collected in Fig. 4.5 where panels in upper row are the reconstructed solution with full Dirichlet/Neumann data and panels in bottom row are the reconstructed solution only with Dirichlet data. In both rows, the reconstructed solutions have similar properties and one can not distinguish the difference clearly. In Fig. 4.6 we present the relative error slopes between the exact solution and the reconstructed one with Dirichlet/Neumann data and only with Dirichlet data respectively. As one can observe, in the low wave number interval, for instance $k \in (1, 20)$, the difference between both settings is indistinguishable. If we further increase the wave number, the difference becomes large but both relative errors are of the same tendency. When we increase the wave number interval to $k \in (1, 200)$, the relative error with Dirichlet/Neumann data is 0.3134 and the one only with Dirichlet data is 0.3357. Numerical evidence of the increasing stability for Dirichlet data can be observed in this numerical example. We mention that similar observation is also obtained when we consider mixed-point sources and details are skipped here.

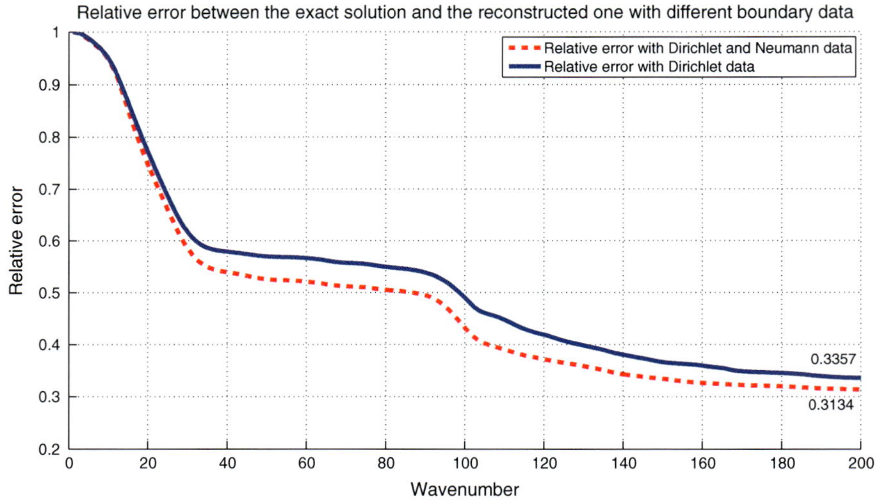

Fig. 4.6 3D domain. Relative error between the exact solution and the reconstructed one with different data. Blue solid line: relative error only with Dirichlet data; Red dashed line: relative error with Dirichlet and Neumann data

4.5 Conclusion

Over the last decade, there has been a rapidly increasing interest in inverse source problems for acoustic Helmholtz and elasticity equations, for instance in [3–5, 7, 10–13, 15, 21, 26]. To overcome the lack of uniqueness induced by non-radiating sources, one seeks to reconstruct the compactly supported source function by boundary measurements of multiple wave numbers in a non-empty interval which brings forth the multi frequency regimes. Then the length of the wave number interval becomes crucial to the (increasing) stability of source recovery. In our work [7], we have considered a novel approach to implement the Fourier transform in time to reduce our inverse source problem to the time domain, which is equivalent to the identification of the initial data in the hyperbolic initial value problem by lateral Cauchy data (observability in control theory) and, as a consequence, obtained most complete increasing (with growing K) stability results. Such an increasing stability, i.e. Theorem 4.2, contains a Lipschitz part of the (weighted) multi frequency boundary measurement and a tail part depending on the highest wave number and a logarithmic factor of the multi frequency boundary measurement. One then can choose the non-empty wave number interval large enough such that the Lipschitz part dominates the tail part and the increasing stability is well preserved.

On the other hand, the high frequency regime covers more than the inverse source problems. In a series of work in [14, 17, 20], the authors have shown both theoretically and numerically that the stability in the Cauchy problem for the Helmholtz equation is enhanced if the wave number increases. Such a setting is different from that of inverse source problems, where the boundary Cauchy data of only one wave number is needed. The high frequency regime means that the chosen wave number shall be large enough. Another important high frequency regime is the inverse potential problem where the Dirichlet-to-Neumann map of the large wave number is considered as the boundary measurements, c.f. [22, 23]. To verify the increasing stability numerically, we have developed a novel reconstruction algorithm based on the linearised inverse potential problem as reported in [16].

Acknowledgements This research is supported in part by the Emylou Keith and Betty Dutcher Distinguished Professorship and the NSF grant DMS 15-14886. S. Lu is supported by National Key Research and Development Program of China (No. 2017YFC1404103), NSFC (No. 11925104), and Program of Shanghai Academic/Technology Research Leader (19XD1420500).

References

1. H. Ammari, G. Bao, J. Fleming, Inverse source problem for Maxwell's equation in magnetoencephalography. SIAM J. Appl. Math. **62**, 1369–1382 (2002)
2. C.A. Balanis, *Antenna Theory-Analysis and Design* (Wiley, Hoboken, NJ, 2005)
3. G. Bao, J. Lin, F. Triki, A multi-frequency inverse source problem. J. Differ. Equ. **249**, 3443–3465 (2010)
4. G. Bao, J. Lin, F. Triki, Numerical solution of the inverse source problem for the Helmholtz equation with multiple frequency data. Contemp. Math. **548**, 45–60 (2011)
5. G. Bao, S. Lu, W. Rundell, B. Xu, A recursive algorithm for multi-frequency acoustic inverse source problems. SIAM J. Numer. Anal. **53**, 1608–1628 (2015)
6. N. Bleistein, J.K. Cohen, Non uniqueness in the inverse source problem in acoustics and electromagnetics. J. Math. Phys. **18**(2), 194–201 (1977)
7. J. Cheng, V. Isakov, S. Lu, Increasing stability in the inverse source problem with many frequencies. J. Differ. Equ. **260**, 4786–4804 (2016)
8. M. Di Cristo, L. Rondi, Examples of exponential instability for inverse inclusion and scattering problems. Inverse Probl. **19**, 685–701 (2003)
9. M. Eller, V. Isakov, G. Nakamura, D. Tataru, Uniqueness and stability in the cauchy problem for Maxwell' and elasticity systems, in *Nonlinear Partial Differential Equations and Their Applications*, eds. by D. Cioranescu, J.-L. Lions (Elsevier Science, North-Holland, 2002), pp. 329–351
10. M. Eller, N.P. Valdivia, Acoustic source identification using multiple frequency information. Inverse Probl. **25**(11), 20 (2009)
11. M. Entekhabi, V. Isakov, Increasing stability in acoustic and elastic inverse source problems, arXiv:1808.10528 (submitted for publication)
12. M. Entekhabi, Increasing stability in the two dimensional inverse source scattering problem with attenuation and many frequencies. Inverse Probl. **34**, 115001 (2018)
13. M. Entekhabi, V. Isakov, On increasing stability in the two dimensional inverse source scattering problem with many frequencies. Inverse Probl. **34**, 05505 (2018)
14. V. Isakov, On increasing stability in the Cauchy problem for general elliptic equations, in *New Prospects in Direct, Inverse, and Control Problems for Evolution Equations*, ed. by A. Favini et al. Springer INdAM Series, vol. 10 (Elsevier Science, North-Holland, 2014)
15. V. Isakov, S. Lu, Inverse source problems without (pseudo) convexity assumptions. Inverse Probl. Imaging **12**, 955–970 (2018)
16. V. Isakov, S. Lu, B. Xu, Linearized inverse Schrödinger potential problem at a large wave number. SIAM J. Appl. Math. (to appear). arXiv:1812.05011

17. V. Isakov, Increased stability in the continuation for the Helmholtz equation with variable coefficient. Contemp. Math. **426**, 255–269 (2007)
18. V. Isakov, *Inverse Problems for Partial Differential Equations* (Springer, New York, 2017)
19. V. Isakov, On increasing stability in the continuation for elliptic equations of second order without (pseudo) convexity assumptions. Inverse Probl. Imaging **13**, 983–1006 (2019)
20. V. Isakov, S. Kindermann, Regions of stability in the Cauchy problem for the Helmholtz equation. Methods Appl. Anal. **18**, 1–30 (2011)
21. V. Isakov, S. Lu, Increasing stability in the inverse source problem with attenuation and many frequencies. SIAM J. Appl. Math. **78**, 1–18 (2018)
22. V. Isakov, J.-N. Wang, Increasing stability for determining the potential in the Schrödinger equation with attenuation from the Dirichlet-to Neumann map. Inverse Probl. Imaging **8**, 1139–1150 (2014)
23. V. Isakov, R.-Y. Lai, J.-N. Wang, Increasing stability for conductivity and attenuation coefficients. SIAM J. Math. Anal. **48**, 569–594 (2016)
24. F. John, Continuous dependence on data for solutions of partial differential equations with a prescribed bound. Comm. Pure Appl. Math. **13**, 551–587 (1960)
25. I. Lasiecka, J.-L. Lions, R. Triggiani, Non homogeneous boundary value problems for second order hyperbolic equations. J. Math. Pures Appl. **65**, 149–192 (1986)
26. P. Li, G. Yuan, Increasing stability for the inverse source scattering problem with multi-frequencies. Inverse Probl. Imaging **11**, 745–759 (2017)
27. D. Tataru, Unique continuation for solutions to PDE's: between Hörmander's Theorem and Holmgren's Theorem. Comm. Part. Diff. Equat. **20**, 855–884 (1995)

Chapter 5
Inverse Moving Source Problem for Fractional Diffusion(-Wave) Equations: Determination of Orbits

Guanghui Hu, Yikan Liu and Masahiro Yamamoto

Abstract This paper is concerned with the inverse problem on determining the orbit of a moving source in fractional diffusion(-wave) equations either in a connected bounded domain of $\mathbb{R}^d$ or in the whole space $\mathbb{R}^d$. Based on a newly established fractional Duhamel's principle, we derive a Lipschitz stability estimate in the case of a localized moving source by the observation data at d interior points. The uniqueness for the general non-localized moving source is verified with additional data of more interior observations.

Keywords Inverse moving source problem · Fractional diffusion(-wave) equation · Fractional Duhamel's principle · Lipschitz stability · Uniqueness

G. Hu
Beijing Computational Science Research Center, Building 9, East Zone, ZPark II, No. 10 Xibeiwang East Road, Haidian District, Beijing 100193, China
e-mail: hu@csrc.ac.cn

Y. Liu (✉)
Research Institute for Electronic Science, Hokkaido University, N12W7, Kita-Ward, Sapporo 060-0812, Japan
e-mail: ykliu@es.hokudai.ac.jp

M. Yamamoto
Graduate School of Mathematical Sciences, The University of Tokyo, 3-8-1 Komaba, Meguro-ku, Tokyo 153-8914, Japan
e-mail: myama@ms.u-tokyo.ac.jp

Honorary Member of Academy of Romanian Scientists, Splaiul Independentei Street, No. 54, 050094 Bucharest, Romania

Peoples' Friendship University of Russia (RUDN University), 6 Miklukho-Maklaya Street, Moscow 117198, Russian Federation

J. Cheng et al. (eds.), *Inverse Problems and Related Topics*, Springer Proceedings in Mathematics & Statistics 310,
https://doi.org/10.1007/978-981-15-1592-7_5

5.1 Introduction

Let $T > 0$ and $\Omega \subset \mathbb{R}^d$, $d = 1, 2, 3, \ldots$, be either a connected bounded domain with a smooth boundary $\partial\Omega$ or $\Omega = \mathbb{R}^d$. For $0 < \alpha \le 2$, consider an initial(-boundary) value problem for a time-fractional diffusion(-wave) equation

$$\begin{cases} (\partial_t^\alpha + \mathcal{L})u(\boldsymbol{x}, t) = g(\boldsymbol{x} - \boldsymbol{\gamma}(t)), & \boldsymbol{x} \in \Omega,\ 0 < t < T, \\ \begin{cases} u(\boldsymbol{x}, 0) = 0 & \text{if } 0 < \alpha \le 1, \\ u(\boldsymbol{x}, 0) = \partial_t u(\boldsymbol{x}, 0) = 0 & \text{if } 1 < \alpha \le 2, \end{cases} & \boldsymbol{x} \in \Omega, \\ u(\boldsymbol{x}, t) = 0 \quad \text{if } \Omega \text{ is bounded}, & \boldsymbol{x} \in \partial\Omega,\ 0 < t < T. \end{cases} \tag{5.1}$$

Here, $\mathcal{L}$ is an elliptic operator with respect to the spatial variable $\boldsymbol{x} \in \mathbb{R}^d$, and ∂_t^α denotes the Caputo derivative with respect to the time variable $t \in \mathbb{R}_+ := (0, \infty)$, which will be precisely defined in Sect. 5.2. The function g is an approximation of Dirac's delta function in $\mathbb{R}^d$, and $\boldsymbol{\gamma} : \mathbb{R}_+ \longrightarrow \mathbb{R}^d$ describes the orbit of a moving source in $\mathbb{R}^d$. The governing equation in (5.1) is called a (time-fractional) diffusion equation when $\alpha \in (0, 1]$, whereas is called a (time-fractional) diffusion-wave equation or a (time-fractional) wave equation when $\alpha \in (1, 2]$. Hence, the system (5.1) approximates a moving point source problem for the (time-fractional) diffusion(-wave) equation.

In this paper, we are interested in the inverse moving source problem of recovering an unknown orbit function $\boldsymbol{\gamma}(t)$ from the solution data detected at a finite number of interior receivers. More precisely, we investigate the following problem.

Problem 5.1 (Inverse moving source problem) *Let u be the solution to* (5.1) *and pick N interior points $\boldsymbol{x}^j \in \Omega$ $(j = 1, \ldots, N)$. Provided that the source profile $g(\boldsymbol{x})$ is suitably given, determine the source orbit $\boldsymbol{\gamma}(t)$ $(0 \le t \le T)$ by the multiple point observations of u at $\{\boldsymbol{x}^j\}_{j=1}^N \times [0, T]$.*

We remark that the relation between the orbit function and the received dynamical signals is nonlinear, whereas the operator which maps the source profile function g to the forward solution u is linear. Hence, the problem considered in this paper is a nonlinear inverse issue.

We refer to Isakov [12] for an overview of uniqueness and stability results on inverse source problems. The approaches of applying Carleman estimates and the unique continuation of evolutionary equations have been widely used in the literature and have led to uniqueness and stability results for both inverse coefficient and inverse source problems with the dynamical data over a finite time (see e.g., [2, 4, 6, 13, 14, 27, 28] as an incomplete list). The concept of increasing stability was explored in [11] and later investigated further for inverse source problems in [5]. For stationary (non-moving) sources, the uniqueness in determining source positions with boundary surface data was deduced in [3], and those for upper and lower estimates of source positions were derived in [15, 16] in one and higher dimensions.

To the best of the authors' knowledge, literature on inverse moving source problems arising in fractional diffusion(-wave) equations is rather limited and even remain open. A logarithmic stability and an iterative inversion scheme were considered in [19] for recovering temporal source terms in fractional diffusion equations. Using the moment theory, a uniqueness result to inverse moving source problems in electromagnetism was proved (see [10]) using boundary surface data. In a series of works [20–22], numerical algorithms were examined for reconstructing a moving orbit from boundary data of solutions of the scalar wave equation.

The aim of this paper is to derive stability and uniqueness results with a finite number of interior monitoring points for the fractional model (5.1) with $\alpha \in (0, 2]$. Our arguments rely on the fixed point theory similarly to [8, 24, 26] but are modified to be applicable to the system (5.1). For this purpose, we have deduced Duhamel's principle for time-fractional partial differential equations of any order (see Lemma 5.2) and a uniform solution representation for the fractional order $\alpha \in (0, 2]$ via the Fourier transform in the whole space (see Lemma 5.3). These lemmas have generalized the corresponding well-known results for equations with integral orders in bounded domains, making new contributions to the theory of fractional equations. The interior observation data are here formulated from the theoretical viewpoint, and in a forthcoming paper, on the basis of the current work, we will discuss the inverse problem with more practical data.

The remaining part of this paper is organized as follows. Preliminary knowledge and our main results for fractional equations in bounded and unbounded domains will be stated in Sect. 5.2. Three auxiliary lemmas will be proved in Sect. 5.3. The proofs of our stability result for localized moving sources (Theorem 5.1) and the uniqueness for non-localized ones (Corollary 5.1) will be carried out in Sect. 5.4.

5.2 Preliminaries and Main Results

Throughout this paper, we set $\mathbb{N} := \{1, 2, 3, \ldots\}$, and by $C > 0$ we denote generic constants which may change from line to line. For $\beta \in \mathbb{R}$, denote the largest integer smaller than or equal to β by the floor function $\lfloor \beta \rfloor$, and the smallest integer larger than or equal to β by the ceiling function $\lceil \beta \rceil$. For $\beta \in [0, 1]$, define the Riemann-Liouville integral operator J^β by

$$J^\beta f(t) := \begin{cases} f(t), & \beta = 0, \\ \dfrac{1}{\Gamma(\beta)} \displaystyle\int_0^t \frac{f(s)}{(t-s)^{1-\beta}}\, \mathrm{d}s, & 0 < \beta \le 1, \end{cases} \qquad f \in C[0, \infty),$$

where $\Gamma(\,\cdot\,)$ is the Gamma function. Then for $\beta \in \mathbb{R}_+$, the Caputo derivative ∂_t^β and the Riemann-Liouville derivative D_t^β are defined as

$$\partial_t^\beta := J^{\lceil \beta \rceil - \beta} \circ \frac{\mathrm{d}^{\lceil \beta \rceil}}{\mathrm{d}t^{\lceil \beta \rceil}}, \quad D_t^\beta := \frac{\mathrm{d}^{\lceil \beta \rceil}}{\mathrm{d}t^{\lceil \beta \rceil}} \circ J^{\lceil \beta \rceil - \beta},$$

where $\circ$ denotes the composition. For the solution representation, we recall the familiar Mittag-Leffler function

$$E_{\beta,\mu}(z) := \sum_{\ell=0}^{\infty} \frac{z^\ell}{\Gamma(\beta\ell+\mu)}, \quad \beta \in \mathbb{R}_+,\ \mu \in \mathbb{R},\ z \in \mathbb{C},$$

which satisfies the frequently used estimate (see Podlubny [23, Theorem 1.5])

$$|E_{\beta,\mu}(z)| \le \frac{C}{1+|z|}, \quad \beta \in (0,2),\ \mu \in \mathbb{R}_+,\ \frac{\pi\beta}{2} < |\arg z| \le \pi. \tag{5.2}$$

For later use, we state the following formula concerning the Riemann-Liouville derivative and Mittag-Leffler functions.

Lemma 5.1 *For $\beta \in \mathbb{R}_+$ and $\lambda \in \mathbb{R}$, we have*

$$D_t^{\lceil\beta\rceil-\beta}\left(t^{\lceil\beta\rceil-1} E_{\beta,\lceil\beta\rceil}(\lambda t^\beta)\right) = t^{\beta-1} E_{\beta,\beta}(\lambda t^\beta).$$

Next, we generalize useful Duhamel's principle to time-fractional evolution equations with arbitrary orders $\beta \in \mathbb{R}_+$.

Lemma 5.2 (Fractional Duhamel's principle) *Let $\beta \in \mathbb{R}_+$ and $\Omega \subset \mathbb{R}^d$ be a domain. Let $F : \Omega \times (0,T) \longrightarrow \mathbb{R}$ be a smooth function, and $\mathcal{P}$ be a linear partial differential operator with respect to $\boldsymbol{x}$ defined in Ω whose coefficients are independent of t. If a smooth function u satisfies*

$$\begin{cases} (\partial_t^\beta + \mathcal{P})u = F & \text{in } \Omega \times (0,T), \\ \partial_t^m u = 0\ (m = 0, \dots, \lceil\beta\rceil - 1) & \text{in } \Omega \times \{0\}, \end{cases} \tag{5.3}$$

then u allows the representation

$$u(\,\cdot\,,t) = \int_0^t D_t^{\lceil\beta\rceil-\beta} v(\,\cdot\,,t;s)\,\mathrm{d}s, \quad 0 < t < T, \tag{5.4}$$

where $v(\,\cdot\,,\,\cdot\,;s)$ is a smooth function satisfying the following homogeneous equation with a parameter $s \in (0,T)$:

$$\begin{cases} (\partial_t^\beta + \mathcal{P})v = 0 & \text{in } \Omega \times (s,T), \\ \partial_t^m v = 0\ (m = 0, \dots, \lceil\beta\rceil - 2),\ \partial_t^{\lceil\beta\rceil-1} v = F(\,\cdot\,,s) & \text{in } \Omega \times \{s\}. \end{cases} \tag{5.5}$$

Here we automatically interpret

$$D_t^\beta v(\,\cdot\,,t;s)=\frac{\partial_t^{\lceil\beta\rceil}}{\Gamma(\lceil\beta\rceil-\beta)}\int_s^t\frac{v(\,\cdot\,,r;s)}{(t-r)^{\beta-\lfloor\beta\rfloor}}\,\mathrm{d}r,$$
$$\partial_t^\beta v(\,\cdot\,,t;s)=\frac{1}{\Gamma(\lceil\beta\rceil-\beta)}\int_s^t\frac{\partial_r^{\lceil\beta\rceil}v(\,\cdot\,,r;s)}{(t-r)^{\beta-\lfloor\beta\rfloor}}\,\mathrm{d}r$$

for $\beta\notin\mathbb{N}$, since $v(\,\cdot\,,t;s)$ is only defined for $t<s$.

The above lemma generalizes similar results in [17, 18], where the source term was assumed to take the form of separated variables. For other fractional varieties of Duhamel's principle, we refer to Umarov and Saidamatov [25], Zhang and Xu [29].

In the sequel, all vectors are by default column vectors unless specified otherwise. For instance, we write $\boldsymbol{x}=(x_1,\dots,x_d)^{\mathrm{T}}\in\mathbb{R}^d$ and $\nabla f=(\partial_1 f,\dots,\partial_d f)^{\mathrm{T}}$, where $(\,\cdot\,)^{\mathrm{T}}$ stands for the transpose and $\partial_k=\frac{\partial}{\partial x_k}$ $(k=1,\dots,d)$. The inner product in $\mathbb{R}^d$ is denoted by $\boldsymbol{x}\cdot\boldsymbol{y}$, and the Euclidean distance $|\cdot|$ is induced as $|\boldsymbol{x}|=(\boldsymbol{x}\cdot\boldsymbol{x})^{1/2}$. Given a matrix $\boldsymbol{\Psi}=(\psi_{jk})\in\mathbb{R}^{d\times d}$, the ℓ^2 norm $|\cdot|$ and the Frobenius norm $\|\cdot\|_{\mathrm{F}}$ of $\boldsymbol{\Psi}$ are defined as

$$|\boldsymbol{\Psi}|:=\max_{|\boldsymbol{x}|=1}|\boldsymbol{\Psi}\boldsymbol{x}|,\quad \|\boldsymbol{\Psi}\|_{\mathrm{F}}:=\left(\sum_{j,k=1}^d\psi_{jk}^2\right)^{1/2}.$$

By the norm equivalence in finite dimensional vector spaces, there exists a constant $C>0$ such that

$$C^{-1}|\boldsymbol{\Psi}|\le\|\boldsymbol{\Psi}\|_{\mathrm{F}}\le C|\boldsymbol{\Psi}|,\quad\forall\,\boldsymbol{\Psi}\in\mathbb{R}^{d\times d}. \tag{5.6}$$

The open ball centered at $\boldsymbol{x}\in\mathbb{R}^d$ with radius $r>0$ is denoted by $B_r(\boldsymbol{x}):=\{\boldsymbol{y}\in\mathbb{R}^d\mid|\boldsymbol{y}-\boldsymbol{x}|<r\}$, and especially we abbreviate $B_r(\boldsymbol{0})=B_r$.

For a domain $\Omega\subset\mathbb{R}^d$, denote the usual L^2 inner product by $(\,\cdot\,,\,\cdot\,)$, and let $H^p(\Omega)$ $(p\in\mathbb{R})$ be the L^2-based Sobolev spaces (see Adams [1]). If Ω is a connected bounded domain, then the elliptic operator $\mathcal{L}$ in the initial-boundary value problem (5.1) is defined as

$$\mathcal{L}:H^2(\Omega)\cap H_0^1(\Omega)\longrightarrow L^2(\Omega),\quad f\longmapsto-\mathrm{div}(\boldsymbol{A}\nabla f)+c\,f,$$

where the matrix $\boldsymbol{A}=(a_{jk})_{1\le j,k\le d}\in(C^1(\overline{\Omega}))^{d\times d}$ is symmetric and strictly positive definite uniformly on $\overline{\Omega}$, and $c\in L^\infty(\Omega)$ is non-negative. In this case, the operator $\mathcal{L}$ generates an eigensystem $\{(\lambda_n,\varphi_n)\}_{n\in\mathbb{N}}$ such that $\mathcal{L}\varphi_n=\lambda_n\varphi_n$ in Ω. Moreover, it is well known that $0<\lambda_1<\lambda_2\le\cdots$, $\lambda_n\to\infty$ as $n\to\infty$, and $\{\varphi_n\}\subset H^2(\Omega)\cap H_0^1(\Omega)$ forms a complete orthonormal system of $L^2(\Omega)$.

For a square integrable function $f\in L^2(\mathbb{R}^d)$, denote its Fourier transform by

$$\widetilde{f}(\boldsymbol{\xi})=(\mathcal{F}f)(\boldsymbol{\xi}):=\frac{1}{(2\pi)^{d/2}}\int_{\mathbb{R}^d}f(\boldsymbol{x})\,\mathrm{e}^{-\mathrm{i}\boldsymbol{x}\cdot\boldsymbol{\xi}}\,\mathrm{d}\boldsymbol{x}.$$

For $p \in \mathbb{R}$, the $H^p(\mathbb{R}^d)$ norm can be represented by using Fourier transform as

$$\|f\|_{H^p(\mathbb{R}^d)} = \left(\int_{\mathbb{R}^d} (1 + |\boldsymbol{\xi}|^2)^p |\widetilde{f}(\boldsymbol{\xi})|^2 \, \mathrm{d}\boldsymbol{\xi} \right)^{1/2}, \quad f \in H^p(\mathbb{R}^d).$$

If $\Omega = \mathbb{R}^d$, then we define the elliptic operator $\mathcal{L}$ in the initial value problem (5.1) as

$$\mathcal{L} : H^2(\mathbb{R}^d) \longrightarrow L^2(\mathbb{R}^d), \quad f \longmapsto -\mathrm{div}(\boldsymbol{A}\nabla f) + \boldsymbol{b} \cdot \nabla f + c\, f,$$

where we assume that the matrix $\boldsymbol{A} = (a_{jk})_{1\le j,k\le d}$, the vector $\boldsymbol{b} = (b_1, \dots, b_d)^{\mathrm{T}}$ and the scalar c are constants, and $\boldsymbol{A}$ is strictly positive definite. Especially, only in the case of $\alpha = 2$ we additionally assume $\boldsymbol{b} = \boldsymbol{0}$ and $c \ge 0$.

Regarding the initial(-boundary) value problem

$$\begin{cases} (\partial_t^\alpha + \mathcal{L})V = 0 & \text{in } \Omega \times \mathbb{R}_+, \\ \begin{cases} V = v_0 & \text{if } 0 < \alpha \le 1, \\ V = 0,\ \partial_t V = v_1 & \text{if } 1 < \alpha \le 2 \end{cases} & \text{in } \Omega \times \{0\}, \\ V = 0 \quad \text{if } \Omega \text{ is bounded} & \text{on } \partial\Omega \times \mathbb{R}_+, \end{cases} \tag{5.7}$$

we provide the well-posedness results in the following lemma.

Lemma 5.3 *Let $\Omega \subset \mathbb{R}^d$ be either a connected bounded domain or $\Omega = \mathbb{R}^d$, and $v_0 \in H^p(\Omega)$, $v_1 \in H^{p-2/\alpha}(\Omega)$ for some fixed $p \in \mathbb{R}$. Then there exists a unique solution $V \in C([0, T]; H^p(\Omega))$ to* (5.7). *Moreover, if Ω is bounded, then the solution to* (5.7) *takes the form*

$$V(\boldsymbol{x}, t) = \sum_{n=1}^{\infty} (v_{\lceil\alpha\rceil-1}, \varphi_n)\, t^{\lceil\alpha\rceil-1} E_{\alpha,\lceil\alpha\rceil}(-\lambda_n t^\alpha) \varphi_n(\boldsymbol{x}). \tag{5.8}$$

If $\Omega = \mathbb{R}^d$, then the solution to (5.7) *takes the form*

$$V(\boldsymbol{x}, t) = \begin{cases} \dfrac{t^{\lceil\alpha\rceil-1}}{(2\pi)^{d/2}} \displaystyle\int_{\mathbb{R}^d} \widetilde{v}_{\lceil\alpha\rceil-1}(\boldsymbol{\xi}) E_{\alpha,\lceil\alpha\rceil}(-S(\boldsymbol{\xi})t^\alpha)\, \mathrm{e}^{\mathrm{i}\boldsymbol{\xi}\cdot\boldsymbol{x}}\, \mathrm{d}\boldsymbol{\xi}, & 0 < \alpha < 2, \\ \dfrac{1}{(2\pi)^{d/2}} \displaystyle\int_{\mathbb{R}^d} \widetilde{v}_1(\boldsymbol{\xi}) S(\boldsymbol{\xi})^{-1/2} \sin(S(\boldsymbol{\xi})^{1/2} t)\, \mathrm{e}^{\mathrm{i}\boldsymbol{\xi}\cdot\boldsymbol{x}}\, \mathrm{d}\boldsymbol{\xi}, & \alpha = 2, \end{cases} \tag{5.9}$$

where $S(\boldsymbol{\xi}) := \boldsymbol{A}\boldsymbol{\xi} \cdot \boldsymbol{\xi} + \mathrm{i}\, \boldsymbol{b} \cdot \boldsymbol{\xi} + c$.

In a bounded domain, the solution representation by the eigensystem is well known (see Sakamoto and Yamamoto [24]). However, solutions to (5.7) in the whole space $\mathbb{R}^d$ seem not well investigated to the best of our knowledge, and we refer to Eidelman and Kochubei [7] for the fundamental solution. In such a sense, formula (5.9) in Lemma 5.3 gives a novel solution representation via the Fourier transform.

Now we are well prepared to discuss Problem 5.1. We begin with specifying the choices of the source profile $g(\boldsymbol{x})$ and the orbit $\boldsymbol{\gamma}(t)$. Assume that g is smooth and compactly supported, i.e., $g \in C_0^\infty(\Omega)$ and there exists a constant $\delta > 0$ such that $\operatorname{supp} g \subset B_\delta$. A typical choice of g can be the following bell-shaped function

$$g(\boldsymbol{x}) = \begin{cases} C \exp\left(\dfrac{1}{|\boldsymbol{x}|^2 - \delta^2}\right), & |\boldsymbol{x}| < \delta, \\ 0, & |\boldsymbol{x}| \ge \delta. \end{cases} \tag{5.10}$$

For the unknown $\boldsymbol{\gamma}$, basically we restrict it in the admissible set

$$\begin{aligned} \mathcal{U}_0 := \{\boldsymbol{\gamma} \in (C^\infty[0,T])^d \mid \boldsymbol{\gamma}(0) = \boldsymbol{0},\ \|\boldsymbol{\gamma}'\|_{C[0,T]} \le K, \\ \boldsymbol{\gamma}(t) + \operatorname{supp} g \subset \Omega,\ 0 \le t \le T\}, \end{aligned} \tag{5.11}$$

where $K > 0$ is a constant. In other words, we restrict our consideration in such orbits that they are smooth and start from the origin with a maximum velocity.

First we investigate a special case of a localized moving source. More precisely, for a sufficiently small $\varepsilon > 0$, we further restrict the unknown orbit in

$$\mathcal{U}_1 := \{\boldsymbol{\gamma} \in \mathcal{U}_0 \mid \|\boldsymbol{\gamma}\|_{C[0,T]} \le \varepsilon\}, \tag{5.12}$$

which means $\{\boldsymbol{\gamma}(t)\}_{0 \le t \le T} \subset B_\varepsilon$ for all $\boldsymbol{\gamma} \in \mathcal{U}_1$.

Since there are d components in the orbit, it is natural to take at least d observation points for the unique identification. Within the admissible set $\mathcal{U}_1$, we pick the minimum necessary d observation points $\boldsymbol{x}^j$ $(j = 1, \dots, d)$ and make the following key assumption: there exists a constant $C > 0$ depending on g, $\{\boldsymbol{x}^j\}_{j=1}^d$ and ε such that

$$\left|\left(\nabla g(\boldsymbol{y}^1)\ \nabla g(\boldsymbol{y}^2)\ \cdots\ \nabla g(\boldsymbol{y}^d)\right)^{-1}\right| \le C, \quad \forall\, \boldsymbol{y}^j \in \overline{B_\varepsilon(\boldsymbol{x}^j)},\ j = 1, \dots, d. \tag{5.13}$$

In other words, we assume that the matrix $(\nabla g(\boldsymbol{y}^1)\ \cdots\ \nabla g(\boldsymbol{y}^d))$ is invertible for all $(\boldsymbol{y}^1, \dots, \boldsymbol{y}^d) \in \prod_{j=1}^d \overline{B_\varepsilon(\boldsymbol{x}^j)}$.

Example 5.1 We rephrase assumptions (5.12) and (5.13) in the case of $d = 1$. For any $\gamma \in \mathcal{U}_1$, we have $\gamma(0) = 0$ and $|\gamma(t)| \le \varepsilon$, $|\gamma'(t)| \le K$ for $0 \le t \le T$. As for the observation point x^1, the assumption (5.13) means

$$|g'(y)| \ge C^{-1} > 0, \quad \forall\, y \in [x^1 - \varepsilon, x^1 + \varepsilon],$$

which implies $[x^1 - \varepsilon, x^1 + \varepsilon] \subset \operatorname{supp} g'$.

Now we can state Lipschitz stability and uniqueness results for Problem 5.1 with the observation data taken at $\{\boldsymbol{x}^j\}_{j=1}^d \times [0, T]$.

Theorem 5.1 *Fix $\boldsymbol{\gamma}_1, \boldsymbol{\gamma}_2 \in \mathcal{U}_1$, where $\mathcal{U}_1$ was defined by* (5.12)*. Denote by u_1 and u_2 the solutions to* (5.1) *with $\boldsymbol{\gamma} = \boldsymbol{\gamma}_1$ and $\boldsymbol{\gamma} = \boldsymbol{\gamma}_2$, respectively. If the set of observation points $\{\boldsymbol{x}^j\}_{j=1}^d$ satisfies* (5.13)*, then there exists a constant $C > 0$ depending on g, $\{\boldsymbol{x}^j\}_{j=1}^d$ and $\mathcal{U}_1$ such that*

$$\|\boldsymbol{\gamma}_1 - \boldsymbol{\gamma}_2\|_{C[0,T]} \le C \sum_{j=1}^{d} \|\partial_t^\alpha (u_1 - u_2)(\boldsymbol{x}^j, \cdot)\|_{C[0,T]}.$$

Especially, $u_1(\boldsymbol{x}^j, \cdot) = u_2(\boldsymbol{x}^j, \cdot)$ $(j = 1, \dots, d)$ on $[0, T]$ implies $\boldsymbol{\gamma}_1 = \boldsymbol{\gamma}_2$ on $[0, T]$.

Our main result Theorem 5.1 requires condition (5.13) for $\boldsymbol{x}^j$, $j = 1, 2, \dots, N$, and especially the number N of the monitoring points $\boldsymbol{x}^j$ should be at least d which is the spatial dimensions. This is reasonable because as unknowns we have to determine d components of $\boldsymbol{\gamma}(t)$, and our data are N functions in $t \in [0, T]$.

The key to proving the above theorem is reducing the original problem to a vector-valued Volterra integral equation of the second kind with respect to the difference $\boldsymbol{\gamma}_1 - \boldsymbol{\gamma}_2$. To this end, the representations of solutions to (5.1) are essential, where Lemmas 5.2 and 5.3 play important roles. Such an argument is also witnessed in [8, 24, 26] which also rely on similar non-vanishing assumptions as (5.13). Nevertheless, due to the nonlinearity of our problem with respect to the orbits, assumption (5.13) looks more complicated than that in [8, 24, 26].

Remarkably, the constant C in the stability estimate of Theorem 5.1 does not depend on the order $\alpha \in (0, 2]$. Indeed, such a uniform estimate of Lipschitz type is achieved at the cost of accessing the αth order derivative of the observation data. Meanwhile, one can also see from the proof that the ill-posedness resulted from α is overwhelmed by the key assumption (5.13) along with the admissible set $\mathcal{U}_1$.

In Theorem 5.1, the Lipschitz stability with minimum possible observation points is achieved within the admissible set $\mathcal{U}_1$ in (5.12), which is rather restrictive. Moreover, since (5.13) implies $\boldsymbol{x}^j \in \operatorname{supp}(\nabla g)$ $(j = 1, \dots, d)$, the required observation condition seems also strict in practice. On the opposite direction, we can remove the localization assumption $\|\boldsymbol{\gamma}\|_{C[0,T]} \le \varepsilon$ in (5.12) and obtain a uniqueness result at the cost of very dense observation points.

Corollary 5.1 *Fix $\boldsymbol{\gamma}_1, \boldsymbol{\gamma}_2 \in \mathcal{U}_0$, where $\mathcal{U}_0$ was defined by* (5.11)*. Denote by u_1 and u_2 the solutions to* (5.1) *with $\boldsymbol{\gamma} = \boldsymbol{\gamma}_1$ and $\boldsymbol{\gamma} = \boldsymbol{\gamma}_2$, respectively. Assume that there exist a finite set of observation points $X := \{\boldsymbol{x}^j\}_{j=1}^N$ and a constant $\varepsilon > 0$ such that for any $\boldsymbol{y} \in \Omega \cap \overline{B_{KT}}$, there exist d observation points $\{\boldsymbol{x}^j(\boldsymbol{y})\}_{j=1}^d \subset X \cap B_\delta(\boldsymbol{y})$ and a constant $C > 0$ such that*

$$\left| \left(\nabla g(z^1) \cdots \nabla g(z^d)\right)^{-1} \right| \le C, \quad \forall\, z^j \in \overline{B_\varepsilon(\boldsymbol{x}^j(\boldsymbol{y}) - \boldsymbol{y})}, \ j = 1, \dots, d. \quad (5.14)$$

Then the relation $u_1(\boldsymbol{x}^j, \cdot) = u_2(\boldsymbol{x}^j, \cdot)$ $(j = 1, \dots, N)$ on $[0, T]$ implies $\boldsymbol{\gamma}_1 = \boldsymbol{\gamma}_2$ on $[0, T]$.

As one can imagine, the above corollary follows from the repeated application of Theorem 5.1, where the invertibility assumption (5.14) is a generalization of (5.13). It suffices to restrict $\boldsymbol{y} \in \Omega$ in the ball $\overline{B_{KT}}$ because $\|\boldsymbol{\gamma}\|_{C[0,T]} \le KT$ for any $\boldsymbol{\gamma} \in \mathcal{U}_0$ by the definition (5.11) of $\mathcal{U}_0$. Since $\Omega \cap \overline{B_{KT}}$ is bounded, the number N of observation points can definitely be finite.

Example 5.2 In the one-dimensional case, if g takes the form of a bell-shaped function (5.10), then it is readily seen that a choice of ε and X in Corollary 5.1 can be

$$\varepsilon = \frac{\delta}{9}, \quad \boldsymbol{x}^j = \frac{(-1)^j \lfloor j/2 \rfloor \delta}{4}, \quad N = \lceil 4(KT+\delta)/\delta \rceil.$$

5.3 Proofs of Lemmas 5.1–5.3

Proof (*Proof of Lemma* 5.1) By the definitions of Mittag-Leffler functions and the Riemann-Liouville derivative, we direct calculate

$$\begin{aligned}
&D_t^{\lceil\beta\rceil-\beta}\left(t^{\lceil\beta\rceil-1}E_{\beta,\lceil\beta\rceil}(\lambda t^\beta)\right)\\
&= \frac{1}{\Gamma(\beta-\lfloor\beta\rfloor)}\sum_{\ell=0}^{\infty}\frac{\lambda^\ell}{\Gamma(\beta\ell+\lceil\beta\rceil)}\frac{\mathrm{d}}{\mathrm{d}t}\int_0^t \frac{s^{\beta\ell+\lceil\beta\rceil-1}}{(t-s)^{\lceil\beta\rceil-\beta}}\,\mathrm{d}s\\
&= \frac{1}{\Gamma(\beta-\lfloor\beta\rfloor)}\sum_{\ell=0}^{\infty}\frac{\Gamma(\beta\ell+\lceil\beta\rceil)\Gamma(\beta-\lfloor\beta\rfloor)\lambda^\ell(t^{\beta(\ell+1)})'}{\Gamma(\beta\ell+\lceil\beta\rceil)\Gamma(\beta\ell+\lceil\beta\rceil+\beta-\lfloor\beta\rfloor)}\\
&= \sum_{\ell=0}^{\infty}\frac{\beta(\ell+1)\lambda^\ell t^{\beta(\ell+1)-1}}{\Gamma(\beta(\ell+1)+1)} = t^{\beta-1}\sum_{\ell=0}^{\infty}\frac{(\lambda t^\beta)^\ell}{\Gamma(\beta\ell+\beta)} = t^{\beta-1}E_{\beta,\beta}(\lambda t^\beta),
\end{aligned}$$

where we have used the formula $\Gamma(\beta+1) = \beta\,\Gamma(\beta)$. □

Proof (*Proof of Lemma* 5.2) The case of $\beta \in \mathbb{N}$ is straightforward and we only give a proof for the case of $\beta \notin \mathbb{N}$. Actually, it suffices to verify that the function u defined by (5.4) satisfies (5.3). Since F, u, v are assumed to be smooth, we can take any derivatives when needed.

First, it follows from the definition of the Riemann-Liouville derivative that

$$\begin{aligned}
D_t^{\lceil\beta\rceil-\beta}F(\,\cdot\,,t) &= \frac{1}{\Gamma(\beta-\lfloor\beta\rfloor)}\frac{\partial}{\partial t}\int_0^t \frac{F(\,\cdot\,,t-s)}{s^{\lceil\beta\rceil-\beta}}\,\mathrm{d}s\\
&= \frac{1}{\Gamma(\beta-\lfloor\beta\rfloor)}\left(\frac{F(\,\cdot\,,0)}{t^{\lceil\beta\rceil-\beta}} + \int_0^t \frac{\partial_s F(\,\cdot\,,s)}{(t-s)^{\lceil\beta\rceil-\beta}}\,\mathrm{d}s\right). \qquad (5.15)
\end{aligned}$$

Next, from (5.3) we calculate

$$
\begin{aligned}
\partial_t u(\cdot,t) &= D_t^{\lceil\beta\rceil-\beta} v(\cdot,t;t) + \int_0^t \partial_t D_t^{\lceil\beta\rceil-\beta} v(\cdot,t;s)\,\mathrm{d}s \\
&= \begin{cases} D_t^{1-\beta} F(\cdot,t) + \int_0^t D_t^{1-\beta}\partial_t v(\cdot,t;s)\,\mathrm{d}s, & 0<\beta<1, \\ \int_0^t D_t^{\lceil\beta\rceil-\beta}\partial_t v(\cdot,t;s)\,\mathrm{d}s, & \beta>1, \end{cases}
\end{aligned}
$$

where we used the initial condition at $t = s$ in (5.5). Inductively, we obtain

$$
\partial_t^m u(\cdot,t) = \begin{cases} \int_0^t D_t^{\lceil\beta\rceil-\beta}\partial_t^m v(\cdot,t;s)\,\mathrm{d}s, & m<\lceil\beta\rceil, \\ D_t^{\lceil\beta\rceil-\beta} F(\cdot,t) + \int_0^t D_t^{\lceil\beta\rceil-\beta}\partial_t^{\lceil\beta\rceil} v(\cdot,t;s)\,\mathrm{d}s, & m=\lceil\beta\rceil. \end{cases} \tag{5.16}
$$

Since v is sufficiently smooth, for $m < \lceil\beta\rceil$ we pass $t \to 0$ in (5.16) to find

$$
\partial_t^m u(\cdot,0) = 0, \quad m = 0,\ldots,\lceil\beta\rceil-1,
$$

i.e., the function u satisfies the initial condition in (5.3). Meanwhile, substituting (5.16) with $m = \lceil\beta\rceil$ into the definition of the Caputo derivative gives

$$
\partial_t^\beta u(\cdot,t) = \frac{1}{\Gamma(\lceil\beta\rceil-\beta)} \int_0^t \frac{\partial_s^{\lceil\beta\rceil} u(\cdot,s)}{(t-s)^{\beta-\lfloor\beta\rfloor}}\,\mathrm{d}s =: \frac{\mathrm{I}_1+\mathrm{I}_2}{\Gamma(\lceil\beta\rceil-\beta)}, \tag{5.17}
$$

where

$$
\begin{aligned}
\mathrm{I}_1 &:= \int_0^t \frac{D_s^{\lceil\beta\rceil-\beta} F(\cdot,s)}{(t-s)^{\beta-\lfloor\beta\rfloor}}\,\mathrm{d}s, \\
\mathrm{I}_2 &:= \int_0^t \frac{1}{(t-s)^{\beta-\lfloor\beta\rfloor}} \int_0^s D_s^{\lceil\beta\rceil-\beta}\partial_s^{\lceil\beta\rceil} v(\cdot,s;r)\,\mathrm{d}r\mathrm{d}s.
\end{aligned}
$$

For I_1, the application of (5.15) yields

$$
\begin{aligned}
\mathrm{I}_1 &= \frac{1}{\Gamma(\beta-\lfloor\beta\rfloor)} \int_0^t \frac{1}{(t-s)^{\beta-\lfloor\beta\rfloor}} \left(\frac{F(\cdot,0)}{s^{\lceil\beta\rceil-\beta}} + \int_0^s \frac{\partial_r F(\cdot,r)}{(s-r)^{\lceil\beta\rceil-\beta}}\,\mathrm{d}r \right) \mathrm{d}s \\
&= \frac{1}{\Gamma(\beta-\lfloor\beta\rfloor)} \left(F(\cdot,0) \int_0^t \frac{\mathrm{d}s}{(t-s)^{\beta-\lfloor\beta\rfloor} s^{\lceil\beta\rceil-\beta}} \right. \\
&\qquad \left. + \int_0^t \partial_r F(\cdot,r) \int_r^t \frac{\mathrm{d}s}{(t-s)^{\beta-\lfloor\beta\rfloor}(s-r)^{\lceil\beta\rceil-\beta}}\,\mathrm{d}r \right) \\
&= \Gamma(\lceil\beta\rceil-\beta) \left(F(\cdot,0) + \int_0^t \partial_r F(\cdot,r)\,\mathrm{d}r \right) = \Gamma(\lceil\beta\rceil-\beta)\, F(\cdot,t).
\end{aligned} \tag{5.18}
$$

For I_2, by suitably exchanging the order of integration, we utilize the definition of Caputo and Riemann-Liouville derivatives to calculate

$$
\begin{aligned}
\mathrm{I}_2 &= \int_0^t \frac{1}{(t-s)^{\beta-\lfloor\beta\rfloor}} \int_0^s D_s^{\lceil\beta\rceil-\beta} \partial_s^{\lceil\beta\rceil} v(\,\cdot\,, s; r)\,\mathrm{d}r\mathrm{d}s \\
&= \frac{1}{\Gamma(\beta-\lfloor\beta\rfloor)} \int_0^t \frac{1}{(t-s)^{\beta-\lceil\beta\rceil}} \int_0^s \partial_s \int_0^{s-r} \frac{\partial_s^{\lceil\beta\rceil} v(\,\cdot\,, s-\tau; r)}{\tau^{\lceil\beta\rceil-\beta}}\,\mathrm{d}\tau\mathrm{d}r\mathrm{d}s \\
&= \frac{1}{\Gamma(\beta-\lfloor\beta\rfloor)} \int_0^t \frac{1}{(t-s)^{\beta-\lceil\beta\rceil}} \int_0^s \frac{\partial_r^{\lceil\beta\rceil} v(\,\cdot\,, r; r)}{(s-r)^{\lceil\beta\rceil-\beta}}\,\mathrm{d}r\mathrm{d}s \\
&\quad + \frac{1}{\Gamma(\beta-\lfloor\beta\rfloor)} \int_0^t \frac{1}{(t-s)^{\beta-\lceil\beta\rceil}} \int_0^s\!\!\int_r^s \frac{\partial_\tau^{\lceil\beta\rceil+1} v(\,\cdot\,, \tau; r)}{(s-\tau)^{\lceil\beta\rceil-\beta}}\,\mathrm{d}\tau\mathrm{d}r\mathrm{d}s \\
&= \frac{1}{\Gamma(\beta-\lfloor\beta\rfloor)} \int_0^t \partial_r^{\lceil\beta\rceil} v(\,\cdot\,, r; r) \int_r^t \frac{\mathrm{d}s}{(t-s)^{\beta-\lfloor\beta\rfloor}(s-r)^{\lceil\beta\rceil-\beta}}\,\mathrm{d}r \\
&\quad + \frac{1}{\Gamma(\beta-\lfloor\beta\rfloor)} \int_0^t\!\!\int_0^\tau \partial_\tau^{\lceil\beta\rceil+1} v(\,\cdot\,, \tau; r) \int_\tau^t \frac{\mathrm{d}s}{(t-s)^{\beta-\lfloor\beta\rfloor}(s-\tau)^{\lceil\beta\rceil-\beta}}\,\mathrm{d}r\mathrm{d}\tau \\
&= \Gamma(\lceil\beta\rceil-\beta)\left(\int_0^t \partial_r^{\lceil\beta\rceil} v(\,\cdot\,, r; r)\,\mathrm{d}r + \int_0^t\!\!\int_0^\tau \partial_\tau^{\lceil\beta\rceil+1} v(\,\cdot\,, \tau; r)\,\mathrm{d}r\mathrm{d}\tau\right). \qquad (5.19)
\end{aligned}
$$

On the other hand, since the operator $\mathcal{P}$ is independent of t, we calculate $-\mathcal{P}u$ as

$$
\begin{aligned}
&-\mathcal{P}u(\,\cdot\,, t) = -\int_0^t D_t^{\lceil\beta\rceil-\beta} \mathcal{P} v(\,\cdot\,, t; s)\,\mathrm{d}s = \int_0^t D_t^{\lceil\beta\rceil-\beta} \partial_t^{\beta} v(\,\cdot\,, t; s)\,\mathrm{d}s \\
&= \int_0^t \frac{\partial_t}{\Gamma(\beta-\lfloor\beta\rfloor)} \int_s^t \frac{1}{(t-r)^{\lceil\beta\rceil-\beta}} \frac{1}{\Gamma(\lceil\beta\rceil-\beta)} \int_s^r \frac{\partial_\tau^{\lceil\beta\rceil} v(\,\cdot\,, \tau; s)}{(r-\tau)^{\beta-\lfloor\beta\rfloor}}\,\mathrm{d}\tau\mathrm{d}r\mathrm{d}s \\
&= \frac{1}{\Gamma(\beta-\lfloor\beta\rfloor)\Gamma(\lceil\beta\rceil-\beta)} \int_0^t \partial_t \int_0^{t-s} \frac{1}{r^{\lceil\beta\rceil-\beta}} \int_s^{t-r} \frac{\partial_\tau^{\lceil\beta\rceil} v(\,\cdot\,, \tau; s)}{(t-r-\tau)^{\beta-\lfloor\beta\rfloor}}\,\mathrm{d}\tau\mathrm{d}r\mathrm{d}s \\
&= \frac{1}{\Gamma(\beta-\lfloor\beta\rfloor)\Gamma(\lceil\beta\rceil-\beta)} \int_0^t\!\!\int_0^{t-s} \frac{\partial_t}{r^{\lceil\beta\rceil-\beta}} \int_s^{t-r} \frac{\partial_\tau^{\lceil\beta\rceil} v(\,\cdot\,, \tau; s)}{(t-r-\tau)^{\beta-\lfloor\beta\rfloor}}\,\mathrm{d}\tau\mathrm{d}r\mathrm{d}s \\
&= \frac{1}{\Gamma(\beta-\lfloor\beta\rfloor)\Gamma(\lceil\beta\rceil-\beta)} \left\{\int_0^t\!\!\int_0^{t-s} \frac{\partial_s^{\lceil\beta\rceil} v(\,\cdot\,, s; s)}{r^{\lceil\beta\rceil-\beta}(t-s-r)^{\beta-\lfloor\beta\rfloor}}\,\mathrm{d}r\mathrm{d}s + \mathrm{I}_3\right\} \\
&= \int_0^t \partial_s^{\lceil\beta\rceil} v(\,\cdot\,, s; s)\,\mathrm{d}s + \frac{\mathrm{I}_3}{\Gamma(\beta-\lfloor\beta\rfloor)\Gamma(\lceil\beta\rceil-\beta)}, \qquad (5.20)
\end{aligned}
$$

where

$$
\begin{aligned}
\mathrm{I}_3 &:= \int_0^t\!\!\int_0^{t-s} \frac{1}{r^{\lceil\beta\rceil-\beta}} \int_0^{t-s-r} \frac{\partial_t^{\lceil\beta\rceil+1} v(\,\cdot\,, t-r-\tau; s)}{\tau^{\beta-\lfloor\beta\rfloor}}\,\mathrm{d}\tau\mathrm{d}r\mathrm{d}s \\
&= \int_0^t\!\!\int_s^t \frac{1}{(r-s)^{\lceil\beta\rceil-\beta}} \int_r^t \frac{\partial_t^{\lceil\beta\rceil+1} v(\,\cdot\,, t-\tau+s; s)}{(\tau-r)^{\beta-\lfloor\beta\rfloor}}\,\mathrm{d}\tau\mathrm{d}r\mathrm{d}s \\
&= \int_0^t\!\!\int_0^\tau \partial_t^{\lceil\beta\rceil+1} v(\,\cdot\,, t-\tau+s; s) \int_s^\tau \frac{\mathrm{d}r}{(\tau-r)^{\beta-\lfloor\beta\rfloor}(r-s)^{\lceil\beta\rceil-\beta}}\,\mathrm{d}s\mathrm{d}\tau
\end{aligned}
$$

$$
\begin{aligned}
&= \Gamma(\beta-\lfloor\beta\rfloor)\Gamma(\lceil\beta\rceil-\beta)\int_0^t\int_0^\tau \partial_t^{\lceil\beta\rceil+1}v(\,\cdot\,,t-\tau+s;s)\,\mathrm{d}s\mathrm{d}\tau\\
&= \Gamma(\beta-\lfloor\beta\rfloor)\Gamma(\lceil\beta\rceil-\beta)\int_0^t\int_0^{t-s} \partial_\tau^{\lceil\beta\rceil+1}v(\,\cdot\,,\tau+s;s)\,\mathrm{d}\tau\mathrm{d}s\\
&= \Gamma(\beta-\lfloor\beta\rfloor)\Gamma(\lceil\beta\rceil-\beta)\int_0^t\int_0^{\tau} \partial_\tau^{\lceil\beta\rceil+1}v(\,\cdot\,,\tau;s)\,\mathrm{d}s\mathrm{d}\tau. \qquad (5.21)
\end{aligned}
$$

The combination of (5.17)–(5.21) immediately indicates

$$
\begin{aligned}
&(\partial_t^\beta+\mathcal{P})u(\,\cdot\,,t)\\
&= \frac{\mathrm{I}_1+\mathrm{I}_2}{\Gamma(\lceil\beta\rceil-\beta)} - \int_0^t \partial_s^{\lceil\beta\rceil}v(\,\cdot\,,s;s)\,\mathrm{d}s - \frac{\mathrm{I}_3}{\Gamma(\beta-\lfloor\beta\rfloor)\Gamma(\lceil\beta\rceil-\beta)}\\
&= F(\,\cdot\,,t)+\int_0^t\int_0^\tau \partial_\tau^{\lceil\beta\rceil+1}v(\,\cdot\,,\tau;r)\,\mathrm{d}r\mathrm{d}\tau - \frac{\mathrm{I}_3}{\Gamma(\beta-\lfloor\beta\rfloor)\Gamma(\lceil\beta\rceil-\beta)} = F(\,\cdot\,,t).
\end{aligned}
$$

Therefore, it is verified that the function u defined by (5.4) indeed satisfies (5.3), and the proof of Lemma 5.2 is completed. □

Proof (*Proof of Lemma* 5.3) If Ω is a bounded domain, the results follow immediately by the same argument as that in Sakamoto and Yamamoto [24]. Henceforth we only deal with the unbounded case of $\Omega=\mathbb{R}^d$.

Recalling the definition of $S(\boldsymbol{\xi})$ in Lemma 5.3, formally we have $\mathcal{F}(\mathcal{L}V(\,\cdot\,,t))(\boldsymbol{\xi}) = S(\boldsymbol{\xi})\widetilde{V}(\boldsymbol{\xi},t)$. Then taking Fourier transform in (5.7) with respect to the spatial variables yields a fractional ordinary differential equation with a parameter $\boldsymbol{\xi}$:

$$
\begin{cases}
(\partial_t^\alpha+S(\boldsymbol{\xi}))\widetilde{V}(\boldsymbol{\xi},t)=0, & t>0,\\
\begin{cases}\widetilde{V}(\boldsymbol{\xi},0)=\widetilde{v}_0(\boldsymbol{\xi}) & \text{if } 0<\alpha\le 1,\\ \widetilde{V}(\boldsymbol{\xi},0)=0,\ \partial_t\widetilde{V}(\boldsymbol{\xi},0)=\widetilde{v}_1(\boldsymbol{\xi}) & \text{if } 1<\alpha\le 2.\end{cases}
\end{cases}
$$

The solution to the above equation turns out to be

$$
\widetilde{V}(\boldsymbol{\xi},t)=\begin{cases}\widetilde{v}_{\lceil\alpha\rceil-1}(\boldsymbol{\xi})t^{\lceil\alpha\rceil-1}E_{\alpha,\lceil\alpha\rceil}(-S(\boldsymbol{\xi})t^\alpha), & 0<\alpha<2,\\ \widetilde{v}_1(\boldsymbol{\xi})S(\boldsymbol{\xi})^{-1/2}\sin(S(\boldsymbol{\xi})^{1/2}t), & \alpha=2,\end{cases} \qquad (5.22)
$$

where $S(\boldsymbol{\xi})\ge 0$ for $\alpha=2$ because we assumed $\boldsymbol{b}=\mathbf{0}$ and $c\ge 0$ in this case.

For any fixed $t\ge 0$, our aim is to verify the boundedness of $\|V(\,\cdot\,,t)\|_{H^p(\mathbb{R}^d)}$. In the case of $0<\alpha<2$, we have to estimate $|E_{\alpha,\lceil\alpha\rceil}(-S(\boldsymbol{\xi})t^\alpha)|$. Denoting by $\kappa>0$ the smallest eigenvalue of the strict positive definite matrix $\boldsymbol{A}$, we see

$$
\operatorname{Re} S(\boldsymbol{\xi}) = \boldsymbol{A}\boldsymbol{\xi}\cdot\boldsymbol{\xi}+c \ge \frac{\kappa}{2}|\boldsymbol{\xi}|^2 \text{ for } |\boldsymbol{\xi}|\gg 1,\quad |\operatorname{Im} S(\boldsymbol{\xi})| = |\boldsymbol{b}\cdot\boldsymbol{\xi}|\le|\boldsymbol{b}||\boldsymbol{\xi}|.
$$

Hence, there exists a constant $R=R(\alpha)>0$ such that $\frac{\pi\alpha}{2}<|\arg(-S(\boldsymbol{\xi})t^\alpha)|\le\pi$ for $|\boldsymbol{\xi}|\ge R$. Then we can employ (5.2) to estimate

$$|E_{\alpha,\lceil\alpha\rceil}(-S(\boldsymbol{\xi})t^\alpha)| \le \frac{C}{1+|S(\boldsymbol{\xi})|t^\alpha} \le \frac{C}{1+|\boldsymbol{\xi}|^2 t^\alpha}, \quad |\boldsymbol{\xi}| \ge R.$$

For $|\boldsymbol{\xi}| < R$, it is readily seen that $|E_{\alpha,\lceil\alpha\rceil}(-S(\boldsymbol{\xi})t^\alpha)|$ is uniformly bounded.

For $0 < \alpha \le 1$, we divide $\mathbb{R}^d = B_R \cup (\mathbb{R}^d \setminus B_R)$ and use (5.22) to estimate

$$\begin{aligned}
\|V(\,\cdot\,,t)\|_{H^p(\mathbb{R}^d)}^2 &= \left(\int_{B_R} + \int_{\mathbb{R}^d\setminus B_R}\right)(1+|\boldsymbol{\xi}|^2)^p |\widetilde{v}_0(\boldsymbol{\xi})|^2 |E_{\alpha,1}(-S(\boldsymbol{\xi})t^\alpha)|^2 \,\mathrm{d}\boldsymbol{\xi} \\
&\le C\int_{B_R}(1+|\boldsymbol{\xi}|^2)^p |\widetilde{v}_0(\boldsymbol{\xi})|^2 \,\mathrm{d}\boldsymbol{\xi} \\
&\quad + \int_{\mathbb{R}^d\setminus B_R}(1+|\boldsymbol{\xi}|^2)^p |\widetilde{v}_0(\boldsymbol{\xi})|^2 \left(\frac{C}{1+|\boldsymbol{\xi}|^2 t^\alpha}\right)^2 \mathrm{d}\boldsymbol{\xi} \\
&\le C\int_{\mathbb{R}^d}(1+|\boldsymbol{\xi}|^2)^p |\widetilde{v}_0(\boldsymbol{\xi})|^2 \,\mathrm{d}\boldsymbol{\xi} = C\|v_0\|_{H^p(\mathbb{R}^d)}^2.
\end{aligned}$$

For $1 < \alpha < 2$, we utilize the same argument as above and the uniform boundedness of $\frac{\zeta^{1/\alpha}}{1+\zeta}$ for $\zeta \ge 0$ to deduce

$$\begin{aligned}
\|V(\,\cdot\,,t)\|_{H^p(\mathbb{R}^d)}^2 &= t^2\left(\int_{B_R} + \int_{\mathbb{R}^d\setminus B_R}\right)(1+|\boldsymbol{\xi}|^2)^p |\widetilde{v}_1(\boldsymbol{\xi})|^2 |E_{\alpha,2}(-S(\boldsymbol{\xi})t^\alpha)|^2 \,\mathrm{d}\boldsymbol{\xi} \\
&\le C t^2\int_{B_R}(1+|\boldsymbol{\xi}|^2)^{2/\alpha}(1+|\boldsymbol{\xi}|^2)^{p-2/\alpha} |\widetilde{v}_1(\boldsymbol{\xi})|^2 \,\mathrm{d}\boldsymbol{\xi} \\
&\quad + \int_{\mathbb{R}^d\setminus B_R}(1+|\boldsymbol{\xi}|^2)^{p-2/\alpha} |\widetilde{v}_1(\boldsymbol{\xi})|^2 \left(\frac{C((1+|\boldsymbol{\xi}|^2)t^\alpha)^{1/\alpha}}{1+|\boldsymbol{\xi}|^2 t^\alpha}\right)^2 \mathrm{d}\boldsymbol{\xi} \\
&\le C(t^2+1)\int_{\mathbb{R}^d}(1+|\boldsymbol{\xi}|^2)^{p-2/\alpha} |\widetilde{v}_1(\boldsymbol{\xi})|^2 \,\mathrm{d}\boldsymbol{\xi} \\
&= C(t^2+1)\|v_1\|_{H^{p-2/\alpha}(\mathbb{R}^d)}^2.
\end{aligned}$$

In the case of $\alpha = 2$, thanks to the assumption $\boldsymbol{b} = \boldsymbol{0}$ and $c \ge 0$, we have $S(\boldsymbol{\xi}) \ge \kappa|\boldsymbol{\xi}|^2$. Then we estimate (5.22) as

$$\begin{aligned}
\|V(\,\cdot\,,t)\|_{H^p(\mathbb{R}^d)}^2 &= \left(\int_{B_1} + \int_{\mathbb{R}^d\setminus B_1}\right)\frac{(1+|\boldsymbol{\xi}|^2)^p |\widetilde{v}_1(\boldsymbol{\xi})||\sin(S(\boldsymbol{\xi})^{1/2}t)|}{S(\boldsymbol{\xi})}\,\mathrm{d}\boldsymbol{\xi} \\
&\le \int_{B_1}(1+|\boldsymbol{\xi}|^2)^{p-1} |\widetilde{v}_1(\boldsymbol{\xi})|^2 (1+|\boldsymbol{\xi}|^2)\frac{S(\boldsymbol{\xi})t^2}{S(\boldsymbol{\xi})}\,\mathrm{d}\boldsymbol{\xi} \\
&\quad + \int_{\mathbb{R}^d\setminus B_1}(1+|\boldsymbol{\xi}|^2)^{p-1} |\widetilde{v}_1(\boldsymbol{\xi})|^2 \frac{1+|\boldsymbol{\xi}|^2}{\kappa|\boldsymbol{\xi}|^2}\,\mathrm{d}\boldsymbol{\xi} \\
&\le C t^2\int_{B_1}(1+|\boldsymbol{\xi}|^2)^{p-1} |\widetilde{v}_1(\boldsymbol{\xi})|^2 \,\mathrm{d}\boldsymbol{\xi}
\end{aligned}$$

$$
\begin{aligned}
&+ C \int_{\mathbb{R}^d \setminus B_1} (1+|\boldsymbol{\xi}|^2)^{p-1} |\widetilde{v}_1(\boldsymbol{\xi})|^2 \, \mathrm{d}\boldsymbol{\xi} \\
&\le C(t^2+1) \|v_1\|^2_{H^{p-1}(\mathbb{R}^d)}.
\end{aligned}
$$

The proof of Lemma 5.3 is completed. □

5.4 Proofs of the Main Results

Proof (*Proof of Theorem* 5.1) Let u_1, u_2 be the solutions to (5.1) with orbits $\boldsymbol{\gamma}_1, \boldsymbol{\gamma}_2 \in \mathcal{U}_1$, respectively. Setting $w := u_1 - u_2$, it is easy to observe that w satisfies the following initial(-boundary) value problem

$$
\begin{cases}
(\partial_t^\alpha + \mathcal{L})w = G & \text{in } \Omega \times (0,T), \\
\begin{cases} w = 0 & \text{if } 0 < \alpha \le 1, \\ w = \partial_t w = 0 & \text{if } 1 < \alpha \le 2 \end{cases} & \text{in } \Omega \times \{0\}, \\
w = 0 \quad \text{if } \Omega \text{ is bounded} & \text{on } \partial\Omega \times (0,T),
\end{cases}
\tag{5.23}
$$

where $G(\boldsymbol{x}, t) := g(\boldsymbol{x} - \boldsymbol{\gamma}_1(t)) - g(\boldsymbol{x} - \boldsymbol{\gamma}_2(t))$. According to the mean value theorem, there exists a smooth function $\boldsymbol{\eta} : \Omega \times (0, T) \longrightarrow \mathbb{R}^d$ such that

$$
G(\boldsymbol{x}, t) = \nabla g(\boldsymbol{\eta}(\boldsymbol{x}, t)) \cdot \boldsymbol{\rho}(t) = \sum_{k=1}^{d} G_k(\boldsymbol{x}, t) \rho_k(t),
$$

where $\boldsymbol{\eta}(\boldsymbol{x}, t)$ is a point lying on the segment between $\boldsymbol{x} - \boldsymbol{\gamma}_1(t)$ and $\boldsymbol{x} - \boldsymbol{\gamma}_2(t)$, and

$$
\boldsymbol{\rho} := \boldsymbol{\gamma}_2 - \boldsymbol{\gamma}_1 = (\rho_1, \dots, \rho_d)^{\mathrm{T}}, \quad G_k(\boldsymbol{x}, t) := \partial_k g(\boldsymbol{\eta}(\boldsymbol{x}, t)) \; (k = 1, \dots, d).
$$

Substituting the observation points $\boldsymbol{x} = \boldsymbol{x}^j$ $(j = 1, \dots, d)$ into the governing equation of (5.23), we obtain

$$
\sum_{k=1}^{d} G_k(\boldsymbol{x}^j, t) \rho_k(t) = \partial_t^\alpha w(\boldsymbol{x}^j, t) + \mathcal{L} w(\boldsymbol{x}^j, t), \quad j = 1, \dots, d. \tag{5.24}
$$

In order to give a representation of $\mathcal{L} w(\boldsymbol{x}^j, t)$, we take advantage of Lemma 5.2 to write $\mathcal{L} w$ as

$$
\mathcal{L} w(\,\cdot\,, t) = \int_0^t D_t^{\lceil \alpha \rceil - \alpha} \mathcal{L} v(\,\cdot\,, t; s) \, \mathrm{d}s, \quad 0 < t \le T, \tag{5.25}
$$

where v satisfies the following homogeneous initial(-boundary) value problem with a parameter $s \in (0, T)$:

$$\begin{cases} (\partial_t^\alpha+\mathcal{L})v=0 & \text{in } \Omega\times(s,T),\\ \begin{cases} v=G(\,\cdot\,,s) & \text{if } 0<\alpha\le 1,\\ v=0,\ \partial_t v=G(\,\cdot\,,s) & \text{if } 1<\alpha\le 2 \end{cases} & \text{in } \Omega\times\{s\},\\ v=0 \quad \text{if } \Omega \text{ is bounded} & \text{on } \partial\Omega\times(s,T). \end{cases}$$

In the case of a bounded domain Ω, it follows from (5.8) that

$$\mathcal{L}v(\,\cdot\,,t;s)=\sum_{n=1}^\infty \lambda_n(G(\,\cdot\,,s),\varphi_n)(t-s)^{\lceil\alpha\rceil-1}E_{\alpha,\lceil\alpha\rceil}(-\lambda_n(t-s)^\alpha)\varphi_n.$$

Using Lemma 5.1, we substitute the above equality into (5.25) with $\boldsymbol{x}=\boldsymbol{x}^j$ to represent

$$\begin{aligned} \mathcal{L}w(\boldsymbol{x}^j,t)&=\int_0^t\sum_{n=1}^\infty \lambda_n(G(\,\cdot\,,s),\varphi_n)(t-s)^{\alpha-1}E_{\alpha,\alpha}(-\lambda_n(t-s)^\alpha)\varphi_n(\boldsymbol{x}^j)\,\mathrm{d}s\\ &=\int_0^t\sum_{k=1}^d q_{jk}(t,s)\rho_k(s)\,\mathrm{d}s,\quad j=1,\dots,d, \end{aligned} \tag{5.26}$$

where

$$q_{jk}(t,s):=(t-s)^{\alpha-1}\sum_{n=1}^\infty \lambda_n(G_k(\,\cdot\,,s),\varphi_n)\varphi_n(\boldsymbol{x}^j)E_{\alpha,\alpha}(-\lambda_n(t-s)^\alpha). \tag{5.27}$$

In the case of $\Omega=\mathbb{R}^d$, we turn to the Fourier transform to see $\mathcal{F}(\mathcal{L}v(\,\cdot\,,t;s))=S\widetilde{v}(\,\cdot\,,t;s)$, where we recall $S(\boldsymbol{\xi})=\boldsymbol{A}\boldsymbol{\xi}\cdot\boldsymbol{\xi}+\mathrm{i}\,\boldsymbol{b}\cdot\boldsymbol{\xi}+c$. Then it follows from (5.9) that

$$\mathcal{F}(\mathcal{L}v(\,\cdot\,,t;s))(\boldsymbol{\xi})=\begin{cases} S(\boldsymbol{\xi})\widetilde{G}(\boldsymbol{\xi},s)(t-s)^{\lceil\alpha\rceil-1}E_{\alpha,\lceil\alpha\rceil}(-S(\boldsymbol{\xi})t^\alpha), & 0<\alpha<2,\\ S(\boldsymbol{\xi})^{1/2}\widetilde{G}(\boldsymbol{\xi},s)\sin(S(\boldsymbol{\xi})^{1/2}(t-s)), & \alpha=2. \end{cases}$$

Taking the inverse Fourier transform in the above equality and applying Lemma 5.1 to (5.25) again, we obtain

$$\begin{aligned} \mathcal{L}w(\,\cdot\,,t)&=\mathcal{F}^{-1}\left(\int_0^t D_t^{\lceil\alpha\rceil-\alpha}\mathcal{F}(\mathcal{L}v(\,\cdot\,,t;s))\,\mathrm{d}s\right)\\ &=\begin{cases} \mathcal{F}^{-1}\left(\int_0^t S(\boldsymbol{\xi})\widetilde{G}(\boldsymbol{\xi},s)(t-s)^{\alpha-1}E_{\alpha,\alpha}(-S(\boldsymbol{\xi})(t-s)^\alpha)\,\mathrm{d}s\right), & 0<\alpha<2,\\ \mathcal{F}^{-1}\left(\int_0^t S(\boldsymbol{\xi})^{1/2}\widetilde{G}(\boldsymbol{\xi},s)\sin(S(\boldsymbol{\xi})^{1/2}(t-s))\,\mathrm{d}s\right), & \alpha=2. \end{cases} \end{aligned}$$

By $\widetilde{G}(\boldsymbol{\xi},s)=\sum_{k=1}^d\widetilde{G}_k(\boldsymbol{\xi},s)\rho_k(s)$ and taking $\boldsymbol{x}=\boldsymbol{x}^j$ $(j=1,\dots,d)$, again we arrive at the expression (5.26), where $q_{jk}(t,s)$ is defined by

$$q_{jk}(t,s) := \begin{cases} \mathcal{F}^{-1}(S(\boldsymbol{\xi})\widetilde{G}_k(\boldsymbol{\xi},s)E_{\alpha,\alpha}(-S(\boldsymbol{\xi})(t-s)^\alpha))(\boldsymbol{x}^j) \\ \times(t-s)^{\alpha-1}, & 0<\alpha<2, \\ \mathcal{F}^{-1}(S(\boldsymbol{\xi})^{1/2}\widetilde{G}_k(\boldsymbol{\xi},s)\sin(S(\boldsymbol{\xi})^{1/2}(t-s)))(\boldsymbol{x}^j), & \alpha=2. \end{cases} \tag{5.28}$$

Since the expression (5.26) is valid for both bounded and unbounded Ω, we plug (5.26) into (5.24) and rewrite it in form of a linear system as

$$\boldsymbol{P}(t)\boldsymbol{\rho}(t) = \partial_t^\alpha \boldsymbol{h}(t) + \int_0^t \boldsymbol{Q}(t,s)\boldsymbol{\rho}(s)\,\mathrm{d}s, \tag{5.29}$$

where $\boldsymbol{h}(t) := (w(\boldsymbol{x}^1,t),\dots,w(\boldsymbol{x}^d,t))^{\mathrm{T}}$ and

$$\boldsymbol{P}(t) := (G_k(\boldsymbol{x}^j,t))_{1\le j,k\le d}, \quad \boldsymbol{Q}(t,s) := (q_{jk}(t,s))_{1\le j,k\le d}$$

are $d\times d$ matrices. Recalling the admissible set $\mathcal{U}_1$ for $\boldsymbol{\gamma}_1$ and $\boldsymbol{\gamma}_2$, we see that $\boldsymbol{\eta}(\boldsymbol{x}^j,t)\in B_\varepsilon(\boldsymbol{x}^j)$ for all $t\in[0,T]$. Therefore, by $G_k(\boldsymbol{x},t)=\partial_k g(\boldsymbol{\eta}(\boldsymbol{x},t))$, the key assumption (5.13) indicates that the matrix

$$\boldsymbol{P}(t) = \big(\nabla g(\eta(\boldsymbol{x}^1,t))\ \nabla g(\eta(\boldsymbol{x}^2,t))\ \cdots\ \nabla g(\eta(\boldsymbol{x}^d,t))\big)^{\mathrm{T}}$$

is invertible for all $t\in[0,T]$. In other words, there exists a constant $C>0$ such that

$$|\boldsymbol{P}(t)^{-1}| \le C, \quad \forall\, t\in[0,T]. \tag{5.30}$$

As for the matrix $\boldsymbol{Q}(t,s)$, it suffices to estimate $|q_{jk}(t,s)|$ appearing in (5.27) and (5.28) separately. In the case of (5.27), the uniform boundedness of $E_{\alpha,\alpha}(-\zeta)$ for $\zeta\ge 0$ yields

$$\begin{aligned} |q_{jk}(t,s)| &\le C(t-s)^{\alpha-1}\left|\sum_{n=1}^\infty \lambda_n(G_k(\,\cdot\,,s),\varphi_n)\varphi_n(\boldsymbol{x}^j)\right| \\ &= C(t-s)^{\alpha-1}|\mathcal{L}G_k(\boldsymbol{x}^j,s)| = C(t-s)^{\alpha-1}|\mathcal{L}\partial_k g(\boldsymbol{\eta}(\boldsymbol{x}^j,t))| \\ &\le C(t-s)^{\alpha-1}\|\mathcal{L}\partial_k(g\circ\boldsymbol{\eta})(\,\cdot\,,s)\|_{C(\overline{\Omega})}. \end{aligned}$$

Since g is a given smooth function and $\boldsymbol{\eta}$ is also smooth and depends only on the admissible set $\mathcal{U}_1$, it turns out that $\|\mathcal{L}\partial_k(g\circ\boldsymbol{\eta})\|_{C(\overline{\Omega}\times[0,T])}$ are uniformly bounded for $k=1,\dots,d$, implying

$$|q_{jk}(t,s)| \le C(t-s)^{\alpha-1}, \quad 1\le j,k\le d. \tag{5.31}$$

For (5.28), we deal with the cases of $0<\alpha<2$ and $\alpha=2$ separately. For $0<\alpha<2$, the similar argument to that in the proof of Lemma 5.3 guarantees a constant $R=R(\alpha)>0$ such that $\frac{\pi\alpha}{2}<|\arg(-S(\boldsymbol{\xi})t^\alpha)|\le\pi$ for $|\boldsymbol{\xi}|\ge R$. Then we employ (5.2) to estimate

$$
\begin{aligned}
|E_{\alpha,\alpha}(-S(\boldsymbol{\xi})(t-s)^\alpha)| &\le \begin{cases} C, & |\boldsymbol{\xi}| \le R, \\ \dfrac{C}{1+|S(\boldsymbol{\xi})|(t-s)^\alpha}, & |\boldsymbol{\xi}| \ge R \end{cases} \\
&\le C, \quad \boldsymbol{\xi} \in \mathbb{R}^d,\ 0 \le s < t \le T.
\end{aligned}
$$

On the other hand, it is readily seen that $|S(\boldsymbol{\xi})| \le |\boldsymbol{A}\boldsymbol{\xi}\cdot\boldsymbol{\xi}| + |\boldsymbol{b}\cdot\boldsymbol{\xi}| + |c| \le C(1+|\boldsymbol{\xi}|^2)$. Thus, based on the definition of the inverse Fourier transform, we can estimate

$$
\begin{aligned}
|q_{jk}(t,s)| &= (t-s)^{\alpha-1}\left|\mathcal{F}^{-1}\left(S(\boldsymbol{\xi})\widetilde{G}_k(\boldsymbol{\xi},s)E_{\alpha,\alpha}(-S(\boldsymbol{\xi})(t-s)^\alpha)\right)(\boldsymbol{x}^j)\right| \\
&\le \frac{(t-s)^{\alpha-1}}{(2\pi)^{d/2}}\int_{\mathbb{R}^d}|S(\boldsymbol{\xi})||\widetilde{G}_k(\boldsymbol{\xi},s)||E_{\alpha,\alpha}(-S(\boldsymbol{\xi})(t-s)^\alpha)||\mathrm{e}^{\mathrm{i}\,\boldsymbol{\xi}\cdot\boldsymbol{x}^j}|\,\mathrm{d}\boldsymbol{\xi} \\
&\le C(t-s)^{\alpha-1}\int_{\mathbb{R}^d}\left((1+|\boldsymbol{\xi}|^2)^{(d+5)/4}|\widetilde{G}_k(\boldsymbol{\xi},s)|\right)(1+|\boldsymbol{\xi}|^2)^{-(d+1)/4}\,\mathrm{d}\boldsymbol{\xi} \\
&\le C(t-s)^{\alpha-1}\left(\int_{\mathbb{R}^d}(1+|\boldsymbol{\xi}|^2)^{(d+5)/2}|\widetilde{G}_k(\boldsymbol{\xi},s)|^2\right)^{1/2} \qquad (5.32)\\
&\quad\times\left(\int_{\mathbb{R}^d}\frac{\mathrm{d}\boldsymbol{\xi}}{(1+|\boldsymbol{\xi}|^2)^{(d+1)/2}}\right)^{1/2} \\
&\le C\|G_k(\,\cdot\,,s)\|_{H^{(d+5)/2}(\mathbb{R}^d)}(t-s)^{\alpha-1},
\end{aligned}
$$

where we used the Cauchy-Schwarz inequality in (5.32). Since g, $\boldsymbol{\eta}$ are smooth and g is compactly supported, we see that $G_k = \partial_k(g\circ\boldsymbol{\eta})$ is also smooth and compactly supported, indicating the uniform boundedness of $\|G_k(\,\cdot\,,s)\|_{H^{(d+5)/2}(\mathbb{R}^d)}$ for $0<s<T$ and $k=1,\dots,d$. Therefore, again we arrive at (5.31) in the case $\Omega=\mathbb{R}^d$ with $0<\alpha<2$. Finally, for $\alpha=2$ we estimate in the same manner as

$$
\begin{aligned}
|q_{jk}(t,s)| &\le \frac{1}{(2\pi)^{d/2}}\int_{\mathbb{R}^d}S(\boldsymbol{\xi})^{1/2}|\widetilde{G}_k(\boldsymbol{\xi},s)||\sin(S(\boldsymbol{\xi})^{1/2}(t-s))||\mathrm{e}^{\mathrm{i}\,\boldsymbol{\xi}\cdot\boldsymbol{x}^j}|\,\mathrm{d}\boldsymbol{\xi} \\
&\le C(t-s)\int_{\mathbb{R}^d}S(\boldsymbol{\xi})|\widetilde{G}_k(\boldsymbol{\xi},s)|\,\mathrm{d}\boldsymbol{\xi} \le C(t-s)\int_{\mathbb{R}^d}(1+|\boldsymbol{\xi}|^2)|\widetilde{G}_k(\boldsymbol{\xi},s)|\,\mathrm{d}\boldsymbol{\xi} \\
&\le C(t-s)\left(\int_{\mathbb{R}^d}(1+|\boldsymbol{\xi}|^2)^{(d+5)/2}|\widetilde{G}_k(\boldsymbol{\xi},s)|^2\,\mathrm{d}\boldsymbol{\xi}\right)^{1/2} \\
&\le C(t-s)\|G_k(\,\cdot\,,s)\|_{H^{(d+5)/2}(\mathbb{R}^d)} \le C(t-s), \quad 1\le j,k\le d,
\end{aligned}
$$

which is consistent with (5.31). Consequently, it reveals that the upper bound (5.31) holds for both cases of domains and remains valid for any $0<\alpha\le 2$. This together with (5.6) implies the estimate

$$
|\boldsymbol{Q}(t,s)| \le C\|\boldsymbol{Q}(t,s)\|_{\mathrm{F}} = C\left(\sum_{j,k=1}^d|q_{jk}(t,s)|\right)^{1/2} \le C(t-s)^{\alpha-1}.
$$

The combination of (5.29), (5.30) and the above estimate yields

$$
\begin{aligned}
|\boldsymbol{\rho}(t)| &\le |\boldsymbol{P}(t)^{-1}| \left(|\partial_t^\alpha \boldsymbol{h}(t)| + \int_0^t |\boldsymbol{Q}(t,s)||\boldsymbol{\rho}(s)| \, \mathrm{d}s \right) \\
&\le C \left(|\partial_t^\alpha \boldsymbol{h}(t)| + \int_0^t (t-s)^{\alpha-1} |\boldsymbol{\rho}(s)| \, \mathrm{d}s \right).
\end{aligned}
$$

Eventually, we employ Grönwall's inequality with a weakly singular kernel (see Henry [9, Lemma 7.1.1]) to conclude

$$
\begin{aligned}
|\boldsymbol{\rho}(t)| &\le C \left(|\partial_t^\alpha \boldsymbol{h}(t)| + C \int_0^t \frac{\mathrm{d}}{\mathrm{d}\zeta} E_{\alpha,1}(\zeta^\alpha) \Big|_{\zeta = C(t-s)} |\partial_s^\alpha \boldsymbol{h}(s)| \, \mathrm{d}s \right) \\
&\le C \|\partial_t^\alpha \boldsymbol{h}\|_{C[0,T]} \left(1 + \int_0^t s^{\alpha-1} E_{\alpha,\alpha}(C s^\alpha) \, \mathrm{d}s \right) \le C \|\partial_t^\alpha \boldsymbol{h}\|_{C[0,T]}
\end{aligned}
$$

for $0 < t \le T$, and hence

$$
\begin{aligned}
\|\boldsymbol{\gamma}_1 - \boldsymbol{\gamma}_2\|_{C[0,T]} &= \|\boldsymbol{\rho}\|_{C[0,T]} \le C \|\partial_t^\alpha \boldsymbol{h}\|_{C[0,T]} \le C \left(\sum_{j=1}^d \|\partial_t^\alpha w(\boldsymbol{x}^j, \cdot)\|_{C[0,T]}^2 \right)^{1/2} \\
&\le C \sum_{j=1}^d \|\partial_t^\alpha (u_1 - u_2)(\boldsymbol{x}^j, \cdot)\|_{C[0,T]}.
\end{aligned}
$$

This completes the proof of Theorem 5.1. □

Remark 5.1 In three dimensions, if the source moves on the plane $\{x_3 = C\}$ where $C \in \mathbb{R}$ is known, then two observation points are sufficient to imply the stability. Analogously, if d' ($1 \le d' < d$, $d > 1$) components of the orbit function are known, then the number of observation points can be reduced to $d - d'$.

Proof (*Proof of Corollary* 5.1) Fix the set X of observation points and the constant ε in the assumption of Corollary 5.1. For any $\boldsymbol{\gamma} \in \mathcal{U}_0$, by $\|\boldsymbol{\gamma}'\|_{C[0,T]} \le K$ we know that $\boldsymbol{\gamma}(t) \in \overline{B_\varepsilon}$ for $t \le \varepsilon/K$. Then we define

$$
T_\ell = \begin{cases} \varepsilon\ell/K, & \ell = 0, 1, \ldots, \lceil KT/\varepsilon \rceil - 1, \\ T, & \ell = \lceil KT/\varepsilon \rceil \end{cases}
$$

and consider the intervals $[T_{\ell-1}, T_\ell]$ ($\ell = 1, \ldots, \lceil KT/\varepsilon \rceil$) successively.

We adopt an inductive argument and start from $\ell = 1$. On $[T_0, T_1] = [0, T_1]$, the above observation implies $\|\boldsymbol{\gamma}_i\|_{C[0,T_1]} \le \varepsilon$ ($i = 1, 2$). Taking $\boldsymbol{y} = \boldsymbol{0}$ in (5.14), we see that there exist d observation points $\{\boldsymbol{x}^j(\boldsymbol{0})\}_{j=1}^d \subset X$ satisfying (5.13). Observing that all assumptions in Theorem 5.1 are fulfilled, we utilize the uniqueness result in Theorem 5.1 to conclude that the relation $u_1(\boldsymbol{x}^j(\boldsymbol{0}), \cdot) = u_2(\boldsymbol{x}^j(\boldsymbol{0}), \cdot)$ $(j = 1, \dots, d)$ on $[0, T_1]$ implies $\boldsymbol{\gamma}_1 = \boldsymbol{\gamma}_2$ on $[0, T_1]$.

For general $\ell = 2, \dots, \lceil KT/\varepsilon \rceil$, we make the induction hypothesis that the relation $u_1(\boldsymbol{x}^j, \cdot) = u_2(\boldsymbol{x}^j, \cdot)$ $(j = 1, \dots, N)$ on $[0, T_{\ell-1}]$ implies $\boldsymbol{\gamma}_1 = \boldsymbol{\gamma}_2$ on $[0, T_{\ell-1}]$. By the well-posedness of the forward problem, we have $u_1 = u_2$ in $\Omega \times [0, T_{\ell-1}]$. Introducing

$$w_\ell(\boldsymbol{x}, t) := (u_1 - u_2)(\boldsymbol{x}, t + T_{\ell+1}),$$

we immediately see that w_ℓ satisfies the equation

$$(\partial_t^\alpha + \mathcal{L}) w_\ell = g(\boldsymbol{x} - \boldsymbol{\gamma}_1(t + T_{\ell-1})) - g(\boldsymbol{x} - \boldsymbol{\gamma}_2(t + T_{\ell-1}))$$

in $(\boldsymbol{\gamma}_1(T_{\ell-1}) + \Omega) \times (0, T - T_{\ell-1})$ with the homogeneous initial(-boundary) condition. Repeating the same argument as that in the proofs of Theorem 5.1 and the case of $\ell = 1$, we can take $\boldsymbol{y} = \boldsymbol{y}_\ell := \boldsymbol{\gamma}_1(T_{\ell-1})$ in (5.14), so that again we can find $\{\boldsymbol{x}^j(\boldsymbol{y}_\ell)\}_{j=1}^d \subset X$ such that (5.13) is fulfilled with $\boldsymbol{x}^j$ replaced by $\boldsymbol{x}^j(\boldsymbol{y}_\ell) - \boldsymbol{y}_\ell$ $(j = 1, \dots, d)$. Since all assumptions in Theorem 5.1 are satisfied in $(\boldsymbol{\gamma}_1(T_{\ell-1}) + \Omega) \times (0, T_\ell - T_{\ell-1})$, we conclude that $w_\ell(\boldsymbol{x}^j(\boldsymbol{y}_\ell), \cdot) = 0$ $(j = 1, \dots, d)$ on $[0, T_\ell - T_{\ell-1}]$ implies $\boldsymbol{\gamma}_1 - \boldsymbol{\gamma}_2 = \boldsymbol{0}$ on $[T_{\ell-1}, T_\ell]$ or equivalently,

$$\begin{aligned} &u_1(\boldsymbol{x}^j(\boldsymbol{y}_\ell), \cdot) = u_2(\boldsymbol{x}^j(\boldsymbol{y}_\ell), \cdot)\ (j = 1, \dots, d) \text{ on } [T_{\ell-1}, T_\ell] \\ &\text{implies} \quad \boldsymbol{\gamma}_1 = \boldsymbol{\gamma}_2 \text{ on } [T_{\ell-1}, T_\ell]. \end{aligned} \tag{5.33}$$

By the inductive argument, for any $\ell = 1, \dots, \lceil KT/\varepsilon \rceil$ there exists a set of d observation points $\{\boldsymbol{x}^j(\boldsymbol{y}_\ell)\}_{\ell=1}^d \subset X$ $(\boldsymbol{y}_\ell = \boldsymbol{\gamma}_1(T_{\ell-1}))$ such that (5.33) holds. Consequently, the proof is completed by collecting the uniqueness on all intervals $[T_{\ell-1}, T_\ell]$. □

Acknowledgements This work is partly supported by the A3 Foresight Program "Modeling and Computation of Applied Inverse Problems", Japan Society for the Promotion of Science (JSPS) and National Natural Science Foundation of China (NSFC). G. Hu is supported by the NSFC grant (No. 11671028) and NSAF grant (No. U1930402). Y. Liu and M. Yamamoto are supported by JSPS KAKENHI Grant Number JP15H05740. M. Yamamoto is partly supported by NSFC (Nos. 11771270, 91730303) and RUDN University Program 5-100.

References

1. R.A. Adams, *Sobolev Spaces* (Academic Press, New York, 1975)
2. Y.E. Anikonov, J. Cheng, M. Yamamoto, A uniqueness result in an inverse hyperbolic problem with analyticity. Eur. J. Appl. Math. **15**, 533–543 (2004)
3. A.E. Badia, T. Ha-Duong, Determination of point wave sources by boundary measurements. Inverse Probl. **17**, 1127–1139 (2001)
4. A.L. Bukhgeim, M.V. Klibanov, Global uniqueness of a class of multidimensional inverse problems. Sov. Math. Dokl. **24**, 244–247 (1981)

5. J. Cheng, V. Isakov, S. Lu, Increasing stability in the inverse source problem with many frequencies. J. Differ. Equ. **260**, 4786–4804 (2016)
6. M. Choulli, M. Yamamoto, Some stability estimates in determining sources and coefficients. J. Inverse Ill Posed Probl. **14**, 355–373 (2006)
7. S.D. Eidelman, A.N. Kochubei, Cauchy problem for fractional diffusion equations. J. Differ. Equ. **199**, 211–255 (2004)
8. K. Fujishiro, Y. Kian, Determination of time dependent factors of coefficients in fractional diffusion equations. Math. Control Relat. Fields **6**, 251–269 (2016)
9. D. Henry, *Geometric Theory of Semilinear Parabolic Equations* (Springer, Berlin, 1981)
10. G. Hu, Y. Kian, P. Li, Y. Zhao, Inverse moving source problems in electrodynamics. Inverse Probl. **35**, 075001 (2019)
11. V. Isakov, Stability in the continuation for the Helmholtz equation with variable coefficient, in *Control Methods in PDE Dynamical Systems, Contemporary Mathematics*, vol. 426 (AMS, Providence, RI, 2007), pp. 255–269
12. V. Isakov, *Inverse Source Problems* (AMS, Providence, RI, 1989)
13. D. Jiang, Y. Liu, M. Yamamoto, Inverse source problem for the hyperbolic equation with a time-dependent principal part. J. Differ. Equ. **262**, 653–681 (2017)
14. M.V. Klibanov, Inverse problems and Carleman estimates. Inverse Probl. **8**, 575–596 (1992)
15. V. Komornik, M. Yamamoto, Upper and lower estimates in determining point sources in a wave equation. Inverse Probl. **18**, 319–329 (2002)
16. V. Komornik, M. Yamamoto, Estimation of point sources and applications to inverse problems. Inverse Probl. **21**, 2051–2070 (2005)
17. Y. Liu, Strong maximum principle for multi-term time-fractional diffusion equations and its application to an inverse source problem. Comput. Math. Appl. **73**, 96–108 (2017)
18. Y. Liu, W. Rundell, M. Yamamoto, Strong maximum principle for fractional diffusion equations and an application to an inverse source problem. Fract. Calc. Appl. Anal. **19**, 888–906 (2016)
19. Y. Liu, Z. Zhang, Reconstruction of the temporal component in the source term of a (time-fractional) diffusion equation. J. Phys. A **50**, 305–203 (2017)
20. T. Nara, Algebraic reconstruction of the general-order poles of a meromorphic function. Inverse Probl. **28**, 025008 (2012)
21. T. Ohe, Real-time reconstruction of moving point/dipole wave sources from boundary measurements. Inverse Probl. Sci. Eng. (accepted)
22. T. Ohe, H. Inui, K. Ohnaka, Real-time reconstruction of time-varying point sources in a three-dimensional scalar wave equation. Inverse Probl. **27**, 115011 (2011)
23. I. Podlubny, *Fractional Differential Equations* (Academic Press, San Diego, 1999)
24. K. Sakamoto, M. Yamamoto, Initial value/boundary value problems for fractional diffusion-wave equations and applications to some inverse problems. J. Math. Anal. Appl. **382**, 426–447 (2011)
25. S.R. Umarov, E.M. Saidamatov, A generalization of Duhamel's principle for differential equations of fractional order. Dokl. Math. **75**, 94–96 (2007)
26. T. Wei, X.L. Li, Y.S. Li, An inverse time-dependent source problem for a time-fractional diffusion equation. Inverse Probl. **32**, 085003 (2016)
27. M. Yamamoto, Stability reconstruction formula and regularization for an inverse source hyperbolic problem by control method. Inverse Probl. **11**, 481–496 (1995)
28. M. Yamamoto, Uniqueness and stability in multidimensional hyperbolic inverse problems. J. Math. Pure Appl. **78**, 65–98 (1999)
29. Y. Zhang, X. Xu, Inverse source problem for a fractional diffusion equation. Inverse Probl. **27**, 035010 (2011)

Chapter 6
Inverse Problems for a Compressible Fluid System

Oleg Yu. Imanuvilov and Masahiro Yamamoto

Abstract In this article, we discuss the methodology based on Carleman estimates concerning the unique continuation and inverse problems of determining spatially varying coefficients. First as retrospective views we refer to main works by that methodology starting from the pioneering work by Bukhgeim and Klibanov published in 1981. Then as one possible object of the application of this methodology, we start to consider compressible fluid flows and we prove conditional stability estimates for an inverse source problem and the continuation of solutions from a part of lateral boundary. We apply two types of Carleman estimates respectively with a weight function which is quadratic in time and a weight function which rapidly decays at the end points of the time interval.

Keywords Compressible viscous fluid · Inverse source problem · Unique continuation · Conditional stability · Carleman estimate

---

O. Yu. Imanuvilov—Partially supported by NSF grant DMS 1312900.
M. Yamamoto—Supported by Grant-in-Aid for Scientific Research (S) 15H05740 of Japan Society for the Promotion of Science and prepared with the support of the "RUDN University Program 5-100".

O. Yu. Imanuvilov
Department of Mathematics, Colorado State University, 101 Weber Building, Fort Collins, CO 80523-1874, USA
e-mail: oleg@math.colostate.edu

M. Yamamoto (✉)
Graduate School of Mathematical Sciences, The University of Tokyo, 3-8-1 Komaba, Meguro-ku, Tokyo 153-8914, Japan
e-mail: myama@ms.u-tokyo.ac.jp

Honorary Member of Academy of Romanian Scientists, Splaiul Independentei Street, No. 54, 050094 Bucharest, Romania

Peoples' Friendship University of Russia (RUDN University), 6 Miklukho-Maklaya Street, Moscow 117198, Russian Federation

J. Cheng et al. (eds.), *Inverse Problems and Related Topics*, Springer Proceedings in Mathematics & Statistics 310,
https://doi.org/10.1007/978-981-15-1592-7_6

6.1 Introduction and Main Results

6.1.1 Review for the Main Methodology and the Results

It is a typical inverse problem to determine spatially varying coefficients and/or source terms in partial differential equations, and the mathematical analysis is not only an interesting subject but also indispensable from the practical point of view, related for example to qualification of the numerical methods. For the moment, we consider inverse problems for a general partial differential equation:

$$\mathcal{L}u := B(\partial_t u, \partial_t^2 u)(t, x') - A(x, D')u(t, x') = R(t, x')f(x'), \quad t \in I,\ x' \in \Omega. \tag{$*$}$$

Here $x' = (x_1, ..., x_n)$ and t denote the spatial and the time variables respectively, I is an interval, $\Omega \subset \mathbb{R}^n$ is a bounded domain, that is, a bounded connected open set, and u can be a vector-valued function, B is some function and $A(x, D')$ is a partial x'-differential operator, $R(t, x')f(x')$ is an external source term. For example, choosing $B(\eta_1, \eta_2) = \eta_k$, $k = 1, 2$ and an elliptic operator $A(x, D')$, the system is a parabolic equation and a hyperbolic equation respectively for $k = 1$ and $k = 2$.

Our inverse problem is formulated as follows:

Given $t_0 \in I$, a function $R(t, x')$ and a subboundary $\Gamma \subset \partial\Omega$, determine some coefficients of $A(x, D')$ and/or $f(x')$ by boundary data of u on $I \times \Gamma$ and data of u at $t = t_0$.

In other case where u is a solution to an initial boundary value problem for system ($*$), we can consider a finite number $u_1, ..., u_N$ of such solutions by changing initial and boundary values, and as data for the inverse problem, we can adopt data from $u_1, ..., u_N$.

Our formulation requires a finite number of solutions to ($*$). For inverse coefficient problems, we can refer to other typical formulation by Dirichlet-to-Neumann maps, which requires us to adopt data of an infinite number of solutions. In this paper we do not discuss inverse problems which involve the Dirichlet-to-Neumann map and refer for example to Isakov [55] as for more accounts for that formulation.

For our inverse problem, Bukhgeim and Klibanov [19] created a fundamental methodology which is based on Carleman estimates. By Carleman estimates we mean L^2-weighted estimates for solutions to the system ($*$) and the choices of the weight functions are essential and depend on the type of the system such as parabolic, hyperbolic equations. General treatments can be found for example in Hörmander [35] or Isakov [55], and we do not describe the details here. In particular, the method in [19] produced the uniqueness and the conditional stability for inverse problems by data of a finite number of solutions. We can further refer to Beilina and Klibanov [8], Klibanov [62, 63, 65], Klibanov and Timonov [68]. Klibanov [64] describes some backgrounds.

Since [19], many authors have applied the methodology and published important results. Among them, Imanuvilov and Yamamoto [43–45] first established the best possible Lipschitz stability over the whole domain Ω.

Here we list up such publications on the inverse problems similar to one studied in this paper according to the types of partial differential equations. However we do not intend complete lists and one can supplement by the references listed.

Single system:

- Hyperbolic equations:
  Baudouin, Crépeau and Valein [2], Baudouin, Mercado and Osses [4], Baudouin and Yamamoto [6], Beilina, Cristofol, Li and Yamamoto [7], Bellassoued [9], Bellassoued and Yamamoto [13], Doubova and Osses [25], Imanuvilov and Yamamoto [44, 45, 47], Isakov [56], Klibanov and Yamamoto [69], Riahi [77].
  Also see a monograph Bellassoued and Yamamoto [15]. Especially [4, 77] discuss the case where the principal coefficients are not smooth and see also Sect. 5 of Chap. 1 of [15]. The work [6] considers also the inverse problems for the heat and the Schrödinger equations in the case where the spatial domain is some kind of graph.
- Parabolic equations:
  Bellassoued and Yamamoto [12], Benabdallah, Cristofol, Gaitan and Yamamoto [16], Benabdallah, Dermenjian and Le Rousseau [17], Benabdallah, Gaitan and Le Rousseau [18], Cristofol, Gaitan and Ramoul [23], Imanuvilov and Yamamoto [43, 46], Poisson [76], Yamamoto [81], Yamamoto and Zou [83], Yuan and Yamamoto [85].
  For parabolic systems where the terms of lower-order derivatives are coupled, the same argument can produce the corresponding stability and uniqueness results to the case of single equations [16, 23].
- Schrödinger equations:
  Baudouin and Mercado [3], Baudouin and Puel [5], Mercado, Osses and Rosier [75], Yuan and Yamamoto [86].
- Ultrahyperbolic equations:
  Amirov [1], Gölgeleyen and Yamamoto [34],
- First-order equations (transport equations):
  Cannarsa, Floridia, Gölgeleyen and Yamamoto [20], Gaitan and Ouzzane [30], Gölgeleyen and Yamamoto [33]. In particular, as for the radiative transport equation called the Boltzmann equation, see Klibanov and Pamyatnykh [66, 67], Machida and Yamamoto [74].
- Plate equations:
  Yuan and Yamamoto [84].
- Integro-differential equations:
  Cavaterra, Lorenzi and Yamamoto [21], Loreti, Sforza and Yamamoto [73].
  This type of equations appear related to the viscoelasticity, and see de Buhan and Osses [24], Imanuvilov and Yamamoto [53, 54], Romanov and Yamamoto [78], for example.
- Fractional partial differential equations:
  In the system $(*)$, in place of ∂_t^k, one considers the fractional-order derivatives which are defined by

$$\partial_t^\alpha u(x,t) = \frac{1}{\Gamma(m-\alpha)} \int_0^t (t-s)^{-\alpha+m-1} \partial_s^m u(x,s) ds$$

for $m - 1 < \alpha < m$ with $m \in \mathcal{N}$, and such a fractional partial differential equation is a model equation for anomalous diffusion.
The research on inverse problems for fractional differential equations develop very rapidly and widely and we refer to limited literature: Kawamoto [60, 61], Huang, Li and Yamamoto [72], Xu, Cheng and Yamamoto [79], Yamamoto and Zhang [82].

Thus we may be able to conclude that the current state of researches based on the methodology by Carleman estimates for single equations is satisfactory.

Before the review for inverse problems for the system, we add some explanations of Carleman estimates. According to the types of the weight functions, we have two kinds of applicable Carleman estimates to the inverse problems:

- local Carleman estimate: the weight function is quadratic in time.
  see Proposition 6.3 for the compressible fluid equations.
  - We need not assume boundary condition on the whole lateral boundary $I \times \partial\Omega$ and we can localize data to a part of $I \times \partial\Omega$ to establish the local stability in Ω.
  - This Carleman estimate does not directly yield the global Lipschitz stability in Ω, but conditional Hölder stability under some a priori bounds, in general.
  - Local Carleman estimates can be proved for wider classes of partial differential equations including hyperbolic equations and transport equation (i.e., first-order partial differential equations).
- Global Carleman estimate: the weight function rapidly decays at the end points of the time interval.
  see Theorem 6.3 and Example at the end of Sect. 6.2.
  - We need to assume boundary condition on the whole lateral boundary $I \times \partial\Omega$ and we can establish the stability global in Ω.
  - We can prove Lipschitz stability under some a priori bounds.
  - Global Carleman estimates cannot be proved for hyperbolic equations and transport equation. Parabolic equations and Schrödinger equations admit the global Carleman estimates.

As works on Carleman estimates concerning inverse problems, we are restricted to refer to Imanuvilov [38, 40], Imanuvilov, Puel and Yamamoto [42], Imanuvilov and Yamamoto [46, 48].

In contrast with single equations, there are not many results on the inverse problems for systems of partial differential equations. Now we discuss some of them. First of all, Maxwell's equations in isotropic media are relatively easy, but in anisotropic cases or with less data, the inverse problems require more consideration. For example, we can refer to Li [71], Yamamoto [80].

Many systems of partial differential equations appearing in the mathematical physics are strongly coupled, and it is often difficult to establish Carleman estimates.

One reason is that there are no general theories for strongly coupling systems, although a general theory for Carleman estimates is available for single equations (e.g., [35, 55]). As such strongly coupling systems, the isotropic Lamé system is important:

$$\mu(x')\partial_t^2\mathbf{u}(t,x')=\mu(x')\Delta\mathbf{u}+(\mu(x')+\lambda(x'))\nabla'\mathrm{div}\,\mathbf{u}$$
$$+(\mathrm{div}\,\mathbf{u})\nabla'\lambda+(\nabla'\mathbf{u}+(\nabla'\mathbf{u})^T)\nabla'\mu.$$

Here $\nabla' = (\partial_{x_1}, ..., \partial_{x_n})$ with $x' = (x_1, ..., x_n)$ and $\cdot^T$ denotes the transposes of vectors and matrices under consideration. For Carleman estimates and the inverse problems for the isotropic Lamé system, we refer to Bellassoued, Imanuvilov and Yamamoto [10], Bellassoued and Yamamoto [14], Ikehata, Nakamura and Yamamoto [37], Imanuvilov, Isakov and Yamamoto [41], Imanuvilov and Yamamoto [49–52]. Isakov, Wang and Yamamoto [57, 58] discuss inverse problems for the Lamé system with residual stress. Moreover related to the viscoelasticity, we need to study the Lamé system with integral terms, and we refer to [24, 53, 54, 78].

Among the system in mathematical physics, one next main subject is inverse problems for fluid dynamics. The most fundamental model equation is the (incompressible) Navier–Stokes equations, and as for the Carleman estimates and inverse problems, we refer to Bellassoued, Imanuvilov and Yamamoto [11], Choulli, Imanuvilov, Puel and Yamamoto [22], Fan, Di Cristo, Jiang and Nakamura [27], Fan, Jiang and Nakamura [28]. Also see Huang and Yamamoto [36] for a system of Navier–Stokes equations and Maxwell's equations.

There are great variants of governing equations for the fluids, and as one comprehensive source book on the mathematical modeling, we refer to Giga and Novotný [32] which is a 3-volume handbook with the total pages more than 3,000. In view of the physical importance and mathematical wide interests, the inverse problems should be more exploited for fluid equations.

With the above retrospective views, in this article, now we start to discuss inverse problems for compressible fluid equations. As is later shown, the system is coupled with a strongly coupling parabolic system and a first-order partial differential equation. In the sense of choices of weight functions for Carleman estimates, the component equations of the compressible fluid system belong to different classes, as we explained related to the global and local Carleman estimates. Thus we need more careful consideration for establishing Carleman estimates. As we explained also, the local Carleman estimate is more suitable for the continuation problem of the solution where we are given local data on a part of lateral boundary. However we give also the proof for the local result by means of the global Carleman estimate, in order to demonstrate the applicability of the global one. Such an approach by the local Carleman estimate can be found in Imanuvilov and Yamamoto [46] for a parabolic equation.

6.1.2 Main Results for Inverse Problems for Compressible Fluids

We start with the description of the equation of the compressible fluid. Let $x' = (x_1, ..., x_n) \in \mathbb{R}^n$, and Ω be a bounded domain in $\mathbb{R}^n$ with $\partial\Omega \in C^\infty$, let $\boldsymbol{\nu} = \boldsymbol{\nu}(x')$ be the unit outward normal vector at x' to $\partial\Omega$. Here by $x' \in \mathbb{R}^n$ and t we denote the spatial variable and the time variable respectively. For a positive constant T, we introduce a cylindrical domain and its lateral boundary by

$$Q := (-T, T) \times \Omega, \quad \Sigma := (-T, T) \times \partial\Omega.$$

Here and henceforth $i = \sqrt{-1}$ and $\cdot^T$ denotes the transposes of vectors and matrices under consideration, and $D = (D_t, D')$, $D_t = \frac{1}{i}\partial_t$, $D' = (\frac{1}{i}\partial_{x_1}, \ldots, \frac{1}{i}\partial_{x_n})$, $\nabla' = (\partial_{x_1}, \ldots, \partial_{x_n})$, $\nabla = (\partial_t, \nabla')$. By $W_p^m(\Omega)$, $H^m(\Omega) := W_2^m(\Omega)$, we mean usual Sobolev spaces.

In the cylindrical domain Q, we consider an inverse source problem for an isothermal compressible viscous fluid equations:

$$\partial_t\rho + \mathrm{div}(\mathbf{v}\rho) = 0 \quad \text{in } Q, \tag{6.1}$$

$$\rho\partial_t\mathbf{v} - L_{\lambda,\mu}(x', D')\mathbf{v} + \rho(\mathbf{v}, \nabla')\mathbf{v} + h(\rho)\nabla'\rho = \mathbf{F} \quad \text{in } Q, \tag{6.2}$$

$$\mathbf{v}|_\Sigma = 0. \tag{6.3}$$

As for details of the model equations, see e.g., Landau and Lifshitz [70].

Here ρ and $\mathbf{v}$ describe the density and the velocity field of the fluid under consideration, respectively, and $\mathbf{F} = \mathbf{F}(t, x')$ is an external force. In Eq. (6.1), we assume no generation of fluids substances, that is, there does not exist any source term in (6.1). Moreover Eqs. (6.1) and (6.2) correspond to the conservation of mass and the conservation of energy, respectively. As for the mathematical treatments such as the well-posedness for an initial value problem, we refer to Itaya [59] for example.

The partial differential operator $L_{\lambda,\mu}(x, D')$ is defined by

$$\begin{aligned} L_{\lambda,\mu}(x, D')\mathbf{w} &= \mu(x)\Delta\mathbf{w} + (\mu(x) + \lambda(x))\nabla'\mathrm{div}\,\mathbf{w} \\ &+(\mathrm{div}\,\mathbf{w})\nabla'\lambda + (\nabla'\mathbf{w} + (\nabla'\mathbf{w})^T)\nabla'\mu, \qquad (t, x') \in Q. \end{aligned} \tag{6.4}$$

The coefficients ρ, λ, μ are assumed to satisfy

$$\rho, \lambda, \mu \in C^2(\overline{Q}), \quad \rho(t, x') > 0,\ \mu(t, x') > 0,\ \lambda(t, x') + \mu(t, x') > 0 \quad \text{on } \overline{Q}. \tag{6.5}$$

For any function $\mathbf{g} = (g_1, \ldots, g_n)$, we introduce the differential form $\omega_\mathbf{g} = \sum_{j=1}^n g_j dx_j$. Then $d\omega_\mathbf{f} = \sum_{k<j}^n (\partial_{x_j} g_k - \partial_{x_k} g_j) dx_k \wedge dx_j$.

We identify the differential form $d\omega_\mathbf{f}$ with the vector-function:

$$d\omega_{\mathbf{g}}$$
$$=\big(\partial_{x_2}g_1-\partial_{x_1}g_2,\ \ldots,\ \partial_{x_n}g_1-\partial_{x_1}g_n,\ \partial_{x_3}g_2-\partial_{x_2}g_3\ ,\ \ldots,$$
$$\partial_{x_n}g_2-\partial_{x_2}g_n,\ \ldots,\ \partial_{x_n}g_{n-1}-\partial_{x_{n-1}}g_n\big).$$

For the case $n=3$, we note that $d\omega_{\mathbf{g}}$ coincides with rot $\mathbf{g}$, and so we write $d\omega_{\mathbf{g}}=\text{rot}\,\mathbf{g}$ in general.

Henceforth let $\widetilde{\Gamma}$ be an arbitrarily given relatively open subset of $\partial\Omega$.

First we consider

Inverse Source Problem.

We assume that the external force $\mathbf{F}$ *is represented by*

$$\mathbf{F}(x)=R(t,x')\mathbf{f}(x')$$

with known matrix-valued function $R(t,x')$. *Let* $\rho(0,\cdot),\ \mathbf{v}(0,\cdot)$ *in* Ω *and* $\partial_{\boldsymbol{\nu}}\mathbf{v}$ *on* $(-T,T)\times\widetilde{\Gamma}$ *be given. Then determine a function* $\mathbf{f}(x')$.

We have

Theorem 6.1 *Let* $M_1>0,\ \delta_0,\delta_1>0$ *be constants,* $h\in C^2(\mathbb{R}^1)$ *and let the function* R *satisfy*

$$R\in C^1(\overline{Q}),\quad |det\,R(0,x')|\ge\delta_0>0,\quad x'\in\overline{\Omega}. \tag{6.6}$$

Moreover for $k=1,2$, *we assume that* $(\rho_k,\mathbf{v}_k)\in C^3(\overline{Q})\times C^3(\overline{Q})$ *satisfy (6.1)–(6.3) with* $\mathbf{F}(t,x')=R(t,x')\mathbf{f}_k(x')$, $\mathbf{f}_k\in H_0^1(\Omega)$ *and*

$$\|(\rho_k,\mathbf{v}_k)\|_{C^3(\overline{Q})\times C^3(\overline{Q})}\le M_1 \tag{6.7}$$

and

$$\rho_k(0,x')\ge\delta_1>0\quad\forall x'\in\overline{\Omega},\ \forall k\in\{1,2\}. \tag{6.8}$$

Furthermore let $\lambda=\lambda(x')$ *and* $\mu=\mu(x')$ *satisfy (6.5). Then there exists a constant* $C_1>0$ *independent of* $(\rho_k,\mathbf{v}_k)$ *such that*

$$\|\mathbf{f}_1-\mathbf{f}_2\|_{H^1(\Omega)}\le C_1(\|(\rho_1-\rho_2)(0,\cdot)\|_{H^3(\Omega)} \tag{6.9}$$

$$+\|(\mathbf{v}_1-\mathbf{v}_2)(0,\cdot)\|_{H^3(\Omega)}+\sum_{j=0}^{1}(\|\partial_{x_0}^j(d\omega_{\mathbf{v}_1-\mathbf{v}_2},div\,(\mathbf{v}_1-\mathbf{v}_2))\|_{H^{\frac34,\frac32}((-T,T)\times\widetilde{\Gamma})}$$
$$+\|\partial_{x_0}^j\partial_{\boldsymbol{\nu}}(d\omega_{\mathbf{v}_1-\mathbf{v}_2},div\,(\mathbf{v}_1-\mathbf{v}_2))\|_{H^{\frac14,\frac12}((-T,T)\times\widetilde{\Gamma})}$$
$$+\|\partial_{x_0}^{j+1}\partial_{\boldsymbol{\nu}}(\mathbf{v}_1-\mathbf{v}_2)\|_{L^2((-T,T)\times\widetilde{\Gamma})})).$$

The first equation (6.1) is a hyperbolic equation, and so if we do not know the initial value for ρ, then we must assume that T is sufficiently large and the coefficient

$\mathbf{v}$ of the principal term must satisfy some condition. As for it, although the equation is a simple scalar first-order equation, such conditions for $\mathbf{v}$ and T are discussed for example in Cannarsa, Floridia, Gölgeleyen and Yamamoto [20] and Gölgeleyen and Yamamoto [33], and we can expect that we need a similar condition if $\rho(0,\cdot)$ in Ω is not given.

In a forthcoming paper, we will discuss different inverses problems without data of initial value $\rho(0,\cdot)$ but with data of ρ on $(-T,T)\times\partial\Omega$.

Our inverse problem is concerned with a coupling of a strongly coupled parabolic system and a first-order transport equation. As for an inverse problem for a system of a single parabolic equation and a transport equation, Gaitan and Ouzzane [31] give information.

Second we consider

Continuation of solutions.

Let $S\subset\partial\Omega$ be a subboundary and be fixed. We assume that $(\rho,\mathbf{v})$ satisfies (6.1) and (6.2). Then in some neighborhood $\mathcal{O}$ of S, estimate $(\rho,\mathbf{v})$ by data of $(\rho,\mathbf{v},\partial_\nu\mathbf{v})$ on $S\times(-T,T)$.

We can state our main result for the continuation problem.

Theorem 6.2 *Assume that* $(\rho_k,\mathbf{v}_k)\in C^3(\overline{Q})\times C^3(\overline{Q})$, $k=1,2$, *satisfy the equations*

$$\partial_t\rho_k+div(\mathbf{v}_k\rho_k)=0 \quad in\ Q, \tag{6.10}$$

$$\rho_k\partial_t\mathbf{v}_k-L_{\lambda,\mu}(x',D')\mathbf{v}_k+\rho_k(\mathbf{v}_k,\nabla')\mathbf{v}_k+h(\rho_k)\nabla'\rho_k=0 \quad in\ Q. \tag{6.11}$$

Moreover

$$\|(\rho_k,\mathbf{v}_k)\|_{C^3(\overline{Q})\times C^3(\overline{Q})}\le M, \quad \forall k\in\{1,2\} \tag{6.12}$$

with some constant $M>0$, *and there exist a constant* $\beta_0>0$ *and a point* $a'\in\partial\Omega$ *such that*

$$\rho_k(0,a')\ge\beta_0>0 \quad in\ Q, \quad \forall k\in\{1,2\} \tag{6.13}$$

and

$$|(\mathbf{v}_1(0,a'),\boldsymbol{\nu}(a'))|>0. \tag{6.14}$$

Then there exist a subboundary $S\subset\partial\Omega$ *and an open set* U_0 *in* $\mathbb{R}^n$ *satisfying* $a'\in U_0\cap\partial\Omega\subset S$, *and constants* $\varepsilon>0$, $C=C(M)>0$, $\kappa=\kappa(M)\in(0,1)$ *such that*

$$\|\partial_t(\mathbf{v}_1-\mathbf{v}_2)\|_{L^2((-\varepsilon,\varepsilon)\times(U_0\cap\Omega))}+\|\partial_t div\,(\mathbf{v}_1-\mathbf{v}_2)\|_{L^2((-\varepsilon,\varepsilon)\times(U_0\cap\Omega))}$$

$$+\|\partial_t rot\,(\mathbf{v}_1-\mathbf{v}_2)\|_{L^2((-\varepsilon,\varepsilon)\times(U_0\cap\Omega))}$$
$$+\sum_{|\alpha'|=2}(\|\partial_{x'}^{\alpha'} div\,(\mathbf{v}_1-\mathbf{v}_2)\|_{L^2((-\varepsilon,\varepsilon)\times(U_0\cap\Omega))}+\|\partial_{x'}^{\alpha'} rot\,(\mathbf{v}_1-\mathbf{v}_2)\|_{L^2((-\varepsilon,\varepsilon)\times(U_0\cap\Omega))})$$
$$+\sum_{|\alpha'|\le 2}(\|\partial_{x'}^{\alpha'}(\mathbf{v}_1-\mathbf{v}_2)\|_{L^2((-\varepsilon,\varepsilon)\times(U_0\cap\Omega))}+\|\partial_{x'}^{\alpha'}(\rho_1-\rho_2)\|_{L^2((-\varepsilon,\varepsilon)\times(U_0\cap\Omega))})$$
$$\le C(\|\rho_1-\rho_2\|_{L^2(-T,T;H^2(S))}+\|\nabla_{x',t}\nabla(\mathbf{v}_1-\mathbf{v}_2)\|_{L^2(-T,T;L^2(S))})^{\kappa}.$$

Here and henceforth let $\mathcal{N}=\{1,2,3,...\}$, $t=x_0$, $\nabla=(\partial_{x_0},\partial_{x_1},...,\partial_{x_n})=(\partial_{x_0},\nabla')$ and $\partial_{x'}^{\alpha'}=\partial_{x_1}^{\alpha_1}\cdots\partial_{x_n}^{\alpha_n}$ with $\alpha'=(\alpha_1,...,\alpha_n)$ with $\alpha_k\in\mathcal{N}\cup\{0\}$ and $|\alpha'|=\alpha_1+\cdots+\alpha_n$.

To the best knowledge of the authors, Theorems 6.1 and 6.2 are the first results for inverse problems for the compressible viscous fluid equations. The paper [26] by Ervedoza, Glass and Guerrero studies the exact controllability for the compressible fluid case. On the other hand, for the incompressible viscous fluid equations (i.e., the Navier–Stokes equations), see Fernández-Cara, Guerrrero, Imanuvilov and Puel [29], Imanuvilov [39] for the null exact controllability.

This article is composed of five sections. In Sect. 6.2, we state the global Carleman estimate (Theorem 6.3) for the compressible fluid (6.1)–(6.2). Section 6.3 is devoted to the proof of Theorem 6.1. In Sect. 6.4, we prove the local Carleman estimate (Proposition 6.3) and complete another proof of Theorem 6.2. Finally Sect. 6.5 provides the proof of Theorem 6.2 by the global Carleman estimate Theorem 6.3.

6.2 Global Carleman Estimate

Henceforth we set $t=x_0$ and $x=(x_0,x_1,...,x_n)=(x_0,x')\in\mathbb{R}^{n+1}$ for the notation convenience. In this section, as Theorem 6.3 we state a Carleman estimate for a strongly coupling parabolic system:

$$P(x,D)\mathbf{u}\equiv\rho\partial_{x_0}\mathbf{u}-L_{\lambda,\mu}(x,D')\mathbf{u}=\mathbf{F}\quad\text{in } Q,\quad \mathbf{u}|_{\Sigma}=0, \tag{6.15}$$

where $\mathbf{u}=(u_1,\ldots,u_n)$.

Theorem 6.3 does not require data at $t=\pm T$, and is called a global Carleman estimate, which provides a global Lipschitz stability estimate with Dirichlet boundary data on Σ. Theorem 6.1 is based on such a global Carleman estimate. Also for Theorem 6.2, although we have no boundary data on Σ, we can apply the global Carleman estimate.

In order to formulate a Carleman estimate with boundary for system (6.15), we introduce notations and definitions following Imanuvilov and Yamamoto [54].

Let $\overline{\xi}$ denote the complex conjugate of $\xi\in\mathbf{C}$. We set $<a,b>=\sum_{k=0}^n a_k\overline{b}_k$ for $a=(a_0,\ldots,a_n), b=(b_0,\ldots,b_n)\in\mathbf{C}^n$, $\xi=(\xi_0,\ldots,\xi_n)$, $\xi'=(\xi_1,\ldots,\xi_n)$, $\widetilde{D}=(D_0,\ldots,D_{n-1})$.

For $\beta \in C^2(\overline{Q})$, we introduce the symbol:

$$p_{\rho,\beta}(x,\xi) = i\rho(x)\xi_0 + \beta(x)|\xi'|^2.$$

We recall that $\widetilde{\Gamma}$ is an arbitrarily given open subboundary of $\partial\Omega$. We set $\Gamma_0 := \partial\Omega \setminus \widetilde{\Gamma}$ and $\Sigma_0 = (-T,T) \times \Gamma_0$, $\widetilde{\Sigma} = (-T,T) \times \widetilde{\Gamma}$.

In order to prove the Carleman estimate for the strongly coupling parabolic system, we assume the existence of a real-valued function ψ which is pseudoconvex with respect to the symbols $p_{\rho,\mu}(x,\xi)$ and $p_{\rho,\lambda+2\mu}(x,\xi)$. More precisely, we can state as follows.

For functions $f(x,\xi)$ and $g(x,\xi)$, we introduce the Poisson bracket

$$\{f, g\} = \sum_{j=0}^{n} \left(\partial_{\xi_j} f \partial_{x_j} g - \partial_{\xi_j} g \partial_{x_j} f\right).$$

Denote

$$\widetilde{\varphi}(x_0) = \frac{1}{(x_0+T)^3(T-x_0)^3}. \tag{6.16}$$

We introduce

Condition 6.1 *We say that a function $\varphi\widetilde{\varphi}^{-1} \in C^{0,2}(\overline{Q})$ with $\partial_{x_0}(\varphi\widetilde{\varphi}^{-1}) \in L^\infty(Q)$, is pseudoconvex with respect to the symbol $p_{\rho,\beta}(x,\xi)$ if there exists a constant $C_1 > 0$ such that*

$$\frac{Im\{\overline{p}_{\rho,\beta}(x,\xi_0,\overline{\widetilde{\zeta}}), p_{\rho,\beta}(x,\xi_0,\widetilde{\zeta})\}}{|s|} > C_1\widetilde{\varphi}(x_0)M(\xi,\widetilde{\varphi}(x_0)s)^2 \quad \forall (x,\xi,s) \in \mathcal{S}, \tag{6.17}$$

where

$$\mathcal{S} = \{(x,\xi,s);\ x \in \overline{Q},\ \ M(\xi,\widetilde{\varphi}(x_0)s) = 1,\ p_{\rho,\beta}(x,\xi_0,\widetilde{\zeta}) = 0\},$$

$$\widetilde{\zeta} = (\xi_1 + i|s|\partial_{x_1}\varphi, \dots, \xi_n + i|s|\partial_{x_n}\varphi) \tag{6.18}$$

and $M(\xi,s) = (\xi_0^2 + \sum_{i=1}^n \xi_i^4 + s^4)^{\frac{1}{4}}$.

We assume that there exists a positive constant C_2 such that

$$\partial_{\boldsymbol{\nu}}\varphi(x)|_{\overline{\Sigma_0}} < 0 \quad \text{and} \quad |\partial_{\boldsymbol{\nu}}\varphi(x)| \geq C_2\widetilde{\varphi}(x_0) \quad \forall x \in \overline{\Sigma_0} \tag{6.19}$$

and

$$-\partial_{\boldsymbol{\nu}}\varphi(x) > \frac{1}{\sqrt{2}}\sqrt{\frac{\mu(x)}{(\lambda+2\mu)(x)}}|\partial_{\boldsymbol{\tau}}\varphi(x)| \quad \forall x \in \Sigma_0 \tag{6.20}$$

for all the unit tangential vectors $\boldsymbol{\tau}$.

Let us assume

$$\varphi(x) < 0 \quad \text{on } \overline{Q}, \quad \nabla'\varphi(x) \neq 0 \quad \text{on } \overline{Q}. \tag{6.21}$$

For the function β consider the symbol $a_\beta(x, \xi') = \beta(x)|\xi'|^2$.

Furthermore we introduce

Condition 6.2 *We say that a function $\varphi\widetilde{\varphi}^{-1} \in C^{0,2}(\overline{Q})$ is pseudoconvex with respect to the symbol $a_\beta(x, \xi')$ if there exists a constant $C_3 > 0$, which is independents of x_0, s, ξ', such that*

$$\frac{Im\{\overline{a}_\beta(x, \xi' - i|s|\nabla'\varphi), a_\beta(x, \xi' + i|s|\nabla'\varphi)\}}{|s|} > C_3\widetilde{\varphi}(x_0)|(\xi', \widetilde{\varphi}(x_0)s)|^2 \;\forall (x, \xi', s) \in \mathcal{K},$$

where

$$\mathcal{K} = \{(x, \xi', s);\; x \in \overline{Q},\; |(\xi', \varphi(x_0)s)| = 1,\; a_\beta(x, \xi' + i|s|\nabla'\varphi) = 0\}.$$

Let $\alpha = (\alpha_0, \dots, \alpha_n) =: (\alpha_0, \alpha')$ with $\alpha_0, ..., \alpha_n \in \mathcal{N} \cup \{0\}$ and $|\alpha| = 2\alpha_0 + \sum_{j=1}^n \alpha_j$, $\partial_x^\alpha = \partial_{x_0}^{\alpha_0}\partial_{x_1}^{\alpha_1} \cdots \partial_{x_n}^{\alpha_n}$.

Finally we assume that

$$\lim_{x_0 \to -T+0} \varphi(x_0, x') = \lim_{x_0 \to T-0} \varphi(x_0, x') = -\infty, \tag{6.22}$$

and

$$\frac{C_4}{(x_0+T)^3(T-x_0)^3} \le \sum_{|\alpha'| \le 2} |\partial_{x'}^{\alpha'}\varphi(x)| \le \frac{C_5}{(x_0+T)^3(T-x_0)^3} \tag{6.23}$$

for $x \in Q$, and

$$|\partial_{x_0}\varphi(x)| \le \frac{C_6}{(x_0+T)^4(T-x_0)^4} \quad x \in ((0, T) \times \overline{\Omega}) \cup ((-T, 0) \times \overline{\Omega}). \tag{6.24}$$

Denote

$$\|\mathbf{u}\|^2_{\mathcal{B}(\varphi,s,Q)} = \int_Q \Bigg(\sum_{|\alpha|=0, \alpha_0 \le 1}^{3} (s\widetilde{\varphi})^{4-2|\alpha|}|\partial_x^\alpha \mathbf{u}|^2$$
$$+ \sum_{|\alpha'| \le 2} (s\widetilde{\varphi})^{3-2|\alpha|}(|\partial_{x'}^{\alpha'}\nabla' d\omega_{\mathbf{u}}|^2 + |\partial_{x'}^{\alpha'}\nabla' \mathrm{div}\, \mathbf{u}|^2) + s\widetilde{\varphi}|\nabla' d\omega_{\mathbf{u}}|^2$$
$$+ (s\widetilde{\varphi})^3|d\omega_{\mathbf{u}}|^2 + s\widetilde{\varphi}|\nabla' \mathrm{div}\, \mathbf{u}|^2 + (s\widetilde{\varphi})^3|\mathrm{div}\, \mathbf{u}|^2\Bigg) e^{2s\varphi} dx, \tag{6.25}$$

$$\|\mathbf{u}\|^2_{\mathcal{X}(\varphi,s,\Omega)} = \int_\Omega \Bigg(\sum_{|\alpha'|=0}^{2} (s\widetilde{\varphi})^{4-2|\alpha|} |\partial_{x'}^{\alpha'} \mathbf{u}|^2 + s\widetilde{\varphi}|\nabla' d\omega_{\mathbf{u}}|^2$$

$$+(s\widetilde{\varphi})^3 |d\omega_{\mathbf{u}}|^2 + s\widetilde{\varphi}|\nabla' \mathrm{div}\, \mathbf{u}|^2 + (s\widetilde{\varphi})^3 |\mathrm{div}\, \mathbf{u}|^2 \Bigg) e^{2s\varphi} dx'.$$

Finally we introduce norms

$$\|\mathbf{F}e^{s\varphi}\|^2_{\mathcal{Y}(\varphi,s,Q)}$$

$$= \|\mathrm{div}\, \mathbf{F}e^{s\varphi}\|^2_{L^2(Q)} + \|d\omega_{\mathbf{F}} e^{s\varphi}\|^2_{L^2(Q)} + \|(s\widetilde{\varphi})^{\frac{1}{2}} \mathbf{F}e^{s\varphi}\|^2_{H^{\frac14,\frac12,\widetilde{s}}(\Sigma)} + \|\mathbf{F}e^{s\varphi}\|^2_{L^2(Q)},$$

and for any constant $p > 0$, we introduce the norm

$$\|\mathbf{u}\|_{H^{\frac{p}{2},p,\widetilde{s}}(\Sigma)} = (\|\mathbf{u}\|^2_{H^{\frac{p}{2},p}(\Sigma)} + \|(s\widetilde{\varphi})^p \mathbf{u}\|^2_{L^2(\Sigma)})^{\frac12}.$$

The following is proved in [54].

Theorem 6.3 *Let* $\mathbf{F}, div\, \mathbf{F}, d\omega_{\mathbf{F}} \in L^2(Q)$, *and let* $\mathbf{u} \in L^2(0, T; H^3(\Omega))$ *satisfy* $\partial_{x_0}\mathbf{u} \in L^2(0, T; H^1(\Omega))$ *and (6.15). Let (6.5) hold true, and let a function* φ *satisfy (6.22)–(6.24), Conditions 6.1 and 6.2 with* $\beta = \mu$ *and* $\beta = \lambda + 2\mu$. *Then there exists* $s_0 > 0$ *such that*

$$\|\mathbf{u}\|_{\mathcal{B}(\varphi,s,Q)} + \|\partial^2_{\boldsymbol{\nu}} \mathbf{u} e^{s\varphi}\|_{H^{\frac14,\frac12,\widetilde{s}}(\Sigma_0)} + \|\partial_{\boldsymbol{\nu}} \mathbf{u} e^{s\varphi}\|_{H^{\frac34,\frac32,\widetilde{s}}(\Sigma_0)} \tag{6.26}$$
$$\le C_7 \Big(\|\mathbf{F}e^{s\varphi}\|_{\mathcal{Y}(\varphi,s,Q)} + \|\partial_{\boldsymbol{\nu}} \mathbf{u} e^{s\varphi}\|_{H^{\frac34,\frac32,\widetilde{s}}(\widetilde{\Sigma})} + \|\partial^2_{\boldsymbol{\nu}} \mathbf{u} e^{s\varphi}\|_{H^{\frac14,\frac12,\widetilde{s}}(\widetilde{\Sigma})}$$
$$+ \|\widetilde{\varphi} \partial_{x_0} \partial_{\boldsymbol{\nu}} \mathbf{u} e^{s\varphi}\|_{L^2(\widetilde{\Sigma})} \Big)$$

for any $s > s_0$. *Here the constant* $C_7 > 0$ *is independent of* s.

Example of a function φ **which satisfies all conditions of Theorem** 6.3. Let $\eta \in C^3(\overline{\Omega})$ satisfy $\eta|_{\partial\Omega\setminus\widetilde{\Gamma}} = 0$, $|\nabla'\eta(x')| > 0$ on $x' \in \overline{\Omega}$, $\partial_{\boldsymbol{\nu}}\eta < 0$ on $\overline{\partial\Omega \setminus \widetilde{\Gamma}}$. We set

$$\varphi(x) = \frac{e^{\tau\eta(x')} - e^{2\tau\|\eta\|_{C(\overline{\Omega})}}}{\widetilde{\ell}(x_0)},$$

where $\widetilde{\ell} \in C^3[-T, T]$, $\widetilde{\ell}(x_0) > 0$ in $(-T, T)$, $\partial_{x_0}\widetilde{\ell}(x_0) < 0$ on $(0, T)$ and $\partial_{x_0}\widetilde{\ell}(x_0) > 0$ on $(-T, 0)$, $\partial^j_{x_0}\widetilde{\ell}(\pm T) = 0$ for all $j \in \{0, 1, 2\}$, $\partial^3_{x_0}\widetilde{\ell}(\pm T) \neq 0$.

Provided that parameter $\tau > 0$ is sufficiently large, we can prove that Conditions 6.1 and 6.2 hold true (e.g., Hörmander [35], Imanuvilov, Puel and Yamamoto [42]). The normal derivative of the function φ on Σ_0 is strictly negative and so (6.20) holds true. Inequality (6.21) follows from the fact that the function η does not have critical points on $\overline{\Omega}$. The properties of the function $\widetilde{\ell}$ imply (6.21)–(6.24).

6.3 Proof of Theorem 6.1

We recall that we set $x_0 = t$. Let $(\rho_j, \mathbf{v}_j)$, $j \in \{1, 2\}$ satisfy

$$\partial_{x_0}\rho_j + \mathrm{div}(\mathbf{v}_j\rho_j) = 0 \quad \text{in } Q, \tag{6.27}$$

$$\rho_j\partial_{x_0}\mathbf{v}_j - L_{\lambda,\mu}(x', D')\mathbf{v}_j + \rho_j(\mathbf{v}_j, \nabla')\mathbf{v}_j + h(\rho_j)\nabla'\rho_j = R\mathbf{f}_j \quad \text{in } Q, \tag{6.28}$$

$$\mathbf{v}_j|_\Sigma = 0. \tag{6.29}$$

We set $\rho = \rho_1 - \rho_1, \mathbf{v} = \mathbf{v}_1 - \mathbf{v}_2, \mathbf{f} = \mathbf{f}_1 - \mathbf{f}_2$. Then, from Eqs. (6.27)–(6.29) we have

$$\partial_{x_0}\rho + \mathrm{div}(\mathbf{v}_1\rho) = -\mathrm{div}(\mathbf{v}\rho_2) \quad \text{in } Q, \tag{6.30}$$

$$\begin{aligned} &\rho_1\partial_{x_0}\mathbf{v} + \rho\partial_{x_0}\mathbf{v}_2 - L_{\lambda,\mu}(x', D')\mathbf{v} + \rho(\mathbf{v}_1, \nabla')\mathbf{v}_1 + \rho_2(\mathbf{v}_1, \nabla')\mathbf{v} \\ &+\rho_2(\mathbf{v}, \nabla')\mathbf{v}_2 + h(\rho_1)\nabla'\rho + (h(\rho_1) - h(\rho_2))\nabla'\rho_2 = R\mathbf{f} \quad \text{in } Q, \end{aligned} \tag{6.31}$$

$$\mathbf{v}|_\Sigma = 0. \tag{6.32}$$

Let $y(t) = (y_1(t), \dots, y_n(t))$ be a solution to the system of ordinary differential equations

$$\frac{dy}{dt} = \mathbf{v}_1(t, y(t)).$$

The curve $(t, y(t))$ is a characteristic curve for the linear hyperbolic operator

$$K_{\mathbf{v}_1}(x, D)w := \partial_{x_0}w + (\mathbf{v}_1(x), \nabla'w). \tag{6.33}$$

Consider the initial value problem

$$K_{\mathbf{v}_1}(x, D)\psi = 0 \quad \text{on } Q, \quad \psi(0, x') = \psi_0(x') \quad x' \in \Omega \tag{6.34}$$

where the initial value $\psi_0 \in C^3(\overline{\Omega})$ satisfies

$$\partial_\nu\psi_0|_{\partial\Omega\setminus\widetilde{\Gamma}} < -C_1 < 0, \quad |\nabla'\psi_0(x')| > C_2 > 0 \quad \forall x \in \Omega, \tag{6.35}$$

$$-\partial_\nu\psi_0(x') > \frac{1}{\sqrt{2}}\sqrt{\frac{\mu(0, x')}{(\lambda + 2\mu)(0, x')}}|\partial_{\boldsymbol{\tau}}\psi_0(x')|, \quad \forall x' \in \overline{\Gamma}_0$$

for all the unit tangential vectors $\boldsymbol{\tau}$.

The existence of such a function ψ_0 is proved in [38] for example. Thanks to (6.29), there exists a unique solution $\psi \in C^3(\overline{Q})$ to the Cauchy problem (6.34) satisfying the estimate

$$\|\psi\|_{C^3(\overline{Q})} \le C_3\|\psi_0\|_{C^3(\overline{\Omega})}. \tag{6.36}$$

Then, by (6.36) and (6.35) there exist constants $\delta > 0$ and $C_4, C_5 > 0$ such that

$$\partial_{\boldsymbol{\nu}}\psi|_{[-\delta,\delta]\times(\partial\Omega\setminus\widetilde{\Gamma})} < -C_4 < 0, \quad |\nabla'\psi(x)| > C_5 > 0 \quad \forall x \in Q_\delta := (-\delta, \delta)\times\Omega, \tag{6.37}$$

$$-\partial_{\boldsymbol{\nu}}\psi(x) > \frac{1}{\sqrt{2}}\sqrt{\frac{\mu(x)}{(\lambda+2\mu)(x)}}|\partial_{\boldsymbol{\tau}}\psi(x)| \quad \forall x \in [-\delta,\delta]\times\overline{\Gamma}_0 \tag{6.38}$$

for all the unit tangential vectors $\boldsymbol{\tau}$.

In terms of the function ψ, we construct the weight function φ by

$$\varphi(x) = \frac{e^{\tau\psi(x)} - e^{2\tau\|\psi\|_{C(\overline{Q}_\delta)}}}{\widetilde{\ell}(x_0)}, \quad \widetilde{\varphi}(x_0) = \frac{1}{(\delta - x_0)^3(\delta + x_0)^3}, \tag{6.39}$$

where τ is a large positive parameter and $\widetilde{\ell}$ satisfies $\widetilde{\ell}(t) = 1 - |x_0|$ on $[-\delta/2, \delta/2]$, $\partial_{x_0}\widetilde{\ell}(x_0) < 0$ on $(0, \delta)$ and $\partial_{x_0}\widetilde{\ell}(x_0) > 0$ on $(-\delta, 0)$, $\widetilde{\ell}(x_0) > 0$ on $(-\delta, \delta)$, $\partial^k_{x_0}\widetilde{\ell}(\pm\delta) = 0$ for all $k \in \{0, 1, 2\}$, $\partial^3_{x_0}\widetilde{\ell}(\pm\delta) \ne 0$, $\widetilde{\ell} \in W^1_\infty(-\delta, \delta) \cap C^2[0, \delta] \cap C^2[-\delta, 0]$.

We have

Proposition 6.1 *We choose $\tau > 1$ sufficiently large. There exist a constant $\delta_0 > 0$, independent of τ, such that for all $\delta \in (0, \delta_0)$, the function φ given by (6.39) satisfies (6.19)–(6.24), Conditions 6.1 and 6.2 with $\beta = \mu$ and $\beta = \lambda + 2\mu$. Moreover there exist constants $C_6 > 0$, $C_7 > 0$ such that*

$$\frac{d\varphi(t, y(t))}{dt} < -C_6 < 0 \quad \text{for } t \in [0, \delta) \text{ and } \quad \frac{d\varphi(t, y(t))}{dt} > C_7 > 0 \quad \text{for } t \in (-\delta, 0]. \tag{6.40}$$

Moreover there exist constants $\tau_0 > 0$ and $C_8 > 0$ such that

$$\partial_{x_0}\varphi(x) < -C_8 < 0 \quad \text{on } [0, \delta)\times\overline{\Omega}, \quad \partial_{x_0}\varphi(x) > C_8 > 0 \quad \text{on } (-\delta, 0]\times\overline{\Omega} \tag{6.41}$$

for all $\tau > \tau_0$.

Proof By (6.39) and (6.34), short computations imply

$$\frac{d\varphi(t, y(t))}{dt} = \frac{d}{dt}\left(\frac{e^{\tau\psi(t,y(t))} - e^{2\tau\|\psi\|_{C(\overline{Q}_\delta)}}}{\widetilde{\ell}(t)}\right)$$

$$= \frac{e^{\tau\psi(t,y(t))}(K_{\mathbf{v}_1}(x, D)\psi)(t, y(t))}{\widetilde{\ell}(t)} - \frac{e^{\tau\psi(t,y(t))} - e^{2\tau\|\psi\|_{C(\overline{Q}_\delta)}}}{\widetilde{\ell}^2(t)}\widetilde{\ell}'(t)$$

$$= -\frac{e^{\tau\psi(t,y(t))} - e^{2\tau\|\psi\|_{C(\overline{Q}_\delta)}}}{\widetilde{\ell}^2(t)}\widetilde{\ell}'(t).$$

Since $\widetilde{\ell}'(t) < 0$ on $(0, \delta)$ and $\widetilde{\ell}'(t) > 0$ on $(-\delta, 0)$, from the above formula we have (6.40) for sufficiently large $\tau > 0$.

The inequality (6.19) follows from (6.37) and (6.39). Inequality (6.38) implies (6.20). Formula (6.39) immediately implies (6.22)–(6.24). Now we verify Condition 6.1. For simplicity of notations, we denote $p(x, \xi) = i\rho\xi_0 + \sum_{k,j=1}^n a_{k,j}\xi_k\xi_j$. Here we recall that $\widetilde{\zeta}$ is given by formula (6.18). Observe that

$$\partial_{x_m}\overline{p}(x, \xi_0, \overline{\widetilde{\zeta}}) = \overline{p}_m(x, \xi_0, \overline{\widetilde{\zeta}}) - i|s|\sum_{k=1}^n p^{(k)}(x, \xi_0, \overline{\widetilde{\zeta}})\partial^2_{x_k x_m}\varphi.$$

Then

$$\operatorname{Im}\{\overline{p}(x, \xi_0, \overline{\widetilde{\zeta}}), p(x, \xi_0, \widetilde{\zeta})\}$$
$$= \operatorname{Im}\left(\sum_{k=0}^n \overline{p}^{(k)}(x, \xi_0, \overline{\widetilde{\zeta}})\partial_{x_k}p(x, \xi_0, \widetilde{\zeta}) - \partial_{x_k}\overline{p}(x, \xi_0, \overline{\widetilde{\zeta}})p^{(k)}(x, \xi_0, \widetilde{\zeta})\right).$$

Here we set

$$a_0(x, \widetilde{\eta}, \widetilde{\eta}) = \sum_{k,j=1}^n \frac{\partial a_{k,j}(x)}{\partial x_0}\eta_k\eta_j, \quad a(x, \widetilde{\eta}, \widetilde{\eta}) = \sum_{k,j=1}^n a_{k,j}(x)\eta_k\eta_j.$$

Simple computations provide

$$\operatorname{Im}\left(\partial_{\xi_0}\overline{p}(x, \xi_0, \overline{\widetilde{\zeta}})\partial_{x_0}p(x, \xi_0, \widetilde{\zeta}) - \partial_{x_0}\overline{p}(x, \xi_0, \overline{\widetilde{\zeta}})\partial_{\xi_0}p(x, \xi_0, \widetilde{\zeta})\right)$$
$$= \operatorname{Im}\left((-i\rho)(p_0(x, \xi_0, \widetilde{\zeta}) + i|s|\sum_{m=1}^n p^{(m)}(x, \xi_0, \widetilde{\zeta})\partial^2_{x_m x_0}\varphi)\right.$$
$$\left. - i\rho(\overline{p}_0(x, \xi_0, \overline{\widetilde{\zeta}}) - i|s|\sum_{m=1}^n p^{(m)}(x, \xi_0, \overline{\widetilde{\zeta}})\partial^2_{x_m x_0}\varphi)\right\}$$
$$= -2\rho\operatorname{Re} p_0(x, \xi) + 2\rho s^2\sum_{m=1}^n p^{(m)}(x, \nabla'\varphi)\partial^2_{x_m x_0}\varphi + 2s^2\rho a_0(x, \nabla'\varphi, \nabla'\varphi).$$

Moreover

$$\operatorname{Im}\left(\partial_{\xi_k}\overline{p}(x, \xi_0, \overline{\widetilde{\zeta}})\partial_{x_k}p(x, \xi_0, \widetilde{\zeta}) - \partial_{x_k}\overline{p}(x, \xi_0, \overline{\widetilde{\zeta}})\partial_{\xi_k}p(x, \xi_0, \widetilde{\zeta})\right)$$
$$= \operatorname{Im}\left(\overline{p}^{(k)}(x, \xi_0, \overline{\widetilde{\zeta}})(p_k(x, \xi_0, \widetilde{\zeta}) + i|s|\sum_{m=1}^n p^{(m)}(x, \xi_0, \widetilde{\zeta})\partial^2_{x_k x_m}\varphi)\right.$$
$$\left. - p^{(k)}(x, \xi_0, \widetilde{\zeta})(\overline{p}_k(x, \xi_0, \overline{\widetilde{\zeta}}) - i|s|\sum_{m=1}^n p^{(m)}(x, \xi_0, \overline{\widetilde{\zeta}})\partial^2_{x_k x_m}\varphi)\right)$$

$$= -p^{(k)}(x, |s|\nabla'\varphi)\mathrm{Re}\, p_k(x, \xi_0, \overline{\widetilde{\zeta}}) + p^{(k)}(x, \xi)\mathrm{Im}\, p_k(x, \xi_0, \widetilde{\zeta})$$

$$+|s|p^{(k)}(x, \xi)\sum_{m=1}^{n} p^{(m)}(x, \xi)\partial^2_{x_k x_m}\varphi + |s|p^{(k)}(x, |s|\nabla'\varphi)\sum_{m=1}^{n} p^{(m)}(x, |s|\nabla'\varphi)\partial^2_{x_k x_m}\varphi$$

$$-p^{(k)}(x, |s|\nabla'\varphi)\mathrm{Re}\, \overline{p}_k(x, \xi_0, \overline{\widetilde{\zeta}}) - p^{(k)}(x, \xi)\mathrm{Im}\, p_k(x, \xi_0, \overline{\widetilde{\zeta}})$$

$$+|s|p^{(k)}(x, \xi)\sum_{m=1}^{n} p^{(m)}(x, \xi)\partial^2_{x_k x_m}\varphi + |s|p^{(k)}(x, |s|\nabla'\varphi)\sum_{m=1}^{n} p^{(m)}(x, |s|\nabla'\varphi)\partial^2_{x_k x_m}\varphi$$

for any $k \in \{1, \dots, n\}$. Therefore

$$\frac{1}{2}\mathrm{Im}\{\overline{p}(x, \xi_0, \overline{\widetilde{\zeta}}), p(x, \xi_0, \widetilde{\zeta})\} \tag{6.42}$$

$$= -\rho\mathrm{Re}\, p_0(x, \xi) + \rho s^2 \sum_{m=1}^{n} p^{(m)}(x, \nabla\varphi)\partial^2_{x_m x_0}\varphi + s^2\rho a_0(x, \nabla'\varphi, \nabla'\varphi)$$

$$+\sum_{k=1}^{n}\Big(- p^{(k)}(x, |s|\nabla'\varphi)(\mathrm{Re}\, p_k(x, \xi) - p_k(x, |s|\nabla'\varphi)) + p^{(k)}(x, \xi)\mathrm{Im}\, p_k(x, \xi)\Big)$$

$$+ \sum_{m,k=1}^{n} (|s|p^{(k)}(x, \xi)p^{(m)}(x, \xi) + |s|p^{(k)}(x, |s|\nabla'\varphi)p^{(m)}(x, |s|\nabla'\varphi))\partial^2_{x_k x_m}\varphi.$$

Observing that

$$\partial^2_{x_k x_m}\varphi = (\tau^2\partial_{x_k}\psi\partial_{x_m}\psi + \tau\partial^2_{x_k x_m}\psi)\frac{e^{\tau\psi}}{\widetilde{\ell}}$$

for any $k, m \in \{1, \dots, n\}$, we have

$$I = \sum_{m,k=1}^{n} (|s|p^{(k)}(x, \xi)p^{(m)}(x, \xi) + |s|p^{(k)}(x, |s|\nabla'\varphi)p^{(m)}(x, |s|\nabla'\varphi))\partial^2_{x_k x_m}\varphi$$

$$= \tau^2|s|(a(x, \xi', \nabla'\psi)^2 + s^2\frac{e^{2\tau\psi(x)}}{\widetilde{\ell}^2}a(x, \nabla'\psi, \nabla'\psi)^2)\frac{e^{\tau\psi}}{\widetilde{\ell}}$$

$$+ \sum_{m,k=1}^{n} (|s|p^{(k)}(x, \xi)p^{(m)}(x, \xi) + |s|p^{(k)}(x, |s|\nabla'\varphi)p^{(m)}(x, |s|\nabla'\varphi))\tau\partial^2_{x_k x_m}\psi\frac{e^{\tau\psi}}{\widetilde{\ell}}.$$

Since $(x, \xi, s) \in \mathcal{S}$, the following inequality holds:

$$a(x, \xi', \xi') = s^2 a(x, \nabla'\varphi, \nabla'\varphi) \geq C_9\tau^2\left|\left(\xi', s\frac{e^{\tau\psi}}{\widetilde{\ell}}\right)\right|^2,$$

provided that τ is sufficiently large. Taking $\widehat{\tau}$ sufficiently large, we have

$$I \geq \frac{\tau^4}{2} C_{10}|s| \frac{e^{\tau\psi}}{\widetilde{\ell}} \left| \left(\xi', s\frac{e^{\tau\psi}}{\widetilde{\ell}} \right) \right|^2 \quad \forall (x, \xi, s) \in \mathcal{S} \tag{6.43}$$

for all $\tau \geq \widehat{\tau}$, where the positive constant C_{10} is independent of (τ, x, ξ, s). Finally observing that

$$|\xi_0| \leq \frac{|a(x, \xi', |s|\nabla'\varphi)|}{\rho(x)} \quad \forall (x, \xi, s) \in \mathcal{S},$$

from (6.43) we find

$$I \geq C_{11} \frac{\tau^4 e^{\tau\psi}}{\widetilde{\ell}} |s| M^2 \left(\xi, s\frac{e^{\tau\psi}}{\widetilde{\ell}} \right) \quad \forall (x, \xi, s) \in \mathcal{S}, \tag{6.44}$$

where the positive constant C_{11} is independent of (τ, x, ξ, s). On the other hand

$$\begin{aligned} &\Bigg| -\rho \mathrm{Re}\, p_0(x, \xi) + \rho s^2 a_0(x, \nabla'\varphi, \nabla'\varphi) + \rho s^2 \sum_{m=1}^{n} p^{(m)}(x, \nabla'\varphi)\partial^2_{x_k x_0}\varphi \\ &+ \sum_{m=1}^{n} -p^{(m)}(x, |s|\nabla'\varphi)(p_m(x, \xi) - p_m(x, |s|\nabla'\varphi)) + p^{(m)}(x, \xi)\mathrm{Im}\, p_m(x, \xi) \Bigg| \\ &\leq C_{12}|s|\tau^2 \frac{e^{\tau\psi}}{\widetilde{\ell}} M^2 \left(\xi, s\frac{e^{\tau\psi}}{\widetilde{\ell}} \right). \end{aligned} \tag{6.45}$$

Inequalities (6.45) and (6.44) imply (6.17). In the same way, we can prove that the function φ satisfies Condition 6.2. In order to prove the inequalities in (6.41), we differentiate φ on $(0, \delta/2)$:

$$\begin{aligned} \partial_{x_0}\varphi &= \frac{e^{\tau\psi} - e^{2\tau\|\psi\|_{C(\overline{\mathcal{Q}}_\delta)}}}{\widetilde{\ell}^2(x_0)} + \frac{\tau e^{\tau\psi}\partial_{x_0}\psi}{\widetilde{\ell}(x_0)} = \frac{e^{\tau\psi} - e^{2\tau\|\psi\|_{C(\overline{\mathcal{Q}}_\delta)}}}{\widetilde{\ell}^2(x_0)} - \frac{\tau(\mathbf{v}_1, \nabla'\psi)e^{\tau\psi}}{\widetilde{\ell}(x_0)} \\ &\leq \frac{e^{\tau\psi} - e^{2\tau\|\psi\|_{C(\overline{\mathcal{Q}}_\delta)}}}{\widetilde{\ell}^2(x_0)} - \frac{\tau\|\mathbf{v}_1\|_{C(\overline{\mathcal{Q}}_\delta)}\|\psi_0\|_{C^3(\overline{\Omega})}e^{\tau\psi}}{\widetilde{\ell}(x_0)}. \end{aligned}$$

Hence for all τ sufficiently large, we proved the first inequality in (6.41) on $(0, \delta/2)$. Taking the derivative of φ on $(\delta/2, \delta]$, we have

$$\partial_{x_0}\varphi = \frac{\widetilde{\ell}'(x_0)}{\widetilde{\ell}^2(x_0)}(e^{\tau\psi} - e^{2\tau\|\psi\|_{C(\overline{\mathcal{Q}}_\delta)}}) + \frac{\tau e^{\tau\psi}\partial_{x_0}\psi}{\widetilde{\ell}(x_0)}$$

$$= \frac{-\widetilde{\ell}'(x_0)}{\widetilde{\ell}^2(x_0)}(e^{2\tau\|\psi\|_{C(\overline{Q}_\delta)}} - e^{\tau\psi}) - \frac{\tau(\mathbf{v}_1, \nabla'\psi)e^{\tau\psi}}{\widetilde{\ell}(x_0)}$$

$$\leq \frac{\inf_{x_0\in[\delta/2,\delta]}(-\widetilde{\ell}'(x_0))}{\widetilde{\ell}^2(x_0)}(e^{2\tau\|\psi\|_{C(\overline{Q}_\delta)}} - e^{\tau\psi}) - \frac{\tau\|\mathbf{v}_1\|_{C(\overline{Q}_\delta)}\|\psi_0\|_{C^3(\overline{\Omega})}e^{\tau\psi}}{\widetilde{\ell}(x_0)}.$$

Hence for all τ sufficiently large, we proved the first inequality in (6.41) on $(\delta/2, \delta]$. The proof of the second inequality in (6.41) is the same. Thus the proof of the proposition is complete. ■

Next we prove

Proposition 6.2 *For $p \in L^2(Q_\delta)$ and $\rho_0 \in L^2(\Omega)$, let $\widetilde{\rho} \in C([-\delta, \delta]; L^2(\Omega))$ be a solution to the initial value problem*

$$-K_{\mathbf{v}_1}(x, D)\widetilde{\rho} = p \quad in\ Q_\delta, \quad \widetilde{\rho}(0, \cdot) = \rho_0. \tag{6.46}$$

Then there exists a constant $C_{13} > 0$ independent of s such that

$$\|e^{s\varphi}\widetilde{\rho}\|_{L^2(Q_\delta)} \leq C_{13}\left(\|e^{s\varphi(0,\cdot)}\rho_0\|_{L^2(\Omega)} + \left\|\frac{p}{s\widetilde{\varphi}}e^{s\varphi}\right\|_{L^2(Q_\delta)}\right) \tag{6.47}$$

for all $s \geq 1$.

Proof For $x_0 > 0$, using the method of characteristics, we solve Eq. (6.46):

$$\widetilde{\rho}(x) = \rho_0(y(0))e^{\int_0^{x_0} \operatorname{div} \mathbf{v}_1(t,y(t))dt}$$
$$+e^{\int_0^{x_0} div\mathbf{v}_1(t,y(t))dt}\int_0^{x_0} p(t, y(t))e^{-\int_0^t \operatorname{div} \mathbf{v}_1(s,y(s))ds}dt. \tag{6.48}$$

We define the diffeomorphism $\mathcal{F}_{x_0}$ of domain Ω into Ω by

$$\mathcal{F}_{x_0}(x') = y(0),$$

where the function y solves the Cauchy problem

$$\frac{dy}{dt} = \mathbf{v}_1(t, y), \quad y(x_0) = x'.$$

By (6.40), we can prove

$$\|e^{s\varphi(x_0,y(x_0))}\rho_0(y(0))e^{\int_0^{x_0} \operatorname{div} \mathbf{v}_1(t,y(t))dt}\|_{L^2(\Omega)}$$

$$\leq \|e^{s\varphi(0,y(0))}\rho_0(y(0))e^{\int_0^{x_0} \operatorname{div} \mathbf{v}_1(t,y(t))dt}\|_{L^2(\Omega)}.$$

Short computations imply

$$\|e^{s\varphi(0,y(0))}\rho_0(y(0))e^{\int_0^{x_0}\operatorname{div}\mathbf{v}_1(t,y(t))dt}\|_{L^2(\Omega)} \le C_{14}\|e^{s\varphi(0,y(0))}\rho_0(y(0))\|_{L^2(\Omega)}$$

$$= \|e^{s\varphi(0,\mathcal{F}_{x_0}(x'))}\rho_0(\mathcal{F}_{x_0}(x'))\|_{L^2(\Omega)}.$$

Making the change of variables $\mathcal{F}_{x_0}(x') = z = (z_1, \ldots, z_n)$, we have

$$\|e^{s\varphi(0,\mathcal{F}_{x_0}(x'))}\rho_0(\mathcal{F}_{x_0}(x'))\|_{L^2(\Omega)} = \|e^{s\varphi(0,z)}\rho_0(z)|\det(\mathcal{F}_{x_0}^{-1})'|^{\frac{1}{2}}\|_{L^2(\Omega)}$$

$$\le C_{15}\|e^{s\varphi(0,\cdot)}\rho_0\|_{L^2(\Omega)}.$$

Therefore

$$\int_{-\delta}^{\delta}\|e^{s\varphi(x)}\rho_0(y(0))e^{\int_0^{x_0}\operatorname{div}\mathbf{v}_1(t,y(t))dt}\|^2_{L^2(\Omega)}dx_0 \le 2\delta C_{16}\|e^{s\varphi(0,\cdot)}\rho_0\|^2_{L^2(\Omega)}. \quad (6.49)$$

Next we estimate the L^2-norm of the second term on the right-hand side of (6.48). Then

$$\int_{-\delta}^{\delta}\int_{\Omega}\left(e^{\int_0^{x_0}\operatorname{div}\mathbf{v}_1(t,y(t))dt}e^{s\varphi(x)}\int_0^{x_0}|p(t,y(t))|e^{-\int_0^t\operatorname{div}\mathbf{v}_1(s,y(s))ds}dt\right)^2 dx_0dx'$$

$$\le C_{17}\int_{\Omega}\int_{-\delta}^{\delta}\left(e^{s\varphi(x)}\int_0^{x_0}p(t,y(t))dt\right)^2 dx_0dx'$$

$$\le C_{18}\int_{\Omega}\int_{-\delta}^{\delta}\frac{(K_{\mathbf{v}_1}(x,D)e^{2s\varphi(x)})}{2sK_{\mathbf{v}_1}(x,D)\varphi}\left(\int_0^{x_0}p(t,y(t))dt\right)^2 dx_0dx'.$$

Observe that $\frac{1}{K_{\mathbf{v}_1}(x,D)\varphi} \in W^1_\infty((0,\delta)\times\Omega)\cap W^1_\infty((-\delta,0)\times\Omega)$. Therefore

$$\mathcal{I} = \int_{Q_\delta}\frac{(K_{\mathbf{v}_1}(x,D)e^{2s\varphi})}{2sK_{\mathbf{v}_1}(x,D)\varphi}\left(\int_0^{x_0}p(t,y(t))dt\right)^2 dx_0dx'$$

$$= -\int_{Q_\delta}\frac{p}{sK_{\mathbf{v}_1}(x,D)\varphi}\left(\int_0^{x_0}p(t,y(t))dt\right)e^{2s\varphi}dx$$

$$+\frac{1}{2s}\int_{Q_\delta}e^{2s\varphi}K_{\mathbf{v}_1}(x,D)^*\left(\frac{1}{K_{\mathbf{v}_1}(x,D)\varphi}\right)\left(\int_0^{x_0}p(t,y(t))dt\right)^2 dx_0dx'$$

$$-\frac{1}{2s}\int_{Q_\delta}e^{2s\varphi}\operatorname{div}\mathbf{v}_1\left(\frac{1}{K_{\mathbf{v}_1}(x,D)\varphi}\right)\left(\int_0^{x_0}p(t,y(t))dt\right)^2 dx_0dx'$$

$$=: \sum_{k=1}^{3}\mathcal{I}_k.$$

Simple computations imply

$$\left|\frac{1}{K_{\mathbf{v}_1}(x,D)\varphi}\right| \le \frac{C_{19}}{\widetilde{\varphi}} \quad \text{and} \quad \left|K_{\mathbf{v}_1}(x,D)\left(\frac{1}{K_{\mathbf{v}_1}(x,D)\varphi}\right)\right| \le \frac{C_{20}}{\widetilde{\varphi}} \quad \text{in } Q_\delta.$$

We set $K_{\mathbf{v}_1}(x,D)^*\psi = -\partial_{x_0}\psi - \operatorname{div}(\mathbf{v}_1\psi)$, which is the formal adjoint operator to $K_{\mathbf{v}_1}(x,D)$. Therefore

$$\begin{aligned}
|\mathcal{I}_2| + |\mathcal{I}_3| &= \left|\frac{1}{2s}\int_{Q_\delta} e^{2s\varphi}\operatorname{div}\mathbf{v}_1\left(\frac{1}{K_{\mathbf{v}_1}(x,D)\varphi}\right)\left(\int_0^{x_0} p(t,y(t))dt\right)^2 dx_0dx'\right| \\
&+ \left|\frac{1}{2s}\int_{Q_\delta} e^{2s\varphi}K_{\mathbf{v}_1}(x,D)^*\left(\frac{1}{K_{\mathbf{v}_1}(x,D)\varphi}\right)\left(\int_0^{x_0} p(t,y(t))dt\right)^2 dx_0dx'\right| \\
&\le C_{21}\int_{Q_\delta} e^{2s\varphi}\frac{1}{s\widetilde{\varphi}}\left(\int_0^{x_0} p(t,y(t))dt\right)^2 dx_0dx'.
\end{aligned} \tag{6.50}$$

By (6.40), there exists a constant $C_{22} > 0$ independent of s such that

$$|\mathcal{I}_1| \le \frac{|\mathcal{I}|}{2} + C_{22}\left\|\frac{pe^{s\varphi}}{s\widetilde{\varphi}}\right\|^2_{L^2(Q)}.$$

This inequality with (6.49) and (6.50), implies (6.47). Thus the proof of the proposition is complete. ■

Applying Proposition 6.2 to Eq. (6.27), we have

$$\begin{aligned}
\|e^{s\varphi}\rho\|_{L^2(Q_\delta)} \le C_{23}\Bigg(\|e^{s\varphi(0,\cdot)}\rho(0,\cdot)\|_{L^2(\Omega)} + \left\|\frac{1}{s\widetilde{\varphi}}\operatorname{div}\mathbf{v}e^{s\varphi}\right\|_{L^2(Q_\delta)} \\
+ \left\|\frac{1}{s\widetilde{\varphi}}\mathbf{v}e^{s\varphi}\right\|_{L^2(Q_\delta)}\Bigg), \quad \forall s \ge s_0,
\end{aligned} \tag{6.51}$$

where s_0 is sufficiently large. From (6.30) we have

$$K_{\mathbf{v}_1}(x,D)\partial_{x'}^{\alpha'}\rho = f_{\alpha'} \quad \text{in } Q, \tag{6.52}$$

where

$$f_{\alpha'} = [K_{\mathbf{v}_1}, \partial_{x'}^{\alpha'}]\rho - \partial_{x'}^{\alpha'}\operatorname{div}(\mathbf{v}\rho_2) - \partial_{x'}^{\alpha'}(\operatorname{div}\mathbf{v}_1\rho).$$

Let $|\alpha'| = 1$. Applying Proposition 6.2 to Eq. (6.52) and using (6.51), we obtain

$$\|e^{s\varphi}\nabla'\rho\|_{L^2(Q_\delta)} \le C_{24}\Bigg(\sum_{|\alpha'|\le 1}\|e^{s\varphi(0,\cdot)}\partial_{x'}^{\alpha'}\rho(0,\cdot)\|_{L^2(\Omega)}$$
$$+\left\|e^{s\varphi}\frac{1}{s\widetilde{\varphi}}\nabla'\mathrm{div}\,\mathbf{v}\right\|_{L^2(Q_\delta)}+\sum_{|\alpha'|\le 1}\left\|\frac{1}{s\widetilde{\varphi}}e^{s\varphi}\partial_{x'}^{\alpha'}\mathbf{v}\right\|_{L^2(Q_\delta)}\Bigg) \tag{6.53}$$

for all $s \ge s_1$. Let $|\alpha| = 2$ with $\alpha_0 = 0$. Applying Proposition 6.2 to Eq. (6.52) and using (6.51) and (6.53), we obtain

$$\sum_{|\alpha'|\le 2}\|e^{s\varphi}\partial_{x'}^{\alpha'}\rho\|_{L^2(Q_\delta)} \le C_{25}\Bigg(\sum_{|\alpha'|\le 2}\|e^{s\varphi(0,\cdot)}\partial_{x'}^{\alpha'}\rho(0,\cdot)\|_{L^2(\Omega)}$$
$$+\sum_{|\alpha'|\le 2}\left\|e^{s\varphi}\frac{1}{s\widetilde{\varphi}}\partial_{x'}^{\alpha'}\mathrm{div}\,\mathbf{v}\right\|_{L^2(Q_\delta)}+\sum_{|\alpha'|\le 2}\left\|e^{s\varphi}\frac{1}{s\widetilde{\varphi}}\partial_{x'}^{\alpha'}\mathbf{v}\right\|_{L^2(Q_\delta)}\Bigg) \tag{6.54}$$

for all $s \ge s_2$. From (6.27) and (6.54), we have

$$\|\nabla'\partial_{x_0}\rho e^{s\varphi}\|_{L^2(Q_\delta)} + \|\partial_{x_0}\rho e^{s\varphi}\|_{L^2(Q_\delta)} \tag{6.55}$$

$$\le C_{26}\Bigg(\sum_{|\alpha'|\le 2}\|e^{s\varphi}\partial_{x'}^{\alpha'}\rho\|_{L^2(Q_\delta)}+\sum_{|\alpha'|\le 2}\left\|e^{s\varphi}(s\widetilde{\varphi})^{-1}\partial_{x'}^{\alpha'}\mathrm{div}\,\mathbf{v}\right\|_{L^2(Q_\delta)}$$
$$+\sum_{|\alpha'|\le 2}\left\|e^{s\varphi}(s\widetilde{\varphi})^{-1}\partial_{x'}^{\alpha'}\mathbf{v}\right\|_{L^2(Q_\delta)}\Bigg)$$
$$\le C_{27}\Bigg(\|e^{s\varphi(0,\cdot)}\rho(0,\cdot)\|_{H^{2,s}(\Omega)}+\sum_{|\alpha'|\le 2}\left\|e^{s\varphi}(s\widetilde{\varphi})^{-1}\partial_{x'}^{\alpha'}\mathrm{div}\,\mathbf{v}\right\|_{L^2(Q_\delta)}$$
$$+\sum_{|\alpha'|\le 2}\left\|e^{s\varphi}(s\widetilde{\varphi})^{-1}\partial_{x'}^{\alpha'}\mathbf{v}\right\|_{L^2(Q_\delta)}\Bigg)$$

for all $s \ge s_3$. We differentiate (6.30) with respect to variable x_0 to have

$$K_{\mathbf{v}_1}(x,D)\partial_{x'}^{\alpha'}\partial_{x_0}\rho = \widetilde{f}_{\alpha'} \quad \text{in } Q, \tag{6.56}$$

where

$$\widetilde{f}_{\alpha'} = [K_{\mathbf{v}_1}, \partial_{x'}^{\alpha'}\partial_{x_0}]\rho - \partial_{x'}^{\alpha'}\partial_{x_0}\mathrm{div}(\mathbf{v}\rho_2) - \partial_{x'}^{\alpha'}\partial_{x_0}(\mathrm{div}\,\mathbf{v}_1\rho).$$

From (6.56) and (6.55), we obtain

$$\sum_{|\alpha'|\le 2} \|\partial_{x'}^{\alpha'}\partial_{x_0}\rho e^{s\varphi}\|_{L^2(Q_\delta)} \tag{6.57}$$

$$\le C_{28}\left(\sum_{|\alpha'|\le 2}\|e^{s\varphi(0,\cdot)}\partial_{x'}^{\alpha'}\partial_{x_0}\rho(0,\cdot)\|_{L^2(\Omega)} + \sum_{|\alpha'|\le 2}\left\|e^{s\varphi}\frac{1}{s\widetilde{\varphi}}\partial_{x'}^{\alpha'}\widetilde{f}_{\alpha'}\right\|_{L^2(Q_\delta)}\right)$$

$$\le C_{29}\Bigg(\|e^{s\varphi(0,\cdot)}\rho(0,\cdot)\|_{H^{3,s}(\Omega)} + \|e^{s\varphi(0,\cdot)}\mathbf{v}(0,\cdot)\|_{H^{3,s}(\Omega)}$$

$$+\sum_{|\alpha'|\le 2}\left\|e^{s\varphi}\frac{1}{s\widetilde{\varphi}}\partial_{x'}^{\alpha'}\operatorname{div}\mathbf{v}\right\|_{L^2(Q_\delta)} + \sum_{|\alpha'|\le 2}\left\|e^{s\varphi}\frac{1}{s\widetilde{\varphi}}\partial_{x'}^{\alpha'}\mathbf{v}\right\|_{L^2(Q_\delta)}\Bigg)$$

for all $s \ge s_4$. Observe that ψ satisfies (6.34)–(6.38) and φ defined by (6.39) satisfies (6.21)–(6.24), Conditions 6.1 and 6.2 provided that parameter τ is large enough.

Applying Carleman estimate (6.26) to Eq. (6.31), we have

$$\begin{aligned}\|\mathbf{v}\|_{\mathcal{B}(\varphi,s,Q_\delta)} + \|\partial_{\boldsymbol{\nu}}^2\mathbf{v}e^{s\varphi}\|_{H^{\frac14,\frac12,\widetilde{s}}(\Sigma_{0,\delta})} + \|\partial_{\boldsymbol{\nu}}\mathbf{v}e^{s\varphi}\|_{H^{\frac34,\frac32,\widetilde{s}}(\Sigma_{0,\delta})} \\ \le C_{30}(\|\mathbf{F}e^{s\varphi}\|_{\mathcal{Y}(\varphi,s,Q_\delta)} + \|(d\omega_{\mathbf{v}},\operatorname{div}\mathbf{v})e^{s\varphi}\|_{H^{\frac34,\frac32,\widetilde{s}}(\widetilde{\Sigma}_\delta)} \\ +\|\partial_{\boldsymbol{\nu}}(d\omega_{\mathbf{v}},\operatorname{div}\mathbf{v})e^{s\varphi}\|_{H^{\frac14,\frac12,\widetilde{s}}(\widetilde{\Sigma}_\delta)} + \|\widetilde{\varphi}\partial_{x_0}\partial_{\boldsymbol{\nu}}\mathbf{v}e^{s\varphi}\|_{L^2(\widetilde{\Sigma}_\delta)})\end{aligned} \tag{6.58}$$

for all $s \ge s_5$, where we set

$$\mathbf{F} = -\rho\partial_{x_0}\mathbf{v}_2 - (\mathbf{v},\nabla')\mathbf{v}_2 - h(\rho_1)\nabla'\rho - (h(\rho_1)-h(\rho_2))\nabla'\rho_2 + R\mathbf{f},$$

and $\widetilde{\Sigma}_\delta = (-\delta,\delta)\times\widetilde{\Gamma}$, $\widetilde{\Sigma}_{0,\delta} = (-\delta,\delta)\times(\partial\Omega\setminus\overline{\widetilde{\Gamma}})$.

From (6.54) and (6.58) we obtain

$$\begin{aligned}\|\mathbf{v}\|_{\mathcal{B}(\varphi,s,Q_\delta)} + \|\partial_{\boldsymbol{\nu}}^2\mathbf{v}e^{s\varphi}\|_{H^{\frac14,\frac12,\widetilde{s}}(\Sigma_{0,\delta})} + \|\partial_{\boldsymbol{\nu}}\mathbf{v}e^{s\varphi}\|_{H^{\frac34,\frac32,\widetilde{s}}(\Sigma_{0,\delta})} \\ \le C_{31}(\|R\mathbf{f}e^{s\varphi}\|_{\mathcal{Y}(\varphi,s,Q_\delta)} + \|(d\omega_{\mathbf{v}},\operatorname{div}\mathbf{v})e^{s\varphi}\|_{H^{\frac34,\frac32,\widetilde{s}}(\widetilde{\Sigma}_\delta)} \\ +\|\partial_{\boldsymbol{\nu}}(d\omega_{\mathbf{v}},\operatorname{div}\mathbf{v})e^{s\varphi}\|_{H^{\frac14,\frac12,\widetilde{s}}(\widetilde{\Sigma}_\delta)} + \|\widetilde{\varphi}\partial_{x_0}\partial_{\boldsymbol{\nu}}\mathbf{v}e^{s\varphi}\|_{L^2(\widetilde{\Sigma}_\delta)} \\ +\|e^{s\varphi(0,\cdot)}\rho(0,\cdot)\|_{H^{2,s}(\Omega)})\end{aligned} \tag{6.59}$$

for all $s \ge s_6$.

Next we differentiate Eqs. (6.27)–(6.29) with respect to x_0. Setting $\widetilde{\rho} = \partial_{x_0}\rho$ and $\mathbf{u} = \partial_{x_0}\mathbf{v}$, we obtain

$$\begin{cases} \rho_1\partial_{x_0}\mathbf{u} + \widetilde{\rho}\partial_{x_0}\mathbf{v}_2 + L_{\lambda,\mu}(x',D')\mathbf{u} + (\mathbf{v}_1,\nabla')\mathbf{u} + (\mathbf{u},\nabla')\mathbf{v}_2 + h(\rho_1)\nabla'\widetilde{\rho} \\ +\ \partial_{x_0}(h(\rho_1)-h(\rho_2))\nabla'\rho_2 = \partial_{x_0}R(x)\mathbf{f} + \mathbf{G} \quad \text{in } Q, \\ \mathbf{u}|_\Sigma = 0, \end{cases} \tag{6.60}$$

where we set

$$\mathbf{G} = -\partial_{x_0}\rho_1\partial_{x_0}\mathbf{v} - \rho\partial_{x_0}^2\mathbf{v}_2 - (\partial_{x_0}\mathbf{v}_1, \nabla')\mathbf{v} - (\mathbf{v}, \nabla')\partial_{x_0}\mathbf{v}_2$$
$$-\partial_{x_0}h(\rho_1)\nabla'\rho - (h(\rho_1) - h(\rho_2))\nabla'\partial_{x_0}\rho_2.$$

Applying Carleman estimate (6.26) to Eq. (6.60), we have

$$\begin{aligned} &\|\mathbf{u}\|_{\mathcal{B}(\varphi,s,Q_\delta)} + \|\partial_{\boldsymbol{\nu}}^2\mathbf{u}e^{s\varphi}\|_{H^{\frac14,\frac12,\widetilde{s}}(\Sigma_{0,\delta})} + \|\partial_{\boldsymbol{\nu}}\mathbf{u}e^{s\varphi}\|_{H^{\frac34,\frac32,\widetilde{s}}(\Sigma_{0,\delta})} \\ &\le C_{32}(\|\mathbf{G}e^{s\varphi}\|_{\mathcal{Y}(\varphi,s,Q_\delta)} + \|(d\omega_{\mathbf{u}}, \operatorname{div}\mathbf{u})e^{s\varphi}\|_{H^{\frac34,\frac32,\widetilde{s}}(\widetilde{\Sigma}_\delta)} \\ &+\|\partial_{\boldsymbol{\nu}}(d\omega_{\mathbf{u}}, \operatorname{div}\mathbf{u})e^{s\varphi}\|_{H^{\frac14,\frac12,\widetilde{s}}(\widetilde{\Sigma}_\delta)} + \|\widetilde{\varphi}\partial_{x_0}\partial_{\boldsymbol{\nu}}\mathbf{u}e^{s\varphi}\|_{L^2(\widetilde{\Sigma}_\delta)}) \end{aligned} \tag{6.61}$$

for all $s \ge s_7$.

Using (6.57) and (6.59), from (6.61) we obtain

$$\begin{aligned} &\sum_{k=0}^{1}(\|\partial_{x_0}^k\mathbf{v}\|_{\mathcal{B}(\varphi,s,Q_\delta)} + \|\partial_{\boldsymbol{\nu}}^2\partial_{x_0}^k\mathbf{v}e^{s\varphi}\|_{H^{\frac14,\frac12,\widetilde{s}}(\Sigma_{0,\delta})} + \|\partial_{\boldsymbol{\nu}}\partial_{x_0}^k\mathbf{v}e^{s\varphi}\|_{H^{\frac34,\frac32,\widetilde{s}}(\Sigma_{0,\delta})}) \\ &\le C_{33}\Bigg(\|\mathbf{f}e^{s\varphi}\|_{\mathcal{Y}(\varphi,s,Q_\delta)} + \sum_{k=0}^{1}(\|\partial_{x_0}^k(d\omega_{\mathbf{v}}, \operatorname{div}\mathbf{v})e^{s\varphi}\|_{H^{\frac34,\frac32,\widetilde{s}}(\widetilde{\Sigma}_\delta)} \\ &+\|\partial_{x_0}^k\partial_{\boldsymbol{\nu}}(d\omega_{\mathbf{v}}, \operatorname{div}\mathbf{v})e^{s\varphi}\|_{H^{\frac14,\frac12,\widetilde{s}}(\widetilde{\Sigma}_\delta)} \\ &+\|\widetilde{\varphi}\partial_{x_0}^{k+1}\partial_{\boldsymbol{\nu}}\mathbf{v}e^{s\varphi}\|_{L^2(\widetilde{\Sigma}_\delta)} + \|e^{s\varphi(0,\cdot)}\partial_{x_0}^k\rho(0,\cdot)\|_{H^{2,s}(\Omega)}\Bigg) \end{aligned} \tag{6.62}$$

for all $s \ge s_8$. By (6.41), for any $\epsilon > 0$ there exists a constant $s_9(\epsilon)$ such that

$$\|\mathbf{f}e^{s\varphi}\|_{\mathcal{Y}(\varphi,s,Q_\delta)} \le C_{34}\|\mathbf{f}e^{s\varphi}\|_{L^2(-\delta,\delta;H^{1,\widetilde{s}}(\Omega))} \le \epsilon\|\mathbf{f}e^{\varphi(0,\cdot)}\|_{H^{1,s}(\Omega)} \tag{6.63}$$

for all $s \ge s_9(\epsilon)$.

On the other hand, there exists a constant $C_{35} > 0$, independent of s, such that

$$\begin{aligned} &\|e^{s\varphi(0,\cdot)}\partial_{x_0}\mathbf{v}(0,\cdot)\|^2_{H^{1,s}(\Omega)} \\ &\le C_{35}\sum_{k=0}^{1}\int_{Q_\delta}\left(\frac{1}{s^2\widetilde{\varphi}^2}|\partial_{x_0}^{k+1}\nabla'\mathbf{v}|^2 + s^2\widetilde{\varphi}^2|\partial_{x_0}^k\nabla'\mathbf{v}|^2\right)e^{2s\varphi}dx \\ &\le C_{36}\sum_{k=0}^{1}\|\partial_{x_0}^k\mathbf{v}\|^2_{\mathcal{B}(\varphi,s,Q_\delta)} \end{aligned} \tag{6.64}$$

for all large $s \ge 1$. By (6.28) we obtain

$$\|e^{s\varphi(0,\cdot)}\rho_1(0,\cdot)\partial_{x_0}\mathbf{v}(0,\cdot)\|_{H^{1,s}(\Omega)} \ge \|e^{s\varphi(0,\cdot)}R(0,\cdot)\mathbf{f}\|_{H^{1,s}(\Omega)}$$
$$-C_{37}(\|e^{s\varphi(0,\cdot)}(\mathbf{v}_1(0,\cdot)-\mathbf{v}_2(0,\cdot))\|_{H^{3,s}(\Omega)} + \|e^{s\varphi(0,\cdot)}(\rho_1(0,\cdot)-\rho_2(0,\cdot))\|_{H^{2,s}(\Omega)}) \tag{6.65}$$

for all large $s \ge 1$.

By (6.8) we obtain from (6.65) that there exists a constant $C_{38} > 0$ such that

$$\|e^{s\varphi(0,\cdot)}\rho_1(0,\cdot)\partial_{x_0}\mathbf{v}(0,\cdot)\|_{H^{1,s}(\Omega)} \ge C_{38}\|e^{s\varphi(0,\cdot)}\mathbf{f}\|_{H^{1,s}(\Omega)}$$
$$-C_{39}(\|e^{s\varphi(0,\cdot)}(\mathbf{v}_1(0,\cdot)-\mathbf{v}_2(0,\cdot))\|_{H^{3,s}(\Omega)} + \|e^{s\varphi(0,\cdot)}(\rho_1(0,\cdot)-\rho_2(0,\cdot))\|_{H^{2,s}(\Omega)}) \tag{6.66}$$

for all large $s > 1$. Then (6.66) and (6.64) imply the estimate

$$\|e^{s\varphi(0,\cdot)}\mathbf{f}\|_{H^{1,s}(\Omega)} \le C_{40}\left(\sum_{k=0}^{1}\|\partial_{x_0}^k\mathbf{v}\|_{\mathcal{B}(\varphi,s,Q_\delta)} + \|e^{s\varphi(0,\cdot)}(\mathbf{v}_1(0,\cdot)-\mathbf{v}_2(0,\cdot))\|_{H^{3s}(\Omega)}\right.$$
$$\left.+\|e^{s\varphi(0,\cdot)}(\rho_1(0,\cdot)-\rho_2(0,\cdot))\|_{H^{2s}(\Omega)}\right)$$

for all large $s > 1$. This estimate and (6.62) imply

$$\|e^{s\varphi(0,\cdot)}\mathbf{f}\|_{H^{1,s}(\Omega)} \le C_{41}\left(\|e^{s\varphi(0,\cdot)}(\mathbf{v}_1(0,\cdot)-\mathbf{v}_2(0,\cdot))\|_{H^{3,s}(\Omega)}\right.$$
$$+\|e^{s\varphi(0,\cdot)}(\rho_1(0,\cdot)-\rho_2(0,\cdot))\|_{H^{3,\widetilde{s}}(\Omega)} + \|\mathbf{f}e^{s\varphi}\|_{\mathcal{Y}(\varphi,s,Q_\delta)}$$
$$+\sum_{k=0}^{1}(\|\partial_{x_0}^k(d\omega_{\mathbf{v}},\operatorname{div}\mathbf{v})e^{s\varphi}\|_{H^{\frac34,\frac32,\widetilde{s}}(\widetilde{\Sigma}_\delta)} + \|\partial_{x_0}^k\partial_{\boldsymbol{\nu}}(d\omega_{\mathbf{v}},\operatorname{div}\mathbf{v})e^{s\varphi}\|_{H^{\frac14,\frac12,\widetilde{s}}(\widetilde{\Sigma}_\delta)}$$
$$+\|\widetilde{\varphi}\partial_{x_0}^{k+1}\partial_{\boldsymbol{\nu}}\mathbf{v}e^{s\varphi}\|_{L^2(\widetilde{\Sigma}_\delta)} + \|e^{s\varphi(0,\cdot)}\partial_{x_0}^k\rho(0,\cdot)\|_{H^{2,s}(\Omega)})) \tag{6.67}$$

for all large $s > s_{10}$. Using (6.63) to estimate the norm of function $\mathbf{f}e^{s\varphi}$ on the right-hand side of (6.67), we have

$$\|e^{s\varphi(0,\cdot)}\mathbf{f}\|_{H^{1,s}(\Omega)} \le C_{42}\left(\|e^{s\varphi(0,\cdot)}\mathbf{v}(0,\cdot)\|_{H^{3,s}(\Omega)}\right.$$
$$+\|e^{s\varphi(0,\cdot)}\rho(0,\cdot)\|_{H^{3,s}(\Omega)} + \epsilon\|\varphi(0,\cdot)\|_{H^{1,s}(\Omega)}$$
$$+\sum_{k=0}^{1}(\|\partial_{x_0}^k(d\omega_{\mathbf{v}},\operatorname{div}\mathbf{v})e^{s\varphi}\|_{H^{\frac34,\frac32,\widetilde{s}}(\widetilde{\Sigma}_\delta)} + \|\partial_{x_0}^k\partial_{\boldsymbol{\nu}}(d\omega_{\mathbf{v}},\operatorname{div}\mathbf{v})e^{s\varphi}\|_{H^{\frac14,\frac12,\widetilde{s}}(\widetilde{\Sigma}_\delta)}$$
$$\left.+\|\widetilde{\varphi}\partial_{x_0}^{k+1}\partial_{\boldsymbol{\nu}}\mathbf{v}e^{s\varphi}\|_{L^2(\widetilde{\Sigma}_\delta)} + \|e^{s\varphi(0,\cdot)}\partial_{x_0}^k\rho(0,\cdot)\|_{H^{2,s}(\Omega)})\right) \tag{6.68}$$

for all $s \ge s_{11}$. Then taking the parameter ϵ sufficiently small in (6.68), we obtain (6.9). Thus the proof of theorem is complete. ■

6.4 Proof of Theorem 6.2 by Local Carleman Estimate

Step 1. Let Q_0 be a subdomain in the cylindrical domain $Q := (-T, T) \times \Omega$ such that the boundary ∂Q_0 is piecewise smooth. In this section, we will first establish a local Carleman estimate for the linearized system of (6.1)–(6.2) of the compressible fluid.

Let $(\rho_k, \mathbf{v}_k)$ satisfy problem (6.10)–(6.11). We set $\rho = \rho_1 - \rho_1, \mathbf{v} = \mathbf{v}_1 - \mathbf{v}_2$. Then, from Eqs. (6.10)–(6.11), we have

$$\partial_{x_0}\rho + \mathrm{div}(\mathbf{v}_1\rho) = -\mathrm{div}(\mathbf{v}\rho_2) \quad \text{in } Q,$$

$$\begin{aligned}&\rho_1\partial_{x_0}\mathbf{v} + \rho\partial_{x_0}\mathbf{v}_2 - L_{\lambda,\mu}(x', D')\mathbf{v} + \rho(\mathbf{v}_1, \nabla')\mathbf{v}_1 + \rho_2(\mathbf{v}_1, \nabla')\mathbf{v}\\ &+\rho_2(\mathbf{v}, \nabla')\mathbf{v}_2 + h(\rho_1)\nabla'\rho + (h(\rho_1) - h(\rho_2))\nabla'\rho_2 = 0 \quad \text{in } Q.\end{aligned}$$

First we consider the system with non-homogeneous terms on the right-hand sides:

$$\partial_{x_0}\rho + (\mathbf{v}_1, \nabla'\rho) + (\mathrm{div}\,\mathbf{v}_1)\rho = -\rho_2(\mathrm{div}\,\mathbf{v}) - \mathbf{v}\cdot\nabla'\rho_2 + G \tag{6.69}$$

and

$$\rho_1\partial_{x_0}\mathbf{v} - L_{\lambda,\mu}(x', D')\mathbf{v} = \widetilde{P}_1(x, D')\rho + (p_2, \nabla')\mathbf{v} + p_0\mathbf{v} + \mathbf{F}. \tag{6.70}$$

Here $\widetilde{P}_1 = \widetilde{P}_1(x, D')$ is a first-order differential operator in x' with C^1-coefficients, and $p_2(x)$, $p_0(x)$ are smooth vector-valued and matrix-valued functions respectively. We set

$$\psi(x) = \psi(x_0, x') = d(x') - \beta x_0^2, \quad \varphi(x) = e^{\tau\psi(x)},$$

where $|\nabla' d| > 0$ on $\overline{Q_0}$, $\beta > 0$ is a sufficiently small constant and τ is a sufficiently large constant which are chosen later. We assume

$$|K_{\mathbf{v}_1}(x, D)\psi| = |\partial_{x_0}\psi + (\mathbf{v}_1, \nabla'\psi)| \neq 0 \quad \text{on } \overline{Q}. \tag{6.71}$$

Moreover we set

$$\begin{cases} N_1(\rho) = \sum_{|\alpha'|\le 2}\|\partial_{x'}^{\alpha'}\rho\|^2_{L^2(\partial Q_0)},\\ N_2(\mathbf{v}) = \sum_{k=0}^{1}(\|\nabla^k\mathbf{v}\|^2_{L^2(\partial Q_0)} + \|\nabla^k\mathrm{div}\,\mathbf{v}\|^2_{L^2(\partial Q_0)} + \|\nabla^k\mathrm{rot}\,\mathbf{v}\|^2_{L^2(\partial Q_0)}),\\ N(\rho, \mathbf{v}) = N_1(\rho) + N_2(\mathbf{v}).\end{cases}$$

Throughout this section, C_k denote positive constants which are independent of $s > s_0$, while we write $C_k = C_k(\tau)$ if C_k depends on τ. We now recall that $\nabla = (\partial_{x_0}, \partial_{x_1}, ..., \partial_{x_n}) = (\partial_{x_0}, \nabla')$ and $\partial_{x'}^{\alpha'} = \partial_{x_1}^{\alpha_1}\cdots\partial_{x_n}^{\alpha_n}$ with $\alpha' = (\alpha_1, ..., \alpha_n)$ and $|\alpha'| := \alpha_1 + \cdots + \alpha_n$.

Now we are ready to state a Carleman estimate for system (6.69)–(6.70) of compressible fluid, which is of a different type from Theorem 6.3.

Proposition 6.3 *(local Carleman estimate) We fix $\tau > 0$ sufficiently large. Let $(\rho, \mathbf{v}) \in C^1([-T, T]; C^2(\overline{\Omega})) \times C^3([-T, T] \times \overline{\Omega})$ satisfy (6.69) and (6.70). Then there exist positive constants s_0 and $C_1, C_2(\tau), C_3(\tau)$ such that*

$$\int_{Q_0} \Bigg\{ \frac{1}{s\varphi} \Bigg(|\partial_{x_0} \mathbf{v}|^2 + |\partial_{x_0} div\, \mathbf{v}|^2 + |\partial_{x_0} rot\, \mathbf{v}|^2 + \sum_{|\alpha'|=2} (|\partial_{x'}^{\alpha'} (div\, \mathbf{v})|^2 + |\partial_{x'}^{\alpha'} (rot\, \mathbf{v})|^2) \Bigg)$$

$$+ \tau^2 \sum_{|\alpha'|=2} |\partial_{x'}^{\alpha'} \mathbf{v}|^2 + s\tau^2 \varphi \sum_{|\alpha'| \le 2} |\partial_{x'}^{\alpha'} \rho|^2 + s\tau^2 \varphi (|\nabla' \mathbf{v}|^2 + |\nabla' (div\, \mathbf{v})|^2 + |\nabla' (rot\, \mathbf{v})|^2)$$

$$+ s^3 \tau^4 \varphi^3 (|\mathbf{v}|^2 + |div\, \mathbf{v}|^2 + |rot\, \mathbf{v}|^2) \Bigg\} e^{2s\varphi} dx$$

$$\le C_1 \int_{Q_0} \Bigg(|\mathbf{F}|^2 + |\nabla' \mathbf{F}|^2 + \frac{1}{s\varphi} \sum_{|\alpha'| \le 2} |\partial_{x'}^{\alpha'} G|^2 \Bigg) e^{2s\varphi} dx + C_2(\tau) e^{C_3(\tau) s} N(\rho, \mathbf{v})$$

for all large $s > s_0$.

Here we note that the constant $C_1 > 1$ is independent of τ.

Proposition 6.3 requires the whole boundary data on ∂Q_0, and called a local Carleman estimate.

Proof of Proposition 6.3.

The proof is straightforward by decoupling the parabolic system (6.70) by introducing operators div and rot because we assume data on the whole boundary ∂Q_0. For completeness, we provide the proof of Proposition 6.3.

First we prove

Lemma 6.1 *Let $r_0 \in L^\infty(Q_0)$. We assume (6.71) and fix $\tau > 0$ sufficiently large. Then for any $m \in \mathbb{R}$, there exist positive constants $s_0 > 0$ and $C_4, C_5(\tau), C_6(\tau)$ such that*

$$\int_{Q_0} s^{m+2} \tau^{m+2} \varphi^{m+2} |\rho|^2 e^{2s\varphi} dx$$

$$\le C_4 \int_{Q_0} s^m \tau^m \varphi^m |K_{\mathbf{v}_1}(x, D)\rho + r_0(x_0, x')\rho|^2 dx + C_5(\tau) e^{C_6(\tau) s} \int_{\partial Q_0} |\rho|^2 d\sigma$$

for all $s > s_0$.

Here, as long as $\|r_0\|_{L^\infty(Q_0)} \le M$ with some constant $M > 0$, the constants $C_4, C_5(\tau)$, $C_6(\tau)$ and s_0 are dependent on M, but can be chosen uniformly in such r_0.

Proof of Lemma 6.1. We set $h := K_{\mathbf{v}_1}(x, D)\rho$ and $y = \varphi^{\frac{m}{2}} \rho e^{s\varphi}$. Then direct calculations yield

$$e^{s\varphi}K_{\mathbf{v}_1}(x,D)(e^{-s\varphi}y) = \partial_{x_0}y + (\mathbf{v}_1, \nabla' y) - s\tau\varphi(K_{\mathbf{v}_1}(x,D)\psi)y.$$

Moreover, since

$$\begin{aligned}
&\partial_{x_k}y = \partial_{x_k}(\varphi^{\frac{m}{2}}\rho e^{s\varphi})\\
=&\frac{m}{2}\varphi^{\frac{m}{2}-1}(\partial_{x_k}\varphi)\rho e^{s\varphi} + \varphi^{\frac{m}{2}}(\partial_{x_k}\rho)e^{s\varphi} + s(\partial_{x_k}\varphi)\varphi^{\frac{m}{2}}\rho e^{s\varphi}\\
=&\frac{m}{2}\varphi^{\frac{m}{2}}\tau(\partial_{x_k}\psi)\rho e^{s\varphi} + \varphi^{\frac{m}{2}}(\partial_{x_k}\rho)e^{s\varphi} + \varphi^{\frac{m}{2}+1}s\tau\rho(\partial_{x_k}\psi)e^{s\varphi}
\end{aligned}$$

for $k = 0, 1, ..., n$, we can directly verify

$$\partial_{x_0}y + (\mathbf{v}_1, \nabla' y) - s\tau\varphi(K_{\mathbf{v}_1}(x,D)\psi)y = e^{s\varphi}\left(\varphi^{\frac{m}{2}}h + \frac{m}{2}\varphi^{\frac{m}{2}}\tau(K_{\mathbf{v}_1}(x,D)\psi)\rho\right).$$

Therefore

$$\begin{aligned}
&\int_{Q_0}|e^{s\varphi}K_{\mathbf{v}_1}(x,D)(e^{-s\varphi}y)|^2dx = \int_{Q_0}\left|e^{s\varphi}\left(\varphi^{\frac{m}{2}}h + \frac{m}{2}\varphi^{\frac{m}{2}}\tau(K_{\mathbf{v}_1}(x,D)\psi)\rho\right)\right|^2dx\\
\ge&\int_{Q_0}s^2\tau^2\varphi^2|K_{\mathbf{v}_1}(x,D)\psi|^2|y|^2dx - \int_{Q_0}2s\tau\varphi(K_{\mathbf{v}_1}(x,D)\psi)y(\partial_{x_0}y + (\mathbf{v}_1, \nabla' y))dx.
\end{aligned}$$

By (6.71) and integration by parts, there exist constants C_7, C_8 independent of τ such that

$$\begin{aligned}
&\int_{Q_0}s^2\tau^2\varphi^2|y|^2dx - C_7\int_{Q_0}s\tau^2\varphi|y|^2dx - C_7\int_{\partial Q_0}s\tau\varphi|y|^2d\sigma\\
\le&C_8\int_{Q_0}\left|\varphi^{\frac{m}{2}}h + \frac{m}{2}\varphi^{\frac{m}{2}}\tau(K_{\mathbf{v}_1}(x,D)\psi)\rho\right|^2e^{2s\varphi}dx.
\end{aligned}$$

Therefore the substitution of $y = \varphi^{\frac{m}{2}}\rho e^{s\varphi}$ yields

$$\begin{aligned}
&\int_{Q_0}s^2\tau^2\varphi^{m+2}|\rho|^2e^{2s\varphi}dx\\
\le&C_9\int_{Q_0}\varphi^m|h|^2e^{2s\varphi}dx + C_{10}\int_{Q_0}\varphi^m\tau^2|\rho|^2e^{2s\varphi}dx\\
+&C_{11}\int_{Q_0}s\tau^2\varphi^{m+1}|\rho|^2e^{2s\varphi}dx + C_{12}(\tau)e^{C_{13}(\tau)s}\int_{\partial Q_0}|\rho|^2d\sigma.
\end{aligned}$$

Here and henceforth we use

$$\begin{aligned}
&s\tau\varphi^{m+1}e^{2s\varphi} \le s\tau\left(\max_{\overline{Q_0}}\varphi\right)^{m+1}e^{2s\max_{\overline{Q_0}}\varphi}\\
\le&C_{14}(\tau)se^{C_{15}(\tau)s} \le C_{12}(\tau)e^{C_{13}(\tau)s}
\end{aligned}$$

for all large $s > 0$.

Absorbing the second and the third terms on the right-hand side into the left-hand side, we complete the proof of the lemma in the case of $r_0 = 0$. For $r_0 \neq 0$, since $|K_{\mathbf{v}_1}(x, D)\rho + r_0\rho|^2 \le 2|K_{\mathbf{v}_1}(x, D)\rho|^2 + 2\|r_0\|^2_{L^\infty(Q_0)}|\rho|^2$, we further choose $s_0 > 0$ large, so that the term $\int_{Q_0} 2\|r_0\|^2_{L^\infty(Q_0)}|\rho|^2 e^{2s\varphi}dx$ can be absorbed into the left-hand side. Thus the proof of Lemma 6.1 is complete. ■

Applying Lemma 6.1 with $m = -1$ to (6.69) after taking ∂_{x_k} and $\partial_{x_j}\partial_{x_k}$ for $1 \le j, k \le n$, we have

$$\begin{aligned}
&\int_{Q_0} s\tau^2\varphi \sum_{|\alpha'|\le 2} |\partial_{x'}^{\alpha'}\rho|^2 e^{2s\varphi}dx\\
\le & C_{16}\int_{Q_0} \frac{1}{s\varphi}\sum_{|\alpha'|\le 2}(|\partial_{x'}^{\alpha'}\mathrm{div}\,\mathbf{v}|^2 + |\partial_{x'}^{\alpha'}\mathbf{v}|^2)e^{2s\varphi}dx\\
&+ C_{17}\int_{Q} \frac{1}{s\varphi}\sum_{|\alpha'|\le 2}|\partial_{x'}^{\alpha'}G|^2 e^{2s\varphi}dx + C_{18}(\tau)e^{C_{19}(\tau)s}N_1(\rho).
\end{aligned} \tag{6.72}$$

For (6.70), introducing $\mathrm{div}\,\mathbf{v}$ and $\mathrm{rot}\,\mathbf{v}$, by direct calculations we obtain

$$\partial_{x_0}\mathbf{v} - \frac{\mu}{\rho_1}\Delta\mathbf{v} = R_1(x, D')(\mathbf{v}, \mathrm{div}\,\mathbf{v}, \mathrm{rot}\,\mathbf{v}) + H_1(x, D')\rho + \mathbf{F} \quad \text{in } Q_0,$$

$$\partial_{x_0}\mathrm{div}\,\mathbf{v} - \frac{\lambda + 2\mu}{\rho_1}\Delta\mathrm{div}\,\mathbf{v} = R_2(x, D')(\mathbf{v}, \mathrm{div}\,\mathbf{v}, \mathrm{rot}\,\mathbf{v}) + H_2(x, D')\rho + \mathrm{div}\,\mathbf{F} \quad \text{in } Q_0,$$

and

$$\partial_{x_0}\mathrm{rot}\,\mathbf{v} - \frac{\mu}{\rho_1}\Delta\mathrm{rot}\,\mathbf{v} = R_3(x, D')(\mathbf{v}, \mathrm{div}\,\mathbf{v}, \mathrm{rot}\,\mathbf{v}) + H_3(x, D')\rho + \mathrm{rot}\,\mathbf{F} \quad \text{in } Q_0.$$

Here $R_k(x, D')$ are first-order differential operators in x' and $H_k(x, D')$ are second-order differential operators in x' with smooth coefficients for $k = 1, 2, 3$.

We apply a parabolic Carleman estimate to the parabolic operators $\partial_{x_0} - \frac{\mu}{\rho_1}\Delta$ and $\partial_{x_0} - \frac{\lambda+2\mu}{\rho_1}\Delta$ and choose $s > 0$ large to absorb the terms R_1, R_2, R_3 into the left-hand side, and we obtain

$$\begin{aligned}
&\int_{Q_0}\Bigg\{\frac{1}{s\varphi}\Bigg(|\partial_{x_0}\mathbf{v}|^2 + |\partial_{x_0}\mathrm{div}\,\mathbf{v}|^2 + |\partial_{x_0}\mathrm{rot}\,\mathbf{v}|^2 + \sum_{|\alpha'|=2}(|\partial_{x'}^{\alpha'}\mathbf{v}|^2 + |\partial_{x'}^{\alpha'}(\mathrm{div}\,\mathbf{v})|^2\\
&\quad +|\partial_{x'}^{\alpha'}(\mathrm{rot}\,\mathbf{v})|^2)\Bigg) + s\tau^2\varphi(|\nabla'\mathbf{v}|^2 + |\nabla'(\mathrm{div}\,\mathbf{v})|^2 + |\nabla'(\mathrm{rot}\,\mathbf{v})|^2)\\
&\quad +s^3\tau^4\varphi^3(|\mathbf{v}|^2 + |\mathrm{div}\,\mathbf{v}|^2 + |\mathrm{rot}\,\mathbf{v}|^2)\Bigg\}e^{2s\varphi}dx
\end{aligned}$$

$$\leq C_{20}\int_{Q_0}\sum_{|\alpha'|\leq 2}|\partial_{x'}^{\alpha'}\rho|^2 e^{2s\varphi}dx + C_{21}\int_{Q_0}(|\mathbf{F}|^2+|\nabla'\mathbf{F}|^2)e^{2s\varphi}dx + C_{22}(\tau)e^{C_{23}(\tau)s}N_2(\mathbf{v}) \tag{6.73}$$

for all $s > s_0$ (e.g., Lemma 7.2 in Bellassoued and Yamamoto [15], Yamamoto [81]).

Moreover (6.73) implies

$$\int_{Q_0}\frac{1}{s\varphi}\sum_{|\alpha'|\leq 2}(|\partial_{x'}^{\alpha'}(\mathrm{div}\,\mathbf{v})|^2+|\partial_{x'}^{\alpha'}\mathbf{v}|^2)e^{2s\varphi}dx \leq C_{24}\int_{Q_0}\sum_{|\alpha'|\leq 2}|\partial_{x'}^{\alpha'}\rho|^2 e^{2s\varphi}dx$$

$$+ C_{25}\int_{Q_0}(|\mathbf{F}|^2+|\nabla'\mathbf{F}|^2)e^{2s\varphi}dx + C_{26}(\tau)e^{C_{27}(\tau)s}N_2(\mathbf{v}). \tag{6.74}$$

Applying (6.74) to the first term on the right-hand side of (6.72), we obtain

$$\int_{Q_0}s\tau^2\varphi\sum_{|\alpha'|\leq 2}|\partial_{x'}^{\alpha'}\rho|^2 e^{2s\varphi}dx$$

$$\leq C_{28}\int_{Q_0}\sum_{|\alpha'|\leq 2}|\partial_{x'}^{\alpha'}\rho|^2 e^{2s\varphi}dx + C_{29}\int_{Q_0}(|\mathbf{F}|^2+|\nabla'\mathbf{F}|^2)e^{2s\varphi}dx$$

$$+ C_{30}\int_{Q_0}\frac{1}{s\varphi}\sum_{|\alpha'|\leq 2}|\partial_{x'}^{\alpha'}G|^2 e^{2s\varphi}dx + C_{31}(\tau)e^{C_{32}(\tau)s}(N_1(\rho)+N_2(\mathbf{v})).$$

Therefore, absorbing the first term on the right-hand side of this inequality into the left-hand side, we have

$$\int_{Q_0}s\tau^2\varphi\sum_{|\alpha'|\leq 2}|\partial_{x'}^{\alpha'}\rho|^2 e^{2s\varphi}dx \leq C_{33}\int_{Q_0}(|\mathbf{F}|^2+|\nabla'\mathbf{F}|^2)e^{2s\varphi}dx$$

$$+ C_{34}\int_{Q_0}\frac{1}{s\varphi}\sum_{|\alpha'|\leq 2}|\partial_{x'}^{\alpha'}G|^2 e^{2s\varphi}dx + C_{35}(\tau)e^{C_{36}(\tau)s}N(\rho,\mathbf{v}). \tag{6.75}$$

On the other hand, we have $-\Delta\mathbf{v} = \mathrm{rot}\,\mathrm{rot}\,\mathbf{v} - \nabla'\mathrm{div}\,\mathbf{v}$. Applying the Carleman estimate for Δ (e.g., Lemma 7.2 in Bellassoued and Yamamoto [15]) and using (6.73), we obtain

$$\int_{Q_0}\tau^2\sum_{|\alpha'|=2}|\partial_{x'}^{\alpha'}\mathbf{v}|^2 e^{2s\varphi}dx$$

$$\leq C_{37}\int_{Q_0}s\tau^2\varphi(|\nabla'(\mathrm{rot}\,\mathbf{v})|^2+|\nabla'(\mathrm{div}\,\mathbf{v})|^2)e^{2s\varphi}dx + C_{38}(\tau)e^{C_{39}(\tau)s}N_2(\mathbf{v})$$

$$\leq C_{40}\int_{Q_0}\sum_{|\alpha'|\leq 2}|\partial_{x'}^{\alpha'}\rho|^2 e^{2s\varphi}dx + C_{41}\int_{Q_0}(|\mathbf{F}|^2+|\nabla'\mathbf{F}|^2)e^{2s\varphi}dx \tag{6.76}$$

$$+C_{42}(\tau)e^{C_{43}(\tau)s}N_2(\mathbf{v}).$$

By (6.73), (6.75) and (6.76), we can complete the proof of Proposition 6.3. ■

Step 2. We will complete the proof of Theorem 6.2 in a neighbourhood of a'. For it, we will apply Proposition 6.3. Without loss of generality, we can assume that $a' = 0$ after a translation if necessary. Moreover by a suitable rotation, we can choose a coordinate $x' = (\widetilde{x}, x_n) := (x_1, ..., x_{n-1}, x_n)$ such that $\boldsymbol{\nu}(0) = (0, ..., 0, 1)$ where $\boldsymbol{\nu}(0)$ is the unit outward normal vector to $\partial\Omega$ at 0.

We set $\mathbf{v} = \mathbf{v}_1 - \mathbf{v}_2$, $\rho = \rho_1 - \rho_2$.

Since $|(\mathbf{v}_1(0), \boldsymbol{\nu}(0))| \neq 0$, without loss of generality we can assume

$$(\mathbf{v}_1(0), \boldsymbol{\nu}(0)) =: \mu_0 > 0. \tag{6.77}$$

Next we choose a suitable ellipsoid centered at $(0, p) \in \mathbb{R}^n$ with some $p > 0$ whose axes perpendicular to the x_n-axis are sufficiently small, and we consider the intersection U of the interior of the ellipsoid with Ω. We have to verify geometrical properties of U. To this end, we argue as follows. Henceforth let $B_\delta = \{\widetilde{x} \in \mathbb{R}^{n-1};\ |\widetilde{x}| < \delta\}$. For small $\delta > 0$, we choose $\vartheta \in C^2(\overline{B_\delta})$ such that $\{x' \in \partial\Omega;\ |\widetilde{x}| < \delta\}$ is described by $x_n = \vartheta(\widetilde{x})$ and Ω is locally located below $x_n = \vartheta(\widetilde{x})$. We note that $\vartheta(0) = 0$. Therefore

$$|\vartheta(\widetilde{x})| \le \|\vartheta\|_{C^1(\overline{B_\delta})}|\widetilde{x}|, \quad |\widetilde{x}| < \delta. \tag{6.78}$$

Let p be given arbitrarily such that $\frac{1}{2} < p < 1$. Then we choose $b = b(p) > 0$ such that

$$\max\{b,\ b\|\vartheta\|_{C^1(\overline{B_\delta})}\} < \min\{1 - p, \delta\}. \tag{6.79}$$

Then for $p \in \left(\frac{1}{2}, 1\right)$, we set

$$g_p(\widetilde{x}) = p - \sqrt{1 - \frac{|\widetilde{x}|^2}{b(p)^2}}, \quad |\widetilde{x}| < b(p).$$

From (6.78) it follows that

$$|\vartheta(\widetilde{x})| < 1 - p \quad \text{if } |\widetilde{x}| < b(p). \tag{6.80}$$

By the definition of ϑ and $\frac{1}{2} < p < 1$, we see that

$$0 = \vartheta(0) > g_p(0) = p - 1$$

and

$$|\vartheta(\widetilde{x})| < 1 - p < g_p(\widetilde{x}) = p, \quad |\widetilde{x}| = b(p).$$

Therefore the two hypersurfaces $\{x';\ x_n = \vartheta(\widetilde{x}),\ |\widetilde{x}| \le b(p)\}$ and $\{x';\ x_n = g_p(\widetilde{x}),\ |\widetilde{x}| \le b(p)\}$ can surround a non-empty open set whose closure contains 0. By Ω_p, we denote its connected component such that $0 \in \partial\Omega_p$. On the basis of Ω_p, we will construct the open set U_0 described in Theorem 6.2.

By the definition, we have

$$\Omega_p \subset \{x';\ |\widetilde{x}| < b(p),\ g_p(\widetilde{x}) < x_n < \vartheta(\widetilde{x})\}$$

and we see by $g_p(\widetilde{x}) > p - 1$ that

$$|x_n| < \max\{1 - p,\ \max_{|\widetilde{x}|<b(p)} |\vartheta(\widetilde{x})|\}, \quad (\widetilde{x}, x_n) \in \Omega_p.$$

Hence (6.80) yields $|x_n| < 1 - p$. Moreover (6.79) yields $|\widetilde{x}| < b(p) < 1 - p$ for $x \in \Omega_p$. Consequently

$$\Omega_p \subset \{x';\ |x_n| < 1 - p,\ |\widetilde{x}| < 1 - p\}. \tag{6.81}$$

With $b(p)$ depending on $p \in \left(\frac{1}{2}, 1\right)$, we set

$$d_p(x') = 1 - \frac{|\widetilde{x}|^2}{b(p)^2} - |x_n - p|^2, \quad |\widetilde{x}| \le b(p).$$

We use $d_p(x')$ as $d(x)$ in the weight function for the Carleman estimate established in Proposition 6.3.

We need more properties of d_p and the level set created by $d_p(x')$, which are obtained by choosing p close to 1. We set

$$S_p = \partial\Omega \cap \{d_p(x') > 0\}.$$

We note that $\partial\Omega_p \subset \partial\Omega \cup \{x';\ d_p(x') = 0\}$. We set

$$h(1 - p) := \|d_p\|_{C(\overline{\Omega_p})}. \tag{6.82}$$

Since $d_p(0, p - 1) = 0$ and $(0, p - 1) \in \overline{\Omega_p}$, applying the mean value theorem to $d_p(x') = d_p(x') - d_p(0, p - 1)$ for $x' \in \Omega_p$, we obtain from (6.81) that $\lim_{p\uparrow 1} \|d_p\|_{C(\overline{\Omega_p})} = 0$, that is,

$$\lim_{p\uparrow 1} h(1 - p) = 0. \tag{6.83}$$

We have $\nabla' d_p(0) = 2p\boldsymbol{\nu}(0)$ and (6.77) implies

$$(\mathbf{v}_1(0), \nabla' d_p(0)) = 2p\mu_0,$$

and, if we choose small $\varepsilon_1 > 0$ and $p \in (1-\varepsilon_1, 1)$, by (6.81) and the continuity of $\mathbf{v}_1$ and $\nabla' d_p$, we see

$$(\mathbf{v}_1(x), \nabla' d_p(x')) > p\mu_0 > \frac{\mu_0}{2}, \quad x' \in \Omega_p \text{ with } p \in (1-\varepsilon_1, 1),\ |x_0| < \varepsilon_1. \tag{6.84}$$

Choose $N > 0$, $\beta > 0$ and $p \in (1-\varepsilon_1, 1)$ such that

$$\begin{cases} N > \frac{1}{4}\frac{h(1-p)}{1-p^2}, \\ \frac{16N-1}{16N}h(1-p) < \beta\varepsilon_1^2, \\ \frac{\mu_0}{2} - \beta\varepsilon_1^2 > 0. \end{cases} \tag{6.85}$$

Such β, p, N exist. Indeed, first choosing $\beta > 0$ sufficiently small for ε_1, we can satisfy the third inequality of (6.85). Next with such $\beta > 0$, if we choose $p \in (1-\varepsilon_1, 1)$ such that $1-p$ is sufficiently small, then (6.83) implies $h(1-p) < \beta\varepsilon_1^2$, and so $\frac{16N-1}{16N}h(1-p) < \beta\varepsilon_1^2$, that is, the second inequality of (6.85) holds for all $N \ge N_0$, where N_0 is sufficiently large. Finally, with such β and p, we choose $N > \frac{1}{16}$ sufficiently large, so that the first inequality in (6.85) is satisfied.

We set

$$\theta = \frac{h(1-p)}{16N}, \quad \psi_p(x) = d_p(x') - \beta x_0^2, \quad \varphi_p(x) = e^{\tau\psi_p(x)}$$

with large $\tau > 0$, and

$$Q_\theta = \{x;\ x' \in \overline{\Omega_p},\ \psi_p(x) > \theta\}.$$

The second inequality of (6.85) implies

$$Q_\theta \subset (-\varepsilon_1, \varepsilon_1) \times \overline{\Omega_p}. \tag{6.86}$$

Proof of (6.86). For $x \in Q_\theta$, by (6.82) we have $h(1-p) - \beta x_0^2 \ge d_p(x') - \beta x_0^2 > \theta$. Therefore the second inequality of (6.85) yields

$$\beta x_0^2 < h(1-p) - \theta = \frac{16N-1}{16N}h(1-p) < \beta\varepsilon_1^2,$$

which means that $|x_0| < \varepsilon_1$.

Moreover

$$0 \in Q_{4\theta}. \tag{6.87}$$

In other words, $Q_{4\theta}$ is a non-empty neighborhood of 0. Indeed, the first inequality in (6.85) implies $\psi_p(0) = d_p(0) = 1 - p^2 > \frac{1}{4N}h(1-p) = 4\theta$.

Furthermore we can prove

$$\partial Q_\theta \subset \Sigma_1 \cup \Sigma_2, \tag{6.88}$$

where $\Sigma_1 \subset [-\varepsilon_1, \varepsilon_1] \times S_p$ and $\Sigma_2 = \{x;\ x' \in \Omega_p,\ \psi_p(x) = \theta\}$.
Proof of (6.88). Let $x = (x_0, x') \in \partial Q_\theta$.
Case 1: Let $x' \in \Omega_p$. If $\psi_p(x) > \theta$, then x is an interior point of Q_θ, which is impossible by $x \in \partial Q_\theta$. Therefore $\psi_p(x) = \theta$ must hold, that is, $x \in \Sigma_2$.
Case 2: Let $x' \in \partial\Omega_p$. We assume that $x' \in \partial\Omega_p \setminus S_p$. Then $x' \in \partial\Omega_p \subset \partial\Omega \cup \{x';\ d_p(x') = 0\}$, and so $x' \in \partial\Omega$ or $d_p(x') = 0$. By $x \notin S_p = \partial\Omega \cap \{x';\ d_p(x') > 0\}$, we see that $x' \notin \partial\Omega$ or $d_p(x') = 0$. Therefore in any case, we have $d_p(x') = 0$. Hence $d_p(x') - \beta x_0^2 \le 0$, which implies $\psi_p(x) \le 0$ and contradicts $x \in \overline{Q_\theta}$. Hence $x' \in \partial\Omega_p \setminus S_p$ is impossible, that is, $x' \in S_p$. Moreover (6.86) implies $|x_0| \le \varepsilon_1$. Thus (6.88) is verified.

Now, in order to apply in Q_θ, we verify the conditions for Proposition 6.3. First

$$|\nabla' d_p(x')| \ne 0, \quad x' \in \overline{\Omega_p}.$$

Indeed let $x' = (\widetilde{x}, x_n) \in \overline{\Omega_p}$ and $\nabla' d_p(x') = 0$. Since

$$\nabla' d_p(x') = \left(\frac{-2\widetilde{x}}{b(p)^2},\ 2(p - x_n) \right)^T,$$

it follows that $\widetilde{x} = 0$ and $x_n = p$. By (6.81), we have $|x_n| \le 1 - p$, and $x_n = p$ contradicts $\frac{1}{2} < p < 1$.

Second, since $x \in Q_\theta$ implies $|x_0| \le \varepsilon_1$ by (6.86), we see by (6.84) that the third inequality in (6.85) yields

$$K_{\mathbf{v}_1}(x, D)\psi_p > 0 \quad \text{on } \overline{Q_\theta}.$$

Recall that $(\rho_k, \mathbf{v}_k)$, $k = 1, 2$ satisfy (6.10)–(6.12) and $\rho = \rho_1 - \rho_2$ and $\mathbf{v} = \mathbf{v}_1 - \mathbf{v}_2$. Then we see that $(\rho, \mathbf{v}) \in C^3(\overline{Q}) \times C^3(\overline{Q})$ satisfies (6.69) and (6.70) with $\mathbf{F} = 0$ and $G = 0$.

For applying Proposition 6.3 to $(\rho, \mathbf{v})$ in Q_θ, we need the cut-off because we have boundary data only on $(-\varepsilon_1, \varepsilon_1) \times S_p$. The cut-off is a routine argument (e.g., Yamamoto [81]): We choose $\chi \in C^\infty(\mathbb{R}^{n+1})$ such that $0 \le \chi \le 1$ and

$$\chi(x) = \begin{cases} 1, & \psi_p(x) > 3\theta, \\ 0, & \psi_p(x) < 2\theta. \end{cases} \tag{6.89}$$

We set

$$r = \chi\rho, \quad \mathbf{w} = \chi\mathbf{v} \quad \text{in } Q_\theta.$$

Then by (6.88), we see

$$|\partial_x^\alpha r| = |\partial_x^\alpha \mathbf{w}| = 0 \quad \text{on } \Sigma_2 \text{ with } |\alpha| \le 2. \tag{6.90}$$

Then

$$\partial_{x_0} r + \mathbf{v}_1 \cdot \nabla' r + (\operatorname{div} \mathbf{v}_1) r = -\rho_2 (\operatorname{div} \mathbf{w}) - \mathbf{w} \cdot \nabla' \rho_2 + J_1(\rho, \mathbf{v}) \quad \text{in } Q_\theta$$

and

$$\rho_1 \partial_{x_0} \mathbf{w} - L_{\lambda,\mu}(x', D')\mathbf{w} = \widetilde{P}_1(x, D')r + (p_2, \nabla')\mathbf{w} + p_0\mathbf{w} + \mathbf{J}_2(\rho, \mathbf{v}, \nabla'\mathbf{v}) \quad \text{in } Q_\theta.$$

Here $\widetilde{P}_1(x, D')$ is the first-order differential operator in (6.70). Since all the coefficients of the linear functions J_1 and $\mathbf{J}_2$ in ρ, $\mathbf{v}$, $\nabla'\mathbf{v}$, contain some components of $\nabla\chi(x)$ as factors, we see that the terms J_1 and $\mathbf{J}_2$ satisfy

$$\begin{cases} |J_1(\rho, \mathbf{v})(x)| \le C_{44}(|\rho(x)| + |\mathbf{v}(x)|), \\ |\mathbf{J}_2(\rho, \mathbf{v}, \nabla'\mathbf{v})(x)| \le C_{44}(|\rho(x)| + |\mathbf{v}(x)| + |\nabla'\mathbf{v}(x)|), \\ |\nabla'\mathbf{J}_2(\rho, \mathbf{v}, \nabla'\mathbf{v})(x)| \le C_{44}\left(|\rho(x)| + |\nabla'\rho(x)| + \sum_{|\alpha'|\le 2} |\partial_{x'}^{\alpha'}\mathbf{v}(x)|\right), \\ \qquad x \in Q_\theta, \\ J_1(\rho, \mathbf{v})(x) = |\mathbf{J}_2(\rho, \mathbf{v}, \nabla'\mathbf{v})(x)| = |\nabla'\mathbf{J}_2(\rho, \mathbf{v}, \nabla'\mathbf{v})(x)| = 0 \\ \text{if } x \in Q_\theta \text{ and } \partial_{x_k}\chi(x) = 0 \text{ with some } k = 0, 1, 2, ..., n. \end{cases} \tag{6.91}$$

By the assumption, we see that

$$\|(\mathbf{v}, \rho)\|_{C^3([-T,T]\times\overline{\Omega})} \le 2M. \tag{6.92}$$

Applying Proposition 6.3 to $\mathbf{w}$ and r and fixing τ large, we obtain

$$\int_{Q_\theta} \Bigg(|\partial_{x_0}\mathbf{w}|^2 + |\partial_{x_0}\operatorname{div}\mathbf{w}|^2 + |\partial_{x_0}\operatorname{rot}\mathbf{w}|^2 + \sum_{|\alpha'|=2} (|\partial_{x'}^{\alpha'}\operatorname{div}\mathbf{w}|^2 + |\partial_{x'}^{\alpha'}\operatorname{rot}\mathbf{w}|^2)$$
$$+ \sum_{|\alpha'|\le 2} (|\partial_{x'}^{\alpha'}\mathbf{w}|^2 + |\partial_{x'}^{\alpha'} r|^2) \Bigg) e^{2s\varphi_p} dx$$

$$\leq C_{45}e^{C_{46}s}N(r,\mathbf{w}) + C_{47}s\int_{Q_\theta}(|\mathbf{J}_2|^2 + |\nabla'\mathbf{J}_2|^2 + \sum_{|\alpha'|\leq 2}|\partial_{x'}^{\alpha'}J_1|^2)e^{2s\varphi_p}dx. \quad (6.93)$$

Here and henceforth we write C_{45} etc., not $C_{45}(\tau)$, because τ is fixed.
Setting

$$\widetilde{N}(\rho,\mathbf{v}) := \|\rho\|^2_{L^2(-\varepsilon_1,\varepsilon_1;H^2(S_p))} + \sum_{j,k=0}^{1}\|\nabla^k\nabla'^j\mathbf{v}\|^2_{L^2((-\varepsilon_1,\varepsilon_1)\times S_p)},$$

by (6.90), we see $N(r,\mathbf{w}) \leq C_{48}\widetilde{N}(\rho,\mathbf{v})$. By (6.89), (6.91) and (6.92), we obtain

$$\int_{Q_\theta}(|\mathbf{J}_2|^2 + |\nabla'\mathbf{J}_2|^2 + \sum_{|\alpha'|\leq 2}|\partial_{x'}^{\alpha'}J_1|^2)e^{2s\varphi_p}dx$$
$$\leq C_{49}(M)\int_{Q_\theta\cap\{2\theta\leq\psi_p(x_0,x')\leq 3\theta\}}e^{2s\varphi_p}dx \leq C_{50}(M)e^{2s\mu_3}.$$

We set $\mu_k = e^{k\tau\theta}$, $k = 3, 4$. Replacing the left-hand side of (6.93) by the integration over $Q_{4\theta}$ and noting $\chi = 1$ in $Q_{4\theta}$, we reach

$$\int_{Q_{4\theta}}\Bigg(|\partial_{x_0}\mathbf{v}|^2 + |\partial_{x_0}\mathrm{div}\,\mathbf{v}|^2 + |\partial_{x_0}\mathrm{rot}\,\mathbf{v}|^2 + \sum_{|\alpha'|=2}(|\partial_{x'}^{\alpha'}\mathrm{div}\,\mathbf{v}|^2 + |\partial_{x'}^{\alpha'}\mathrm{rot}\,\mathbf{v}|^2)$$
$$+\sum_{|\alpha'|\leq 2}(|\partial_{x'}^{\alpha'}\mathbf{v}|^2 + |\partial_{x'}^{\alpha'}\rho|^2)\Bigg)e^{2s\varphi_p}dx$$
$$\leq sC_{51}(M)e^{2s\mu_3} + C_{52}e^{C_{53}s}\widetilde{N}(\rho,\mathbf{v}) \quad (6.94)$$

for all large $s > 0$. We set

$$\|(\rho,\mathbf{v})\|_B^2 = \|\partial_{x_0}\mathbf{v}\|^2_{L^2(Q_{4\theta})} + \|\partial_{x_0}\mathrm{div}\,\mathbf{v}\|^2_{L^2(Q_{4\theta})} + \|\partial_{x_0}\mathrm{rot}\,\mathbf{v}\|^2_{L^2(Q_{4\theta})}$$
$$+\sum_{|\alpha'|=2}(\|\partial_{x'}^{\alpha'}\mathrm{div}\,\mathbf{v}\|^2_{L^2(Q_{4\theta})} + \|\partial_{x'}^{\alpha'}\mathrm{rot}\,\mathbf{v}\|^2_{L^2(Q_{4\theta})})$$
$$+\sum_{|\alpha'|\leq 2}(\|\partial_{x'}^{\alpha'}\mathbf{v}\|^2_{L^2(Q_{4\theta})} + \|\partial_{x'}^{\alpha'}\rho\|^2_{L^2(Q_{4\theta})}).$$

Since $\sup_{s>0}se^{-s(\mu_4-\mu_3)} < \infty$ and $\varphi_p(x) > e^{4\tau\theta} = \mu_4$ for $x \in Q_{4\theta}$, by (6.94) we obtain

$$\|(\rho,\mathbf{v})\|_B^2 \leq C_{54}(M)(e^{-(\mu_4-\mu_3)s} + e^{C_{55}s}\widetilde{N}(\rho,\mathbf{v})) \quad (6.95)$$

for all $s > s_0$. Replacing $C_{54}(M)$ by $e^{C_{55}s_0}C_{54}(M)$, we have (6.95) for all $s > 0$. We minimize the right-hand side of (6.95), that is, $m(s) := e^{-(\mu_4-\mu_3)s} + e^{C_{55}s}\widetilde{N}(\rho,\mathbf{v})$ by changing $s > 0$. We solve $\frac{dm}{ds}(s) = 0$ to find

$$\frac{\mu_4 - \mu_3}{C_{55}\widetilde{N}(\rho, \mathbf{v})} = e^{(C_{55}+\mu_4-\mu_3)s}. \tag{6.96}$$

We discuss the two cases separately.

Case1 : $\dfrac{\mu_4 - \mu_3}{C_{55}\widetilde{N}(\rho, \mathbf{v})} > 1.$

Then

$$s_* = \frac{1}{C_{55} + \mu_4 - \mu_3} \log \frac{\mu_4 - \mu_3}{C_{55}\widetilde{N}(\rho, \mathbf{v})} > 0$$

solves (6.96), and $m(s_*) = C_{56}\widetilde{N}(\rho, \mathbf{v})^{2\kappa}$, where

$$\kappa = \frac{1}{2}\frac{\mu_4 - \mu_3}{\mu_4 - \mu_3 + C_{55}} \in (0, 1).$$

Hence

$$\|(\rho, \mathbf{v})\|_B \le C_{56}^{\frac{1}{2}}\widetilde{N}(\rho, \mathbf{v})^{\kappa}.$$

Case2 : $\dfrac{\mu_4 - \mu_3}{C_{55}\widetilde{N}(\rho, \mathbf{v})} \le 1.$

Then there does not exist $s > 0$ satisfying (6.96), but we obtain $\widetilde{N}(\rho, \mathbf{v}) \ge \frac{\mu_4-\mu_3}{C_{55}}$. By (6.92), we have $\|(\rho, \mathbf{v})\|_B \le C_{57}M$. Therefore we can write

$$\|(\rho, \mathbf{v})\|_B \le (C_{57}M\widetilde{N}(\rho, \mathbf{v})^{-\kappa})\widetilde{N}(\rho, \mathbf{v})^{\kappa}$$
$$\le \left(C_{57}M\left(\frac{\mu_4 - \mu_3}{C_{58}}\right)^{-\kappa}\right)\widetilde{N}(\rho, \mathbf{v})^{\kappa} \le C_{58}\widetilde{N}(\rho, \mathbf{v})^{\kappa}.$$

In both cases, we obtain $\|(\rho, \mathbf{v})\|_B \le C_{59}\widetilde{N}(\rho, \mathbf{v})^{\kappa}$.

By (6.87) we have $0 \in Q_{4\theta}$, that is, $d_p(0) > 4\theta$, and so we can choose an open subset $U_0 \subset \overline{\Omega_p}$ and a constant $\varepsilon > 0$ such that $0 \in U_0$ and $\mathcal{U} := (-\varepsilon, \varepsilon) \times U_0 \subset Q_{4\theta}$. Thus the proof of Theorem 6.2 is complete by the local Carleman estimate. ■

6.5 Proof of Theorem 6.2 by Global Carleman Estimate

In this section, we describe an essence of the proof of Theorem 6.2 by means of the global Carleman estimate (Theorem 6.3). Similarly to the proof in Sect. 6.4 by the local Carleman estimate, we need an argument by a cut-off function, and the proof involves other technicalities, as will be described below.

Step 1. For the operator $K_{\mathbf{v}_1}(x, D)$ given by (6.33), we introduce the principal symbol $K_{\mathbf{v}_1}(x, \xi) = \xi_0 + (\mathbf{v}_1(x), \xi')$. Let $\mathcal{O} \subset Q$ be a domain with piecewise smooth boundary and $\mathbf{n}(x)$ be the outward unit normal vector to $\mathcal{O}$. We prove a Carleman estimate for operator (6.33) which is similar to Lemma 6.1.

Proposition 6.4 *Let $\phi \in C^2(\overline{\mathcal{O}})$ satisfy*

$$K_{\mathbf{v}_1}(x, D)\phi \neq 0 \quad on\ \overline{\mathcal{O}}. \tag{6.97}$$

There exist constants $s_0 > 0$ and $C_0 > 0$, independent of w and s, such that

$$s^2\|we^{s\phi}\|^2_{L^2(\mathcal{O})} + s\int_{\partial\mathcal{O}} K_{\mathbf{v}_1}(x, \mathbf{n}(x))w^2e^{2s\phi}K_{\mathbf{v}_1}(x, D)\phi d\sigma$$
$$\leq C_0\|e^{s\phi}K_{\mathbf{v}_1}(x, D)w\|^2_{L^2(\mathcal{O})} \tag{6.98}$$

for $s > s_0$.

Proof Let $z = we^{s\phi}$. Then

$$e^{s\phi}K_{\mathbf{v}_1}(x, D)w = K_{\mathbf{v}_1}(x, D)z - sK_{\mathbf{v}_1}(x, D)\phi z \quad \text{in } \mathcal{O}. \tag{6.99}$$

Taking the norms in $L^2(\mathcal{O})$ of both sides of Eq. (6.99), we have

$$\|K_{\mathbf{v}_1}(x, D)z\|^2_{L^2(\mathcal{O})} - 2s(K_{\mathbf{v}_1}(x, D)z, (K_{\mathbf{v}_1}(x, D)\phi)z)_{L^2(\mathcal{O})}$$
$$+s^2\|(K_{\mathbf{v}_1}(x, D)\phi)z\|^2_{L^2(\mathcal{O})} = \|e^{s\phi}K_{\mathbf{v}_1}(x, D)w\|^2_{L^2(\mathcal{O})}. \tag{6.100}$$

By (6.97) there exists a constant $C_1 > 0$ independent of s such that

$$\|(K_{\mathbf{v}_1}(x, D)\phi)z\|^2_{L^2(\mathcal{O})} \geq C_1\|z\|^2_{L^2(\mathcal{O})}. \tag{6.101}$$

Integrating by parts, we obtain

$$(K_{\mathbf{v}_1}(x, D)z, (K_{\mathbf{v}_1}(x, D)\phi)z)_{L^2(\mathcal{O})} = \frac{1}{2}(K_{\mathbf{v}_1}(x, D)z^2, K_{\mathbf{v}_1}(x, D)\phi)_{L^2(\mathcal{O})}$$
$$= \frac{1}{2}\int_{\mathcal{O}} z^2K^*_{\mathbf{v}_1}(x, D)(K_{\mathbf{v}_1}(x, D)\phi)dx + \int_{\partial\mathcal{O}} K_{\mathbf{v}_1}(x, \mathbf{n})w^2e^{2s\phi}(K_{\mathbf{v}_1}(x, D)\phi)d\sigma. \tag{6.102}$$

By (6.101), taking the parameter s_0 sufficiently large, we obtain

$$s^2\|(K_{\mathbf{v}_1}(x,D)\phi)z\|^2_{L^2(\mathcal{O})}+\frac{s}{2}\int_{\mathcal{O}} z^2 K^*_{\mathbf{v}_1}(x,D)K_{\mathbf{v}_1}(x,D)\phi dx \ge \frac{C_2 s^2}{2}\|we^{s\phi}\|^2_{L^2(\mathcal{O})} \tag{6.103}$$

for any $s > s_0$.

This estimate and (6.100), (6.102), (6.103) yield (6.98). Thus the proof of Proposition 6.4 is complete. ■

Step 2: Change of variables and construction of the weight function. We can assume that there exist a constant $\beta_1 > 0$ and subboundaries S and S' of $\partial\Omega$ such that

$$S' \subset S, \quad |(\mathbf{v}_1(0,x'), \boldsymbol{\nu}(x'))| \ge \beta_1 \quad \text{for } x' \in S'.$$

Let $a' \in \overline{S'}$. Without loss of generality, after a possible translation, we can assume that $a' = 0$. The system (6.2) is rotationary invariant. Hence after a possible rotation, we can further assume that $\boldsymbol{\nu}(0) = (0, \dots, 0, 1) =: \mathbf{e}_n$. Then we have

$$|(\mathbf{v}_1(0), \mathbf{e}_n)| \ge \beta_0 > 0. \tag{6.104}$$

Near $(0, \dots, 0)$, the boundary $\partial\Omega$ is locally given by an equation $x_n - \vartheta(x_1, \dots, x_{n-1}) = 0$ and $x_n - \vartheta(x_1, \dots, x_{n-1}) < 0$ if $(x_1, \dots, x_n) \in \Omega$, where $\vartheta \in C^3$ and $\vartheta(0) = 0$. Since $\boldsymbol{\nu}(0) = \mathbf{e}_n$, we have

$$(\partial_{x_1}\vartheta(0), \dots \partial_{x_{n-1}}\vartheta(0)) = 0. \tag{6.105}$$

Denote

$$F(x) = (x_0, x_1, \dots, x_{n-1}, x_n - \vartheta(x_1, \dots, x_{n-1})). \tag{6.106}$$

The diffeomorphism F maps $Q \cap B(0,\delta)$ into $F(Q \cap B(0,\delta)) \subset \{x;\ x_n < 0\}$. It follows from (6.105) that

$$F'(0) = I, \tag{6.107}$$

where I is the unit matrix and $F'(0)$ is the Jacobi matrix of the mapping F at point $x = 0$. Consider the change of variables $y = F(x)$ and set $\widetilde{w}(y) = w \circ F^{-1}(y)$ and

$$\psi_0(y') = y_n - \sum_{k=1}^{n-1} y_k^2, \quad y' = (y_1, \dots, y_n).$$

For any $\tau \ge 1$, we define the function

$$\phi_0(y) = \frac{e^{\tau\psi_0(y')} - e^{2\tau\|\psi_0 \circ F\|_{C(\overline{\Omega})}}}{\widetilde{\ell}(y_0)},$$

where $\widetilde{\ell} \in C^3[-T, T]$, $\widetilde{\ell}(y_0) > 0$ in $(-T, T)$, $\partial_{y_0}\widetilde{\ell}(y_0) < 0$ on $(0, T)$, $\partial^2_{y_0}\widetilde{\ell}(0) < 0$ and $\partial_{y_0}\widetilde{\ell}(y_0) > 0$ on $(-T, 0)$, $\partial^j_{y_0}\widetilde{\ell}(\pm T) = 0$ for all $j \in \{0, 1, 2\}$, $\partial^3_{y_0}\widetilde{\ell}(\pm T) \neq 0$ and

$$\widetilde{\ell}(y_0) = 1 - y_0^2 \quad \text{on} \quad y_0 \in \left(-\frac{T}{4}, \frac{T}{4}\right). \tag{6.108}$$

We set

$$\psi(x') = \psi_0(F(x)), \quad \varphi(x) = \frac{e^{\tau\psi(x')} - e^{2\tau\|\psi\|_{C(\overline{\Omega})}}}{\widetilde{\ell}(x_0)}. \tag{6.109}$$

Since $\widetilde{\ell}'(0) = 0$ and $\nabla'\psi(0) = \mathbf{e}_n$ for each $\tau > 1$, there exists a neighborhood $G'(\tau)$ of $0 \in \mathbb{R}^{n+1}$ depending on τ such that

$$K_{\mathbf{v}_1}(x, D)\phi_0 \neq 0 \quad \forall x \in G'(\tau). \tag{6.110}$$

Then the conditions of Theorem 6.3 for φ given by (6.109), are satisfied for all sufficiently large τ. At this point we fix parameter τ sufficiently large.

Step 3: Level sets of the function φ. We set

$$\kappa = \varphi(0) < 0.$$

Obviously 0 is the maximum of the function φ over $B(0, \delta) \cap Q$. In order to see this fact, we observe that $\varphi = \phi_0 \circ F$. Let $\widehat{x}$ be a point of the maximum of the function φ. Then $\widehat{y} := F(\widehat{x})$ is the maximum of the function ϕ_0 over the set $\overline{F(B(0, \delta) \cap Q)}$. By the construction of the mapping F, we have

$$F(B(0, \delta) \cap Q) \subset \{y;\ y_n < 0\}.$$

The function $\psi_0(y')$ has the maximum on $\overline{F(B(0, \delta) \cap Q)}$ at point $y \in \overline{F(B(0, \delta) \cap Q)}$ such that $y' = 0$. Hence

$$\phi_0(y) \leq \phi_0(y_0, 0) = \frac{e^{\tau\psi_0(0)} - e^{2\tau\|\psi_0\|_{C(\overline{\Omega})}}}{\widetilde{\ell}(y_0)} \quad \text{for all } y \in F(B(0, \delta) \cap Q).$$

Since $e^{\tau\psi_0(0)} - e^{2\tau\|\psi_0\|_{C(\overline{\Omega})}}$ is strictly negative, the function $\phi_0(y_0, 0)$ reaches the maximum at a point where $\widetilde{\ell}(y_0)$ reaches its global maximum. By the construction of $\widetilde{\ell}(y_0)$, it has the maximum at point $y_0 = 0$.

For strictly positive δ, consider the level set:

$$G_\delta = \{x \in B(0, \delta) \cap Q;\ \varphi(x) > \kappa - \delta\}$$

and the image of this set under the diffeomorphism F

$$\mathcal{G}_\delta = \{y \in F(B(0, \delta) \cap Q);\ \phi_0(y) > \kappa - \delta\} = F(G_\delta). \tag{6.111}$$

By $B'(0, R)$ we denote the ball in $\mathbb{R}^n$ centered at 0 with radius R. We have

Proposition 6.5 *There exist $\delta_0 > 0$ and functions $r(\delta)$ and $t(\delta)$ such that $r, t \in C[0, \delta_0]$,*

$$r(\delta) > 0, \quad t(\delta) > 0 \quad \forall \delta \in (0, \delta_0], \quad \lim_{\delta \to +0} r(\delta) = \lim_{\delta \to +0} t(\delta) = 0$$

and

$$\mathcal{G}_\delta \subset [-t(\delta), t(\delta)] \times (B'(0, r(\delta)) \cap \{y';\ y_n < 0\}).$$

Proof Let $y = (y_0, y')$ satisfy $y_0 \neq 0$ and $y_0 \in \left(-\frac{T}{4}, \frac{T}{4}\right)$. By $\kappa < 0$, short computations and (6.108) imply

$$\phi_0(y) \le \frac{\kappa}{\widetilde{\ell}(y_0)} = \frac{\kappa}{1 - y_0^2} = \kappa + \kappa\left(\frac{1}{1 - y_0^2} - 1\right) = \kappa + \frac{\kappa y_0^2}{1 - y_0^2} < \kappa + \kappa y_0^2.$$

Setting $t(\delta) = \sqrt{\delta/|\kappa|} = \sqrt{-\delta/\kappa}$, assuming that $|y_0| \ge t(\delta)$, applying the above inequality and noting that κ is negative, we obtain that

$$\phi_0(y) < \kappa - \delta.$$

Therefore

$$|y_0| \le t(\delta) \quad \forall y \in \mathcal{G}_\delta. \tag{6.112}$$

Next observe that

$$\phi_0(y) \le \frac{e^{\tau\psi_0(y')} - e^{2\tau\|\psi_0\|_{C(\overline{\Omega})}}}{1 - t^2(\delta)} \le e^{\tau\psi_0(y')} - e^{2\tau\|\psi_0\|_{C(\overline{\Omega})}} \le \kappa + e^{\tau\psi_0(y')} - 1.$$

If $\psi_0(y') < \frac{\ln(1-\delta)}{\tau}$, then

$$\psi_0(y') < \kappa - \delta.$$

Therefore if $y \in \mathcal{G}_\delta$, then

$$\psi_0(y') > \frac{\ln(1 - \delta)}{\tau}.$$

Consequently the definition of the function ψ_0 implies

$$y_n - \frac{\ln(1 - \delta)}{\tau} > \sum_{k=1}^{n-1} y_k^2.$$

Since $y_n < 0$ this inequality yields

$$\sum_{k=1}^{n-1} y_k^2 \le -\frac{\ln(1-\delta)}{\tau} \quad \text{and} \quad y_n^2 \le \left(\frac{\ln(1-\delta)}{\tau}\right)^2.$$

Taking

$$r(\delta) = \sqrt{-\frac{\ln(1-\delta)}{\tau} + \left(\frac{\ln(1-\delta)}{\tau}\right)^2},$$

we obtain that

$$|y'| \le r(\delta) \quad \forall y \in \mathcal{G}_\delta.$$

The proof of Proposition 6.5 is complete. ■

From (6.111) and Proposition 6.5, we have

Proposition 6.6 *There exist* $\delta_1 > 0$ *and functions* $\widetilde{r}(\delta)$ *and* $t(\delta)$ *such that* $\widetilde{r}, t \in C[0, \delta_1]$,

$$\widetilde{r}(\delta) > 0, \quad t(\delta) > 0 \quad \forall \delta \in (0, \delta_1],$$

$$\lim_{\delta \to +0} \widetilde{r}(\delta) = \lim_{\delta \to +0} t(\delta) = 0$$

and

$$G_\delta \subset [-t(\delta), t(\delta)] \times (B'(0, \widetilde{r}(\delta)) \cap \Omega).$$

Proof By (6.111) and (6.106), we see that

$$F^{-1}(\mathcal{G}_\delta) = G_\delta, \quad y_0 = x_0.$$

Therefore applying Proposition 6.5, we obtain that

$$|x_0| < t(\delta) \quad \forall x \in G_\delta.$$

Moreover, Proposition 6.5 yields

$$|y| \le d(\delta) = \sqrt{t^2(\delta) + r^2(\delta)} \quad \forall y \in \mathcal{G}_\delta. \tag{6.113}$$

Using Taylor's formula, we have

$$F^{-1}(y) = F(0) + F'^{-1}(0)y + o(y), \quad \|o(y)\| \le C_3|y|^2 \quad \forall y \in \mathcal{G}_\delta.$$

We have $F(0) = 0$ and $F'^{-1}(0) = I$ by (6.107). Hence

$$|F^{-1}(y)| \le |y| + C_4|y|^2 \le (1 + C_5 d(\delta))|y| \le C_6|y| \le C_7 d(\delta).$$

We take $\widetilde{r}(\delta) = C_7 d(\delta)$. Thus the proof of Proposition 6.6 is complete. ■

Step 4: Carleman estimate. Let $\chi \in C^\infty(\mathbb{R}^1_+)$, $\chi \ge 0$ on $\mathbb{R}^1_+$, $\chi|_{[0,\frac{1}{4}]} = 1$ and $\chi|_{[\frac{1}{2},+\infty)} = 0$. Consider the function $\widetilde{\chi} = \chi(2(\kappa - \varphi)/\delta)$, where we fix $\delta \in [0, \widetilde{\delta}_0]$. Observe that

$$\widetilde{\chi}|_{G_{\frac{\delta}{8}}} = 1 \quad \text{and} \quad \widetilde{\chi}|_{G_{\frac{\delta}{2}} \setminus G_{\frac{\delta}{4}}} = 0. \tag{6.114}$$

Indeed on $G_{\frac{\delta}{8}}$, we have $\varphi(x) > \kappa - \delta/8$. Hence $\delta/8 > k - \varphi(x)$ and $(k - \varphi(x))/\delta \le \frac{1}{8}$ or $2(k - \varphi(x))/\delta \le \frac{1}{4}$. The definition of the function $\widetilde{\chi}$ implies the first equality in (6.114). Now we prove the second one. Indeed on $G_{\frac{\delta}{2}} \setminus G_{\frac{\delta}{4}}$, we have $\varphi(x) < \kappa - \delta/4$. Consequently $\delta/8 > k - \varphi(x)$ and $(k - \varphi(x))/\delta \ge \frac{1}{4}$ or $2(k - \varphi(x))/\delta \ge \frac{1}{2}$. Thus by the definition of the function $\widetilde{\chi}$, we have the second equality in (6.114).

We set

$$\widetilde{\rho} = \widetilde{\chi}\rho, \quad \widetilde{\mathbf{v}} = \widetilde{\chi}\mathbf{v}.$$

By Proposition 6.6, for all sufficiently small δ, there exists $\widetilde{r}(\delta)$ such that

$$G_\delta \subset (-t(\delta), t(\delta)) \times B'(0, \widetilde{r}(\delta)), \quad \lim_{\delta \to +0} \widetilde{r}(\delta) = 0 \quad \text{and} \quad \lim_{\delta \to +0} t(\delta) = 0. \tag{6.115}$$

Hence by (6.104), there exists a constant $\delta_1 > 0$ such that

$$|(\mathbf{v}_1(0, x'), \boldsymbol{\nu}(x'))| \ge \frac{\beta_0}{2} > 0 \quad \text{on} \quad (-t(\delta), t(\delta)) \times B'(0, \widetilde{r}(\delta)) \tag{6.116}$$

for all $\delta \in (0, \delta_1)$.

Decreasing the parameter δ_1 if necessary, we obtain

$$K_{\mathbf{v}_1}(x, D)\varphi \ne 0 \quad \forall x \in (-t(\delta), t(\delta)) \times B'(0, \widetilde{r}(\delta)) \tag{6.117}$$

for all $\delta \in (0, \delta_1)$. We set $\boldsymbol{Q}_\delta = (-t(\delta), t(\delta)) \times B'(0, \widetilde{r}(\delta))$.

Taking the difference between Eq. (6.10) with $k = 1, 2$, we can verify

$$-K_{\mathbf{v}_1}(x, D)^* \partial_{x'}^{\alpha'} \widetilde{\rho} = f_\alpha \quad \text{in} \, \boldsymbol{Q}_\delta, \tag{6.118}$$

where

$$f_\alpha = -\partial_{x'}^{\alpha'}([K^*_{\mathbf{v}_1}, \widetilde{\chi}]\rho) - [K^*_{\mathbf{v}_1}, \partial_{x'}^{\alpha'}]\widetilde{\rho} - \partial_{x'}^{\alpha'} \mathrm{div}\,(\widetilde{\mathbf{v}}\rho_2) + \partial_{x'}^{\alpha'}((\mathbf{v}, \nabla'\widetilde{\chi})\rho_2).$$

Applying estimate (6.98) to Eq. (6.118), we obtain

$$\sum_{|\alpha'|\le 2} s^{4-2|\alpha'|}\|e^{s\varphi}\partial_{x'}^{\alpha'}\widetilde{\rho}\|_{L^2(Q_\delta)} \le C_8\Bigg(\frac{1}{\sqrt{s}}\sum_{|\alpha'|\le 2} s^{4-2|\alpha'|}\|e^{s\varphi}\partial_{x'}^{\alpha'}\widetilde{\rho}\|_{L^2(\Sigma_\delta)} \tag{6.119}$$

$$+\sum_{|\alpha'|\le 2} s^{4-2|\alpha'|}\left\|e^{s\varphi}\frac{1}{s}\partial_{x'}^{\alpha'}\operatorname{div}\widetilde{\mathbf{v}}\right\|_{L^2(Q_\delta)} + \sum_{|\alpha'|\le 2} s^{4-2|\alpha'|}\left\|e^{s\varphi}\frac{1}{s}\partial_{x'}^{\alpha'}\widetilde{\mathbf{v}}\right\|_{L^2(Q_\delta)}$$

$$+\sum_{|\alpha'|\le 2} s^{4-2|\alpha'|}\left\|\frac{1}{s}\partial_{x'}^{\alpha'}([K^*_{\mathbf{v}_1},\widetilde{\chi}]\rho)\right\|_{L^2(Q_\delta)} + \sum_{|\alpha'|\le 2} s^{4-2|\alpha'|}\left\|\frac{1}{s}\partial_{x'}^{\alpha'}((\mathbf{v},\nabla'\widetilde{\chi})\rho_2)\right\|_{L^2(Q_\delta)}\Bigg)$$

for all $s \ge s_0$. Consider the differential operator

$$\mathcal{L}(x,D)\mathbf{v} = \rho_1\partial_{x_0}\mathbf{v} - L_{\lambda,\mu}(x',D')\mathbf{v} + \rho_2(\mathbf{v}_1,\nabla')\mathbf{v} + \rho_2(\mathbf{v},\nabla')\mathbf{v}_2.$$

Taking the difference between Eq. (6.11) with $k = 1, 2$, we can verify that $\widetilde{\mathbf{v}}$ satisfies

$$\mathcal{L}(x,D)\widetilde{\mathbf{v}} = \mathbf{F} \quad \text{in } Q_\delta, \quad \widetilde{\mathbf{v}}|_{\Sigma_\delta} = 0, \tag{6.120}$$

where

$$\mathbf{F} = -\widetilde{\rho}\partial_{x_0}\mathbf{v}_2 - (\widetilde{\mathbf{v}},\nabla')\mathbf{v}_2 - \widetilde{\chi}(h(\rho_1)\nabla'\rho - (h(\rho_1)-h(\rho_2))\nabla'\rho_2)$$
$$+\widetilde{\mathbf{f}} + [\mathcal{L}(x,D),\widetilde{\chi}]\mathbf{v} - \widetilde{\rho}(\mathbf{v}_1,\nabla')\mathbf{v}_1 - \widetilde{\rho}\partial_{x_0}\mathbf{v}_2 \quad \text{and} \quad \widetilde{\mathbf{f}} = \widetilde{\chi}\mathbf{f}.$$

Observe that the function φ defined by (6.109) satisfies (6.21)–(6.24), Conditions 6.1 and 6.2, provided that parameter τ is large enough.

Applying Carleman estimate (6.26) to Eq. (6.120), we have

$$\|\widetilde{\mathbf{v}}\|_{\mathcal{B}(\varphi,s,Q_\delta)} \le C_9(\|\mathbf{F}e^{s\varphi}\|_{\mathcal{Y}(\varphi,s,Q_\delta)} + \|(d\omega_{\widetilde{\mathbf{v}}},\operatorname{div}\widetilde{\mathbf{v}})e^{s\varphi}\|_{H^{\frac{3}{4},\frac{3}{2},s}(\widetilde{\Sigma}_\delta)} \tag{6.121}$$
$$+\|\partial_\nu(d\omega_{\widetilde{\mathbf{v}}},\operatorname{div}\widetilde{\mathbf{v}})e^{s\varphi}\|_{H^{\frac{1}{4},\frac{1}{2},s}(\widetilde{\Sigma}_\delta)} + \|\widetilde{\varphi}\partial_{x_0}\partial_\nu\widetilde{\mathbf{v}}e^{s\varphi}\|_{L^2(\widetilde{\Sigma}_\delta)})$$

for all $s \ge s_1$, where

$$\widetilde{\Sigma}_\delta = Q_\delta \cap ([-T,T]\times(S\cap B'(0,\delta))).$$

Short computations imply

$$\|\mathbf{F}e^{s\varphi}\|_{\mathcal{Y}(\varphi,s,Q_\delta)} \le C_{10}\Bigg(\sum_{|\alpha'|\le 2}\|\partial_{x'}^{\alpha'}\widetilde{\rho}e^{s\varphi}\|_{L^2(Q_\delta)} + \sum_{|\alpha'|\le 1}\|\partial_{x'}^{\alpha'}\widetilde{\mathbf{v}}e^{s\varphi}\|_{L^2(Q_\delta)} \tag{6.122}$$
$$+\|(\widetilde{\mathbf{f}} + [\mathcal{L}(x,D),\widetilde{\chi}]\mathbf{v})e^{s\varphi}\|_{\mathcal{Y}(\varphi,s,Q_\delta)}\Bigg).$$

Applying (6.119) to the first term on the right-hand side of (6.122), we obtain

$$\|\mathbf{F}e^{s\varphi}\|_{\mathcal{Y}(\varphi,s,\mathbf{Q}_\delta)} \le C_{11}\Bigg(\sum_{|\alpha'|\le 1}\|\partial_{x'}^{\alpha'}\widetilde{\mathbf{v}}e^{s\varphi}\|_{L^2(\mathbf{Q}_\delta)} + \|(\widetilde{\mathbf{f}} + [\mathcal{L}(x,D),\widetilde{\chi}]\mathbf{v})e^{s\varphi}\|_{\mathcal{Y}(\varphi,s,\mathbf{Q}_\delta)}$$
$$+\frac{1}{\sqrt{s}}\sum_{|\alpha'|\le 2}s^{4-2|\alpha'|}\|e^{s\varphi}\partial_{x'}^{\alpha'}\widetilde{\rho}\|_{L^2(\Sigma_\delta)} \quad (6.123)$$
$$+\sum_{|\alpha'|\le 2}s^{4-2|\alpha'|}\left\|e^{s\varphi}\frac{1}{s}\partial_{x'}^{\alpha'}\operatorname{div}\widetilde{\mathbf{v}}\right\|_{L^2(\mathbf{Q}_\delta)} + \sum_{|\alpha'|\le 2}s^{4-2|\alpha'|}\left\|e^{s\varphi}\frac{1}{s}\partial_{x'}^{\alpha'}\widetilde{\mathbf{v}}\right\|_{L^2(\mathbf{Q}_\delta)}$$
$$+\sum_{|\alpha'|\le 2}s^{4-2|\alpha'|}\left\|\frac{1}{s}\partial_{x'}^{\alpha'}([K^*_{\mathbf{v}_1},\widetilde{\chi}]\rho)\right\|_{L^2(Q_\delta)} + \sum_{|\alpha'|\le 2}s^{4-2|\alpha'|}\left\|\frac{1}{s}\partial_{x'}^{\alpha'}((\mathbf{v},\nabla'\widetilde{\chi})\rho_2)\right\|_{L^2(\mathbf{Q}_\delta)}\Bigg)$$

for all $s \ge s_2$. From (6.123) and (6.121), we obtain

$$\|\widetilde{\mathbf{v}}\|_{\mathcal{B}(\varphi,s,\mathbf{Q}_\delta)} \le C_{12}\Bigg(\|\widetilde{\mathbf{f}}e^{s\varphi}\|_{\mathcal{Y}(\varphi,s,\mathbf{Q}_\delta)} + \|[\mathcal{L}(x,D),\widetilde{\chi}]\mathbf{v}e^{s\varphi}\|_{\mathcal{Y}(\varphi,s,\mathbf{Q}_\delta)}$$
$$+\|(d\omega_{\widetilde{\mathbf{v}}},\operatorname{div}\widetilde{\mathbf{v}})e^{s\varphi}\|_{H^{\frac{3}{4},\frac{3}{2},\widetilde{s}}(\widetilde{\Sigma}_\delta)} + \|\partial_{\boldsymbol{\nu}}(d\omega_{\widetilde{\mathbf{v}}},\operatorname{div}\widetilde{\mathbf{v}})e^{s\varphi}\|_{H^{\frac{1}{4},\frac{1}{2},\widetilde{s}}(\widetilde{\Sigma}_\delta)}$$
$$+\|\widetilde{\varphi}\partial_{x_0}\partial_{\boldsymbol{\nu}}\widetilde{\mathbf{v}}e^{s\varphi}\|_{L^2(\widetilde{\Sigma}_\delta)} + \frac{1}{\sqrt{s}}\sum_{|\alpha'|\le 2}s^{4-2|\alpha'|}\|e^{s\varphi}\partial_{x'}^{\alpha'}\widetilde{\rho}\|_{L^2(\widetilde{\Sigma}_\delta)}$$
$$+\sum_{|\alpha'|\le 2}s^{4-2|\alpha'|}\left\|\frac{1}{s}\partial_{x'}^{\alpha'}([K^*_{\mathbf{v}_1},\widetilde{\chi}]\rho)\right\|_{L^2(\mathbf{Q}_\delta)} + \sum_{|\alpha'|\le 2}s^{4-2|\alpha'|}\left\|\frac{1}{s}\partial_{x'}^{\alpha'}((\mathbf{v},\nabla'\widetilde{\chi})\rho_2)\right\|_{L^2(\mathbf{Q}_\delta)}\Bigg)$$

for all $s \ge s_3$.

Then we can proceed similarly to complete the proof of Theorem 6.2. ■

References

1. A.Kh. Amirov, *Integral Geometry and Inverse Problems for Kinetic Equations* (VSP, Utrecht, 2001)
2. L. Baudouin, E. Crépeau, J. Valein, Global Carleman estimate on a network for the wave equation and application to an inverse problem. Math. Control Relat. Fields **1**, 1–24 (2011)
3. L. Baudouin, A. Mercado, An inverse problem for Schrödinger equations with discontinuous main coefficient. Appl. Anal. **87**, 1145–1165 (2008)
4. L. Baudouin, A. Mercado, A. Osses, A global Carleman estimate in a transmission wave equation and application to a one-measurement inverse problem. Inverse Probl. **23**, 257–258 (2007)
5. L. Baudouin, J.-P. Puel, Uniqueness and stability in an inverse problem for the Schrödinger equation. Inverse Probl. **18**, 1537–1554 (2002)
6. L. Baudouin, M. Yamamoto, Inverse problem on a tree-shaped network: unified approach for uniqueness. Appl. Anal. **94**, 2370–2395 (2015)
7. L. Beilina, M. Cristofol, S. Li, M. Yamamoto, Lipschitz stability for an inverse hyperbolic problem of determining two coefficients by a finite number of observations. Inverse Probl. **34**, 015001 (2018)
8. L. Beilina, M.V. Klibanov, *Approximate Global Convergence and Adaptivity for Coefficient Inverse Problems* (Springer, Berlin, 2012)

9. M. Bellassoued, Uniqueness and stability in determining the speed of propagation of second-order hyperbolic equation with variable coefficients. Appl. Anal. **83**, 983–1014 (2004)
10. M. Bellassoued, O. Imanuvilov, M. Yamamoto, Inverse problem of determining the density and two Lamé coefficients by boundary data. SIAM J. Math. Anal. **40**, 238–265 (2008)
11. M. Bellassoued, O. Imanuvilov, M. Yamamoto, Carleman estimate for the Navier-Stokes equations and an application to a lateral Cauchy problem. Inverse Probl. **32**, 025001 (23pp) (2016)
12. M. Bellassoued, M. Yamamoto, Inverse source problem for a transmission problem for a parabolic equation. J. Inverse Ill-Posed Probl. **14**, 47–56 (2006)
13. M. Bellassoued, M. Yamamoto, Logarithmic stability in determination of a coefficient in an acoustic equation by arbitrary boundary observation. J. Math. Purés Appl. **85**, 193–224 (2006)
14. M. Bellassoued, M. Yamamoto, Lipschitz stability in determining density and two Lamé coefficients. J. Math. Anal. Appl. **329**, 1240–1259 (2007)
15. M. Bellassoued, M. Yamamoto, *Carleman Estimates and Applications to Inverse Problems for Hyperbolic Systems* (Springer, Tokyo, 2017)
16. A. Benabdallah, M. Cristofol, P. Gaitan, M. Yamamoto, Inverse problem for a parabolic system with two components by measurements of one component. Appl. Anal. **88**, 683–709 (2009)
17. A. Benabdallah, Y. Dermenjian, J. Le Rousseau, Carleman estimates for the one-dimensional heat equation with a discontinuous coefficient and applications to controllability and an inverse problem. J. Math. Anal. Appl. **336**, 865–887 (2007)
18. A. Benabdallah, P. Gaitan, J. Le Rousseau, Stability of discontinuous diffusion coefficients and initial conditions in an inverse problem for the heat equation. SIAM J. Control Optim. **46**, 1849–1881 (2007)
19. A.L. Bukhgeim, M.V. Klibanov, Global uniqueness of class of multidimentional inverse problems. Soviet Math. Dokl. **24**, 244–247 (1981)
20. P. Cannarsa, G. Floridia, F. Gölgeleyen, M. Yamamoto, Inverse coefficient problems for a transport equation by local Carleman estimate. Inverse Probl. **35**, 105013 (2019)
21. C. Cavaterra, A. Lorenzi, M. Yamamoto, A stability result via Carleman estimates for an inverse source problem related to a hyperbolic integro-differential equation. Comput. Appl. Math. **25**, 229–250 (2006)
22. M. Choulli, O. Imanuvilov, J.-P. Puel, M. Yamamoto, Inverse source problem for linearized Navier-Stokes equations with data in arbitrary sub-domain. Appl. Anal. **92**, 2127–2143 (2012)
23. M. Cristofol, P. Gaitan, H. Ramoul, Inverse problems for a 2×2 reaction diffusion system using a Carleman estimate with one observation. Inverse Probl. **22**, 1561–1573 (2006)
24. M. de Buhan, A. Osses, Logarithmic stability in determination of a 3D viscoelastic coefficient and a numerical example. Inverse Probl. **26**, 095006 (2010)
25. A. Doubova, A. Osses, Rotated weights in global Carleman estimates applied to an inverse problem for the wave equation. Inverse Probl. **22**, 265–296 (2006)
26. S. Ervedoza, O. Glass, S. Guerrero, Local exact controllability for the two and three-dimensional compressible Navier-Stokes equations, Commun. Partial Differ. Eqn. **41**, 1660–1691 (2016)
27. J. Fan, M. Di Cristo, Y. Jiang, G. Nakamura, Inverse viscosity problem for the Navier–Stokes equation. J. Math. Anal. Appl. **365**, 750–757 (2010)
28. J. Fan, Y. Jiang, G. Nakamura, Inverse problems for the Boussinesq system. Inverse Probl. **25**, 085007 (2009)
29. E. Fernández-Cara, S. Guerrero, O. Imanuvilov, J.-P. Puel, Local exact controllability of the Navier–Stokes system. J. Math. Pures Appl. **83**, 1501–1542 (2004)
30. P. Gaitan, H. Ouzzane, Inverse problem for a free transport equation using Carleman estimates. Appl. Anal. **93**, 1073–1086 (2014)
31. P. Gaitan, H. Ouzzane, Stability result for two coefficients in a coupled hyperbolic-parabolic system. J. Inverse Ill-Posed Probl. **25**, 265–286 (2017)
32. Y. Giga, A. Novotný (eds.), *Handbook of Mathematical Analysis in Mechanics of Viscous Fluids (3 volumes)* (Springer, Berlin, 2018)
33. F. Gölgeleyen, M. Yamamoto, Stability for some inverse problems for transport equations. SIAM J. Math. Anal. **48**, 2319–2344 (2016)

34. F. Gölgeleyen, M. Yamamoto, *Uniqueness of solution of an inverse source problem for ultra-hyperbolic equations*, arXiv:1909.01399
35. L. Hörmander, *Linear Partial Differential Operators* (Springer, Berlin, 1963)
36. X. Huang, M. Yamamoto, *Carleman estimates for a magnetohydrodynamics system and application to inverse source problem* (2018), arXiv:1806.07576
37. M. Ikehata, G. Nakamura, M. Yamamoto, Uniqueness in inverse problems for isotropic Lamé system. J. Math. Sci. Univ. Tokyo **5**, 627–692 (1998)
38. O. Imanuvilov, Controllability of parabolic equations. Math. Sb. **186**, 879–900 (1995)
39. O. Imanuvilov, Remarks on exact controllability for the Navier–Stokes equations. ESAIM Control Optim. Calc. Var. **6**, 39–72 (2001)
40. O. Imanuvilov, Carleman estimates for hyperbolic equations. Asymptot. Anal. **32**, 185–220 (2002)
41. O. Imanuvilov, V. Isakov, M. Yamamoto, An inverse problem for the dynamical Lamé system with two sets of boundary data. Commun. Pure Appl. Math. **56**, 1366–1382 (2003)
42. O. Imanuvilov, J.P. Puel, M. Yamamoto, Carleman estimates for parabolic equations with nonhomogeneous boundary conditions. Chin. Ann. Math. Ser. B. **30**, 333–378 (2009)
43. O. Imanuvilov, M. Yamamoto, Lipschitz stability in inverse parabolic problems by the Carleman estimate. Inverse Probl. **14**, 1229–1245 (1998)
44. O. Imanuvilov, M. Yamamoto, Global Lipschitz stability in an inverse hyperbolic problem by interior observations. Inverse Probl. **17**, 717–728 (2001)
45. O. Imanuvilov, M. Yamamoto, Global uniqueness and stability in determining coefficients of wave equations. Commun. Partial Differ. Equ. **26**, 1409–1425 (2001)
46. O. Imanuvilov, M. Yamamoto, Carleman estimate for a parabolic equation in a Sobolev space of negative order and its applications, in *Control of Nonlinear Distributed Parameter Systems (College Station, TX, 1999)*. Lecture Notes in Pure and Applied Mathematics, vol. 218 (Dekker, New York, 2001), pp. 113–137
47. O. Imanuvilov, M. Yamamoto, Determination of a coefficient in an acoustic equation with a single measurement. Inverse Probl. **19**, 157–171 (2003)
48. O. Imanuvilov, M. Yamamoto, Carleman estimates for parabolic equations in Sobolev spaces of negative order and exact controllability for semilinear parabolic equation. Publ. Res. Inst. Math. Sci. **39**, 227–274 (2003)
49. O. Imanuvilov, M. Yamamoto, Carleman estimates for the non-stationary Lamé system and the application to an inverse problem. ESAIM Control Optim. Calc. Var. **11**, 1–56 (2005)
50. O. Imanuvilov, M. Yamamoto, Carleman estimates for the three-dimensional nonstationary Lamé system and application to an inverse problem, in *Control Theory of Partial Differential Equations*. Lecture Notes in Pure Applied Mathematics vol. 242. (Chapman & Hall/CRC, Boca Raton, FL, 2005), pp. 337–374
51. O. Imanuvilov, M. Yamamoto, Carleman estimates for the Lamé system with stress boundary condition. Publ. Res. Inst. Math. Sci. **43**, 1023–1093 (2007)
52. O. Imanuvilov, M. Yamamoto, An inverse problem and an observability inequality for the Lamé system with stress boundary condition. Appl. Anal. **88**, 711–733 (2009)
53. O. Imanuvilov, M. Yamamoto, *Carleman estimate for linear viscoelasticity equations and an inverse source problem* to appear in SIAM J. Math. Anal. (2019), arXiv:1711.09276
54. O. Imanuvilov, M. Yamamoto, Carleman estimate and an inverse source problem for the Kelvin-Voigt model for viscoelasticity. to appear in Inverse Probl. (2019)
55. V. Isakov, *Inverse Problems for Partial Differential Equations* (Springer, Berlin, 1998)
56. V. Isakov, Carleman estimates and applications to inverse problems. Milan J. Math. **72**, 249–271 (2004)
57. V. Isakov, J.-N. Wang, M. Yamamoto, Uniqueness and stability of determining the residual stress by one measurement. Commun. Partial Differ. Equ. **32**, 833–848 (2007)
58. V. Isakov, J.-N. Wang, M. Yamamoto, *An inverse problem for a dynamical Lame system with residual stress*. SIAM J. Math. Anal. **39**, 1328–1343 (2007/08)
59. N. Itaya, On the Cauchy problem for the system of fundamental equations describing the movement of compressible viscous fluid. Kôdai Math. Sem. Rep. **23**, 60–120 (1971)

60. A. Kawamoto, Hölder stability estimate in an inverse source problem for a first and half-order time fractional diffusion equation. Inverse Probl. Imaging **12**, 315–330 (2018)
61. A. Kawamoto, Lipschitz stability estimates in inverse source problems for a fractional diffusion equation of half order in time by Carleman estimates. J. Inverse Ill-Posed Probl. **26**, 647–672 (2018)
62. M.V. Klibanov, Inverse problems in the "large" and Carleman bounds. Differ. Equ. **20**, 755–760 (1984)
63. M.V. Klibanov, Inverse problems and Carleman estimates. Inverse Probl. **8**, 575–596 (1992)
64. M.V. Klibanov, Carleman estimates and inverse problems in the last two decades, *Surveys on Solutions Methods for Inverse Problems* (Springer, Wien, 2000), pp. 119–146
65. M.V. Klibanov, Carleman estimates for global uniqueness, stability and numerical methods for coefficient inverse problems. J. Inverse Ill-Posed Probl. **21**, 477–560 (2013)
66. M.V. Klibanov, S.E. Pamyatnykh, Lipschitz stability of a non-standard problem for the non-stationary transport equation via a Carleman estimate. Inverse Probl. **22**, 881–890 (2006)
67. M.V. Klibanov, S.E. Pamyatnykh, Global uniqueness for a coefficient inverse problem for the nonstationary transport equation via Carleman estimate. J. Math. Anal. Appl. **343**, 352–365 (2008)
68. M.V. Klibanov, A. Timonov, *Carleman Estimates for Coefficient Inverse Problems and Numerical Applications* (VSP, Utrecht, 2004)
69. M.V. Klibanov, M. Yamamoto, Lipschitz stability of an inverse problem for an accoustic equation. Appl. Anal. **85**, 515–538 (2006)
70. L.D. Landau, E.M. Lifshitz, *Fluid Mechanics*, vol. 6, Course of Theoretical Physics (Pergamon, Oxford, 1959)
71. S. Li, An inverse problem for Maxwell's equations in bi-isotropic media. SIAM J. Math. Anal. **37**, 1027–1043 (2005)
72. X. Huang, Z. Li, M. Yamamoto, Carleman estimates for the time-fractional advection-diffusion equations and applications. Inverse Probl. **35**, 045003 (2019)
73. P. Loreti, D. Sforza, M. Yamamoto, Carleman estimate and application to an inverse source problem for a viscoelasticity model in anisotropic case. Inverse Probl. **33**, 125014 (2017)
74. M. Machida, M. Yamamoto, Global Lipschitz stability in determining coefficients of the radiative transport equation. Inverse Probl. **30**, 035010 (2014)
75. A. Mercado, A. Osses, L. Rosier, Inverse problems for the Schrödinger equation via Carleman inequalities with degenerate weights. Inverse Probl. **24**, 015017 (2008)
76. O. Poisson, Carleman estimates for the heat equation with discontinuous diffusion coefficients. Appl. Anal. **87**, 1129–1144 (2008)
77. B. Riahi, Stability estimates in determination of a coefficient in transmission wave equation by boundary observation. Appl. Anal. **94**, 2478–2516 (2015)
78. V.G. Romanov, M. Yamamoto, Recovering a Lamé kernel in a viscoelastic equation by a single boundary measurement. Appl. Anal. **89**, 377–390 (2010)
79. X. Xu, J. Cheng, M. Yamamoto, Carleman estimate for a fractional diffusion equation with half order and application. Appl. Anal. **90**, 1355–1371 (2011)
80. M. Yamamoto, *On an inverse problem of determining source terms in Maxwell's equations with a single measurement*, in *Inverse Problems, Tomography, and Image Processing* (Plenum Press, New York, 1998), pp. 241–256
81. M. Yamamoto, Carleman estimates for parabolic equations and applications. Inverse Probl. **25**, 123013 (2009)
82. M. Yamamoto, Y. Zhang, Conditional stability in determining a zeroth-order coefficient in a half-order fractional diffusion equation by a Carleman estimate. Inverse Probl. **28**, 105010 (2012)
83. M. Yamamoto, J. Zou, Simultaneous reconstruction of the initial temperature and heat radiative coefficient. Inverse Probl. **17**, 1181–1202 (2001)
84. G. Yuan, M. Yamamoto, Lipschitz stability in inverse problems for a Kirchhoff plate equation. Asymptot. Anal. **53**, 29–60 (2007)

85. G. Yuan, M. Yamamoto, Lipschitz stability in the determination of the principal part of a parabolic equation. ESAIM Control Optim. Calc. Var. **15**, 525–554 (2009)
86. G. Yuan, M. Yamamoto, Carleman estimates for the Schrödinger equation and applications to an inverse problem and an observability inequality. Chin. Ann. Math. Ser. B **31**, 555–578 (2010)

Chapter 7
Carleman Estimate for a General Second-Order Hyperbolic Equation

Xinchi Huang

Abstract In this article, we consider a general second-order hyperbolic equation. We first establish a modified Carleman estimate for this equation by adding some functions of adjustment. Then general conditions imposed on the principal parts, mixed with the weight function and the functions of adjustment are derived. Finally, we give the realizations of the weight functions by choosing suitable adjustments such that the above general conditions are satisfied in some specific cases.

Keywords Carleman estimate · Hyperbolic equation · Time-dependent principal parts

7.1 Introduction and Main Results

7.1.1 Introduction

Let $T > 0$ and Ω be a bounded domain in $\mathbb{R}^n$ $(n \geq 1)$, with sufficiently smooth boundary $\partial\Omega$, for example, of C^2-class. We deal with the following hyperbolic equation:

$$Lu := \partial_t^2 u - \sum_{i,j=1}^{n} \partial_j(a_{ij}(x,t)\partial_i u) = F(x,t) \tag{7.1}$$

where $(x,t) \in Q := \Omega \times (0,T)$ (or $Q := \Omega \times (-T,T)$), and matrix $A = (a_{ij})_{n\times n}$ is symmetric and strictly positive definite. In other words, we assume $a_{ij} = a_{ji} \in C^1(\overline{Q})$, $1 \leq i, j \leq n$, and there exists a constant $\sigma > 0$ such that

Dedicated to Masahiro Yamamoto Sensei for His Sixtieth Birthday.

X. Huang (✉)
Graduate School of Mathematical Sciences, The University of Tokyo, 3-8-1 Komaba, Meguro, Tokyo 153-8914, Japan
e-mail: huangxc@ms.u-tokyo.ac.jp

J. Cheng et al. (eds.), *Inverse Problems and Related Topics*, Springer Proceedings in Mathematics & Statistics 310,
https://doi.org/10.1007/978-981-15-1592-7_7

$$\sigma \sum_{j=1}^{n} \xi_j^2 \le \sum_{i,j=1}^{n} a_{ij}(x,t)\xi_i\xi_j, \quad (x,t) \in \overline{Q},\ \xi \in \mathbb{R}^n.$$

Here and henceforth, we use notations $\partial_i = \frac{\partial}{\partial x_i}$, $\partial_t = \frac{\partial}{\partial t}$, $i = 1, ..., n$. Moreover, we use $\partial_0 = \partial_t$ somewhere for simplicity and define an $(n+1)$-dimensional symmetric matrix $\widetilde{A} := (\widetilde{a}_{ij})_{i,j=0,...,n}$:

$$\widetilde{a}_{00} = -1,\quad \widetilde{a}_{i0} = 0,\ i = 1, ..., n,\ \widetilde{a}_{0j} = 0,\ j = 1, ..., n,\quad \widetilde{a}_{ij} = a_{ij},\ i, j = 1, ..., n.$$

There are many works on the method of Carleman estimate. For the quick introduction and a comprehensive bibliography, we refer to [2, 10] and the references therein. As for the second-order hyperbolic equations, [4, 6, 7] consider Carleman estimates for the operator $p\partial_t^2 - \Delta$ that the principal parts are constant, while [1] investigate a general case $p\partial_t^2 - \sum_{i,j}^n a_{ij}\partial_i\partial_j$ with p strictly positive. Actually there are two ways to derive a Carleman estimate. One is to verify the conditions for general Carleman estimates (e.g., the pseudoconvexity condition in [4, Theorem 3.2.1′]) and the other is to find a good division of the weighted operator and calculate directly by integration by parts. In this article, we apply the latter way to establish the main results in the next subsection.

It is a well-known fact that some conditions on the coefficients are necessary for the Carleman estimate of hyperbolic type. Here we are mainly devoted to the conditions imposed on the principal parts (i.e., a_{ij}) in the Carleman estimate for a general second-order hyperbolic equation. Even for the case of t-independent principal parts, the previous conditions are difficult to check in general. See, for example, [3, 8, 9]. Bearing in mind the applications to inverse problems, it is necessary to find an explicit choice of the weight function. For instance, in [1], the authors resulted in a local Carleman estimate for $p\partial_t^2 - \sum_{i,j}^n a_{ij}\partial_i\partial_j$ with the choice:

$$\psi(x,t) = -\frac{1}{2\kappa}x_1 - \frac{1}{2}|x'|^2 - \frac{1}{2}t^2 + \frac{1}{2\kappa}\delta_0$$

where x_1 is the first component of x and x' denotes the vector of the rest components. δ_0 is some constant and $\kappa > 0$ is some small parameter. They found that this special choice can deal with the case: $a_{ij} = \delta_{ij}$, $i, j = 1, ..., n$ and

$$\frac{\partial_1 p(x)}{p(x)} \le -c_0$$

for a positive constant c_0, which can be interpreted in physics that the wave speed is increasing in the x_1-direction (see also [4, Theorem 3.4.1]). Here we intend to derive more general conditions on the principal parts a_{ij} which are also relatively easy-checking.

7.1.2 Main Theorems

In order to state our main theorems, we introduce the weight function:

$$\varphi(x,t) = e^{\lambda\psi(x,t)} \tag{7.2}$$

for $(x,t) \in \overline{Q}$ with large parameter $\lambda > 0$ and $\psi \in C^2(\overline{Q})$ satisfying

$$|\nabla_{x,t}\psi| > 0 \text{ on } \overline{Q}. \tag{7.3}$$

Now we are ready to state the main theorems.

Theorem 7.1 *Let φ, ψ satisfy* (7.2) *and* (7.3)*. Assume that there exist sufficiently smooth functions $\hat{\theta}$, θ_k, $k = 0, ..., n$ such that $\widetilde{B}_\theta := (\widetilde{b}_{ij,\theta})_{i,j=0,...,n}$ defined as*

$$\begin{aligned}\widetilde{b}_{ij,\theta} =& \sum_{m=0}^{n}\left(\sum_{l=0}^{n}\left(\widetilde{a}_{il}\partial_l(\widetilde{a}_{mj}) + \widetilde{a}_{jl}\partial_l(\widetilde{a}_{mi}) - \widetilde{a}_{ml}\partial_l(\widetilde{a}_{ij})\right) + \theta_i\widetilde{a}_{mj} + \theta_j\widetilde{a}_{mi}\right)\partial_m\psi \\ &+ \sum_{m,l=0}^{n} 2\widetilde{a}_{il}\widetilde{a}_{mj}\partial_m\partial_l\psi - \widetilde{a}_{ij}\hat{\theta}, \qquad i,j = 0,...,n\end{aligned}$$

is strictly positive definite in Q, that is, we can find a constant $r_0 > 0$ satisfying:

$$\sum_{i,j=0}^{n}\widetilde{b}_{ij,\theta}(x,t)\xi_i\xi_j \ge r_0|\xi|^2, \quad \textit{for all } (x,t) \in \overline{Q} \textit{ and } \xi \in R^n \setminus \{0\}.$$

Then there exist constants $s_0 \ge 1$ and $C > 0$ such that

$$\|(s\varphi)^{\frac{1}{2}}e^{s\varphi}\nabla_{x,t}u\|^2_{L^2(Q)} + \|(s\varphi)^{\frac{3}{2}}e^{s\varphi}u\|^2_{L^2(Q)} \le C\|e^{s\varphi}Lu\|^2_{L^2(Q)}$$

for all $s \ge s_0$ and $u \in H^2(\overline{Q})$ compactly supported in Q.

Next, in particular, we choose

$$\psi(x,t) = d(x) - \beta t^2 \tag{7.4}$$

for $(x,t) \in \overline{Q}$ with $d \in C^2(\overline{\Omega})$ satisfying

$$|\nabla d| > 0 \quad \text{on } \overline{\Omega} \tag{7.5}$$

and $\beta > 0$ to be fixed later. Then we have the second theorem:

Theorem 7.2 *Let d satisfy* (7.5)*. Assume $\theta_k \in C^2(\overline{Q})$, $k = 1, ..., n$ and the coefficients a_{ij}, $i, j = 1, ..., n$ are independent of time variable t and satisfy that $B_\theta = (b_{ij,\theta})_{i,j=1,...,n}$ defined as*

$$b_{ij,\theta} = \sum_{m=1}^{n} \Big(\sum_{l=1}^{n} (a_{il}\partial_l(a_{mj}) + a_{jl}\partial_l(a_{mi}) - a_{ml}\partial_l(a_{ij})) + \theta_i a_{mj} + \theta_j a_{mi} \Big) \partial_m d$$
$$+ \sum_{m,l=1}^{n} 2a_{il}a_{mj}\partial_m\partial_l d, \qquad i, j = 1, ..., n$$

is strictly positive definite in Q, that is, there exists a constant $r_0 > 0$ such that

$$\sum_{i,j=1}^{n} b_{ij,\theta}(x,t)\xi_i\xi_j \geq r_0|\xi|^2, \quad \textit{for all } (x,t) \in \overline{Q} \textit{ and } \xi \in R^n \setminus \{0\}.$$

Then there exists a constant $\beta_0 > 0$, that depends on the coefficients, $\|\theta_k\|_{L^\infty(Q)}$, r_0 and T, fulfilling: for any $0 < \beta \leq \beta_0$, there exist constants $s_0 \geq 1$ and $C > 0$ such that

$$\|(s\varphi)^{\frac{1}{2}} e^{s\varphi} \nabla_{x,t} u\|^2_{L^2(Q)} + \|(s\varphi)^{\frac{3}{2}} e^{s\varphi} u\|^2_{L^2(Q)} \leq C\|e^{s\varphi} Lu\|^2_{L^2(Q)}$$

for all $s \geq s_0$ and $u \in H^2(\overline{Q})$ compactly supported in Q with φ, ψ satisfying (7.2) *and* (7.4).

Remark 7.1 Theorem 7.2 is also valid for time-dependent a_{ij} satisfying: there exists function $h \in L^\infty(Q)$ such that

$$\partial_t(a_{ij})(x,t) = h(x,t)a_{ij}(x,t), \quad (x,t) \in \overline{Q}$$

for any $i, j = 1, ..., n$. In this case, β_0 also depends on $\|h\|_{L^\infty(Q)}$.

Remark 7.2 In Theorems 7.1, 7.2, we introduce some functions of adjustment θ, θ_k, $k = 1, \ldots, n$. From the technical point of view, these functions come from the suitable division of the weighted operator Pw (see γ_k, $k = 0, \ldots, n$ in formulas (7.6) and (7.7)). On the other hand, from the applicable viewpoint, the presence of these functions gives us more flexible choices of the weight functions so that we are able to derive weaker assumptions on the coefficients a_{ij} than those in the previous works we mentioned in the former contexts.

As for a simple case, we can fix θ_k, $k = 1, ..., n$ to be zero, which leads to

Proposition 7.1 *Let d satisfy* (7.5) *and φ, ψ satisfy* (7.2), (7.4). *Assume the coefficients a_{ij}, $i, j = 1, ..., n$ are independent of time variable t and satisfy that $B = (b_{ij})_{i,j=1,...,n}$ defined as*

$$b_{ij} = \sum_{m,l=1}^{n} 2a_{il}a_{mj}\partial_m\partial_l d + \sum_{m,l=1}^{n} \Big(a_{il}\partial_l(a_{mj}) + a_{jl}\partial_l(a_{mi}) - a_{ml}\partial_l(a_{ij})\Big)\partial_m d$$

for $i, j = 1, ..., n$, *is strictly positive definite in* Ω, *that is, there exists a constant* $r_0 > 0$ *such that*

$$\sum_{i,j=1}^{n} b_{ij}(x)\xi_i\xi_j \geq r_0|\xi|^2, \quad \text{for all } x \in \overline{\Omega} \text{ and } \xi \in R^n \setminus \{0\}.$$

Then for any $0 < \beta \leq \beta_0$, $\beta_0 > 0$ *is some constant which depends only on the coefficients* a_{ij} *and* r_0, *there exist constants* $s_0 \geq 1$ *and* $C > 0$ *such that*

$$\|(s\varphi)^{\frac{1}{2}} e^{s\varphi} \nabla_{x,t} u\|^2_{L^2(Q)} + \|(s\varphi)^{\frac{3}{2}} e^{s\varphi} u\|^2_{L^2(Q)} \leq C\|e^{s\varphi} Lu\|^2_{L^2(Q)}$$

for all $s \geq s_0$ *and* $u \in H^2(\overline{Q})$ *compactly supported in* Q.

Remark 7.3 Although in this article we only discuss the operator $L = \partial_t^2 - \sum_{i,j=1}^n a_{ij}(x,t)\partial_i\partial_j$ which is the principal part of second-order hyperbolic operator, we can extend immediately the results to a more general hyperbolic operator $\widetilde{L} = p(x,t)\partial_t^2 - \sum_{i,j=1}^n a_{ij}(x,t)\partial_i\partial_j + \sum_{i=1}^n b_i(x,t)\partial_i + c(x,t)$ where p is strictly positive in Q. This can be proved by multiplying p^{-1} and noting the robustness of the Carleman estimate.

Remark 7.4 In the statements of Theorems 7.1, 7.2 and Proposition 7.1, we can actually add the terms $\sum_{k=1}^2 \|P_k(e^{s\varphi}u)\|^2_{L^2(Q)}$ on the left-hand side of the estimations where P_k, $k = 1, 2$ is defined by (7.6) and (7.7) in the following contexts.

The rest part of the article is organized as follows. In Sect. 7.2, we establish the Carleman estimate with the functions of adjustment θ, θ_k, $k = 1, ...n$ by direct calculations. Then in Sect. 7.3, we give the realizations of the weight functions in some specific choices of the principal coefficients a_{ij}, $i, j = 1, ..., n$.

7.2 Proofs of the Main Results

7.2.1 Setting and Start of Carleman Estimate

Let $w = ue^{s\varphi}$ where weight function φ is smooth enough. By $u = we^{-s\varphi}$, we set

$$\begin{aligned} Pw := e^{s\varphi} Lu =& \partial_t^2 w - \sum_{i,j=1}^{n} \partial_j(a_{ij}\partial_i w) + 2s\left(\sum_{i,j=1}^{n} a_{ij}(\partial_i\varphi)\partial_j w - (\partial_t\varphi)\partial_t w\right) \\ &- s^2\left(\sum_{i,j=1}^{n} a_{ij}(\partial_i\varphi)\partial_j\varphi - |\partial_t\varphi|^2\right) w - s(L\varphi)w. \end{aligned}$$

Now we introduce the following three operators:

$$
\begin{aligned}
P_1 w &:= \partial_t^2 w - \sum_{i,j=1}^{n} \partial_j (a_{ij} \partial_i w) - s^2 \left(\sum_{i,j=1}^{n} a_{ij} (\partial_i \varphi) \partial_j \varphi - |\partial_t \varphi|^2 \right) w \\
&\quad + \sum_{k=1}^{n} \gamma_k (\partial_k w) + \gamma_0 (\partial_t w), \qquad (7.6)\\
P_2 w &:= 2s \left(\sum_{i,j=1}^{n} a_{ij} (\partial_i \varphi) \partial_j w - (\partial_t \varphi) \partial_t w \right) + s\gamma w, \qquad (7.7)\\
P_3 w &:= P_1 w + P_2 w = Pw + s(L\varphi) w + s\gamma w + \sum_{k=1}^{n} \gamma_k (\partial_k w) + \gamma_0 (\partial_t w)
\end{aligned}
$$

where the functions $\gamma = \gamma(x,t)$, $\gamma_0 = \gamma_0(x,t)$ and $\gamma_k = \gamma_k(x,t),\ k = 1, ..., n$ are independent of large parameter s and would be fixed later. Then we calculate

$$
\begin{aligned}
&2(P_1 w, P_2 w) = \\
&\int_Q 4s \sum_{i,j=1}^{n} a_{ij} (\partial_i \varphi)(\partial_j w) \partial_t^2 w \, dxdt - \int_Q 4s (\partial_t \varphi)(\partial_t w) \partial_t^2 w \, dxdt \\
&+ \int_Q 2s\gamma w (\partial_t^2 w) \, dxdt - \int_Q 4s \left(\sum_{i,j=1}^{n} a_{ij} (\partial_i \varphi) \partial_j w \right) \sum_{k,l=1}^{n} \partial_l (a_{kl} \partial_k w) dxdt \\
&+ \int_Q 4s (\partial_t \varphi)(\partial_t w) \sum_{i,j=1}^{n} \partial_j (a_{ij} \partial_i w) dxdt - \int_Q 2s\gamma w \sum_{i,j=1}^{n} \partial_j (a_{ij} \partial_i w) dxdt \\
&- \int_Q 4s^3 X \sum_{i,j=1}^{n} a_{ij} (\partial_i \varphi) \partial_j w \, w \, dxdt + \int_Q 4s^3 X (\partial_t \varphi)(\partial_t w) w \, dxdt \\
&- \int_Q 2s^3 \gamma X |w|^2 dxdt + \int_Q 4s \left(\sum_{i,j=1}^{n} a_{ij} (\partial_i \varphi) \partial_j w \right) \sum_{k=1}^{n} \gamma_k (\partial_k w) \, dxdt \\
&- \int_Q 4s (\partial_t \varphi)(\partial_t w) \sum_{k=1}^{n} \gamma_k (\partial_k w) \, dxdt + \int_Q 2s\gamma \sum_{k=1}^{n} \gamma_k (\partial_k w) w \, dxdt \\
&+ \int_Q 4s \left(\sum_{i,j=1}^{n} a_{ij} (\partial_i \varphi) \partial_j w \right) \gamma_0 (\partial_t w) \, dxdt - \int_Q 4s (\partial_t \varphi)(\partial_t w) \gamma_0 (\partial_t w) \, dxdt \\
&+ \int_Q 2s\gamma\gamma_0 (\partial_t w) w \, dxdt =: \sum_{p=1}^{15} I_p
\end{aligned}
$$

where

$$X := \sum_{i,j=1}^{n} a_{ij}(\partial_i\varphi)\partial_j\varphi - |\partial_t\varphi|^2.$$

By applying integration by parts and noting that $w = ue^{s\varphi}$ is compactly supported in Q, we calculate

$$I_1 = \int_Q 4s\sum_{i,j=1}^{n} a_{ij}(\partial_i\varphi)\partial_j w\partial_t^2 w\,dxdt$$

$$= -4s\int_Q\sum_{i,j=1}^{n}\partial_t(a_{ij}\partial_i\varphi)(\partial_j w)\partial_t w\,dxdt + 2s\int_Q\sum_{i,j=1}^{n}\partial_j(a_{ij}\partial_i\varphi)|\partial_t w|^2 dxdt,$$

$$I_2 = -\int_Q 4s(\partial_t\varphi)(\partial_t w)\partial_t^2 w\,dxdt = 2s\int_Q(\partial_t^2\varphi)|\partial_t w|^2 dxdt,$$

$$I_3 = \int_Q 2s\gamma w(\partial_t^2 w)dxdt = -2s\int_Q\gamma|\partial_t w|^2 dxdt + s\int_Q(\partial_t^2\gamma)|w|^2 dxdt,$$

$$I_4 = -\int_Q 4s\left(\sum_{i,j=1}^{n} a_{ij}(\partial_i\varphi)\partial_j w\right)\sum_{k,l=1}^{n}\partial_l(a_{kl}\partial_k w)\,dxdt$$

$$= 2s\sum_{i,j=1}^{n}\int_Q\left(\sum_{m,l=1}^{n}\big(2a_{il}\partial_l(a_{mj}\partial_m\varphi) - \partial_l(a_{ij}a_{ml}\partial_m\varphi)\big)\right)(\partial_i w)\partial_j w\,dxdt,$$

$$I_5 = \int_Q 4s(\partial_t\varphi)(\partial_t w)\sum_{i,j=1}^{n}\partial_j(a_{ij}\partial_i w)\,dxdt$$

$$= -4s\int_Q\sum_{i,j=1}^{n} a_{ij}\partial_j(\partial_t\varphi)(\partial_i w)\partial_t w\,dxdt + 2s\sum_{i,j=1}^{n}\int_Q\partial_t(a_{ij}\partial_t\varphi)(\partial_i w)\partial_j w\,dxdt,$$

$$I_6 = -\int_Q 2s\gamma w\sum_{i,j=1}^{n}\partial_j(a_{ij}\partial_i w)\,dxdt$$

$$= 2s\sum_{i,j=1}^{n}\int_Q\gamma a_{ij}(\partial_i w)(\partial_j w)dxdt - s\int_Q\sum_{i,j=1}^{n}\partial_i(a_{ij}\partial_j\gamma)|w|^2 dxdt,$$

$$I_7 = -\int_Q 4s^3 X\sum_{i,j=1}^{n} a_{ij}(\partial_i\varphi)\partial_j ww\,dxdt = 2s^3\int_Q\sum_{i,j=1}^{n}\partial_j(Xa_{ij}(\partial_i\varphi))|w|^2 dxdt,$$

$$I_8 = \int_Q 4s^3(\partial_t\varphi)(\partial_t w)Xw\,dxdt = -2s^3\int_Q\partial_t(X\partial_t\varphi)|w|^2 dxdt,$$

$$I_9 = -2s^3\int_Q\gamma X|w|^2 dxdt,$$

$$I_{10} = \int_Q 4s \left(\sum_{i,j=1}^n a_{ij}(\partial_i \varphi)\partial_j w \right) \sum_{k=1}^n \gamma_k (\partial_k w)\, dxdt$$

$$= 4s \sum_{i,j=1}^n \int_Q \gamma_i \left(\sum_{k=1}^n a_{kj}\partial_k \varphi \right) (\partial_i w)(\partial_j w) dxdt,$$

$$I_{11} = -\int_Q 4s(\partial_t \varphi)(\partial_t w) \sum_{k=1}^n \gamma_k (\partial_k w)\, dxdt = -4s \int_Q (\partial_t \varphi) \sum_{i=1}^n \gamma_i (\partial_i w)\partial_t w\, dxdt,$$

$$I_{12} = \int_Q 2s\gamma \sum_{k=1}^n \gamma_k (\partial_k w) w\, dxdt = -s \int_Q \sum_{k=1}^n \partial_k(\gamma\gamma_k)|w|^2 dxdt,$$

$$I_{13} = 4s \int_Q \gamma_0 \sum_{i,j=1}^n a_{ij}\partial_j \varphi(\partial_i w)\partial_t w\, dxdt,$$

$$I_{14} = -4s \int_Q \gamma_0(\partial_t \varphi)|\partial_t w|^2 dxdt,$$

$$I_{15} = \int_Q 2s\gamma\gamma_0(\partial_t w)w\, dxdt = -s \int_Q \partial_t(\gamma\gamma_0)|w|^2 dxdt.$$

Thus,

$$2(P_1 w, P_2 w)_{L^2(Q)} = \sum_{p=1}^{15} I_p =: Prin1 + Prin2 + Low$$

with

$$Prin1 = 2s \int_Q \Bigg\{ \Bigg(\sum_{i,j=1}^n \partial_j(a_{ij}\partial_i \varphi) + \partial_t^2 \varphi - \gamma - 2\gamma_0(\partial_t \varphi) \Bigg) |\partial_t w|^2$$

$$+ 2\sum_{i=1}^n \Bigg(-\sum_{k=1}^n \partial_t(a_{ik}\partial_k \varphi) - \sum_{k=1}^n a_{ik}\partial_k(\partial_t \varphi) - \gamma_i(\partial_t \varphi) + \gamma_0 \sum_{k=1}^n a_{ik}(\partial_k \varphi) \Bigg) (\partial_i w)(\partial_t w)$$

$$+ \sum_{i,j=1}^n \Bigg(\sum_{m,l=1}^n (2a_{il}\partial_l(a_{mj}\partial_m \varphi) - \partial_l(a_{ij}a_{ml}\partial_m \varphi)) + \partial_t(a_{ij}\partial_t \varphi)$$

$$+ \gamma a_{ij} + 2\gamma_i \sum_{k=1}^n a_{kj}(\partial_k \varphi) \Bigg) (\partial_i w)(\partial_j w) \Bigg\} dxdt,$$

$$Prin2 = 2s^3 \int_Q \left(\sum_{i,j=1}^n \partial_j(X a_{ij}\partial_i \varphi) - \partial_t(X\partial_t \varphi) - \gamma X \right) |w|^2 dxdt,$$

$$Low = s \int_Q \left((L\gamma) - \sum_{k=1}^n \partial_k(\gamma\gamma_k) - \partial_t(\gamma\gamma_0) \right) |w|^2 dxdt.$$

Next we introduce a second large parameter. Actually we choose

$$\varphi(x,t) = e^{\lambda\psi(x,t)}$$

with $\lambda \ge 1$. Then we divide $\gamma, \gamma_k, \gamma_0$ according to the power of λ:

$$\gamma = \lambda^2\varphi\tau + \lambda\varphi\theta, \quad \gamma_0 = \lambda\tau_0 + \theta_0 \quad \text{and} \quad \gamma_k = \lambda\tau_k + \theta_k, \; k = 1, ..., n.$$

Then $\tau, \theta, \tau_k, \theta_k, \tau_0, \theta_0$ are all independent of both s and λ. Thus we can divide $Prin1$ and $Prin2$ with respect to the power of λ:

$$Prin1 =: Prin1a + Prin1b, \qquad Prin2 =: Prin2a + Prin2b$$

with

$$Prin1a = \int_Q 2s\lambda^2\varphi\Biggl\{\Biggl(\sum_{i,j=1}^n a_{ij}(\partial_i\psi)\partial_j\psi + |\partial_t\psi|^2 - \tau - 2\tau_0\partial_t\psi\Biggr)|\partial_t w|^2$$

$$+ 2\sum_{i=1}^n\Biggl(-\sum_{k=1}^n 2a_{ik}(\partial_k\psi)\partial_t\psi - (\partial_t\psi)\tau_i + \tau_0\sum_{k=1}^n a_{ik}(\partial_k\psi)\Biggr)(\partial_i w)(\partial_t w)$$

$$+ \sum_{i,j=1}^n\Biggl(\sum_{m,l=1}^n (2a_{il}a_{mj}(\partial_m\psi)\partial_l\psi - a_{ij}a_{ml}(\partial_m\psi)\partial_l\psi) + a_{ij}|\partial_t\psi|^2 + a_{ij}\tau$$

$$+ 2\tau_i\sum_{k=1}^n a_{kj}(\partial_k\psi)\Biggr)(\partial_i w)(\partial_j w)\Biggr\}dxdt,$$

$$Prin1b = \int_Q 2s\lambda\varphi\Biggl\{\Biggl(\sum_{i,j=1}^n \partial_j(a_{ij}\partial_i\psi) + \partial_t^2\psi - \theta - 2\theta_0\partial_t\psi\Biggr)|\partial_t w|^2$$

$$+ 2\sum_{i=1}^n\Biggl(-\sum_{k=1}^n \partial_t(a_{ik}\partial_k\psi) - \sum_{k=1}^n a_{ik}\partial_t(\partial_k\psi) - \theta_i\partial_t\psi + \theta_0\sum_{k=1}^n a_{ik}(\partial_k\psi)\Biggr)(\partial_i w)(\partial_t w)$$

$$+ \sum_{i,j=1}^n\Biggl(\sum_{m,l=1}^n (2a_{il}\partial_l(a_{mj}\partial_m\psi) - \partial_l(a_{ij}a_{ml}\partial_m\psi)) + \partial_t(a_{ij}\partial_t\psi)$$

$$+ a_{ij}\theta + 2\theta_i\sum_{k=1}^n a_{kj}(\partial_k\psi)\Biggr)(\partial_i w)(\partial_j w)\Biggr\}dxdt,$$

$$Prin2a = \int_Q 2s^3\lambda^4\varphi^3\Biggl[3Y\sum_{i,j=1}^n a_{ij}(\partial_i\psi)\partial_j\psi - 3Y|\partial_t\psi|^2 - \tau Y\Biggr]|w|^2dxdt$$

$$= \int_Q 2s^3\lambda^4\varphi^3[3Y^2 - \tau Y]|w|^2dxdt,$$

$$Prin2b = \int_Q 2s^3\lambda^3\varphi^3\Biggl[\sum_{i,j=1}^n \partial_j(Ya_{ij}\partial_i\psi) - \partial_t(Y\partial_t\psi) - \theta Y\Biggr]|w|^2dxdt$$

where

$$Y := \sum_{i,j=1}^{n} a_{ij}(\partial_i \psi)(\partial_j \psi) - |\partial_t \psi|^2.$$

In addition,

$$Low = \int_Q O(s\lambda^4 \varphi)|w|^2 dxdt.$$

7.2.2 Completion of the Proofs

Now we fix τ_i, $i = 1, ..., n$ and τ_0. In fact, we can solve

$$\begin{cases} \sum_{m,l=1}^{n} 2a_{il}a_{mj}(\partial_m \psi)\partial_l \psi + \sum_{m=1}^{n} 2\tau_i a_{mj}\partial_m \psi = 0, & 1 \le \forall i, j \le n, \\ \sum_{j=1}^{n} 2a_{ij}(\partial_j \psi)(\tau_0 - 2\partial_t \psi) - 2\tau_i \partial_t \psi = 0, & 1 \le \forall i \le n. \end{cases}$$

From the first equation we have

$$\tau_i = -\sum_{l=1}^{n} a_{il}\partial_l \psi$$

for each $i = 1, ..., n$. Then inserting this into the second equation leads to

$$\tau_0 = \partial_t \psi.$$

Therefore, we obtain

$$Prin1a = \int_Q \left\{ (Y - \tau)|\partial_t w|^2 + (-Y + \tau) \sum_{i,j=1}^{n} a_{ij}(\partial_i w)(\partial_j w) \right\} dxdt,$$

$$Prin2a = \int_Q 2s^3\lambda^4\varphi^3(3Y^2 - \tau Y)|w|^2 dxdt.$$

Since the signs of the two principal term in $Prin1a$ are opposite, we take

$$\tau = Y = \sum_{i,j=1}^{n} a_{ij}(\partial_i \psi)(\partial_j \psi) - |\partial_t \psi|^2$$

which yields

$$
\begin{aligned}
&Prin1a = 0\\
&Prin2a = \int_Q 4s^3\lambda^4\varphi^3 Y^2|w|^2 dxdt \ge 0.
\end{aligned}
\tag{7.8}
$$

This implies that we need to discuss the positivities of $Prin1b$ and $Prin2b$. Recall that we set

$$
\widetilde{A} = \begin{pmatrix} -1 & 0 \\ 0 & A \end{pmatrix},
$$

$\widetilde{A} = (\widetilde{a}_{ij})_{i,j=0,\dots,n}$. Then we can rewrite $Prin1b$ and $Prin2b$ by noting $\partial_0 = \partial_t$ that

$$
\begin{aligned}
Prin1b = \sum_{i,j=0}^{n} \int_Q 2s\lambda\varphi \Bigg(\sum_{m,l=0}^{n} &(2\widetilde{a}_{il}\partial_l(\widetilde{a}_{mj}\partial_m\psi) - \partial_l(\widetilde{a}_{ij}\widetilde{a}_{ml}\partial_m\psi))\\
&+ \widetilde{a}_{ij}\theta + 2\theta_i \sum_{k=0}^{n} \widetilde{a}_{kj}(\partial_k\psi)\Bigg)(\partial_i w)(\partial_j w)dxdt
\end{aligned}
$$

$$
Prin2b = \int_Q 2s^3\lambda^3\varphi^3 \left(\sum_{i,j=0}^{n} \partial_j(Y\widetilde{a}_{ij}\partial_i\psi) - \theta Y \right)|w|^2 dxdt.
$$

Moreover, we express $Prin1b$ in terms of the symmetric matrix $\widetilde{B}_\theta = (\widetilde{b}_{ij,\theta})_{i,j=0,\dots,n}$ defined in the statement of Theorem 7.1 with

$$
\hat{\theta} = \sum_{m,l=0}^{n} \partial_l(\widetilde{a}_{ml}(\partial_m\psi)) - \theta
$$

and obtain

$$
Prin1b = \sum_{i,j=0}^{n} \int_Q 2s\lambda\varphi\widetilde{b}_{ij,\theta}(\partial_i w)(\partial_j w)dxdt. \tag{7.9}
$$

Furthermore, by calculating

$$
\begin{aligned}
&\sum_{i,j=0}^{n} \partial_j(Y\widetilde{a}_{ij}\partial_i\psi) - \theta Y\\
&= Y\sum_{i,j=0}^{n} \partial_j(\widetilde{a}_{ij}\partial_i\psi) + \sum_{i,j=0}^{n} \widetilde{a}_{ij}(\partial_i\psi)\partial_j\left(\sum_{m,l=0}^{n} \widetilde{a}_{ml}(\partial_m\psi)\partial_l\psi \right) - \theta Y\\
&= \sum_{i,j,m,l=0}^{n} \Big(\widetilde{a}_{ij}\partial_j(\widetilde{a}_{ml})(\partial_i\psi)(\partial_m\psi)\partial_l\psi + \widetilde{a}_{ij}\widetilde{a}_{ml}(\partial_i\psi)(\partial_l\psi)\partial_j\partial_m\psi\\
&\qquad + \widetilde{a}_{ij}\widetilde{a}_{ml}(\partial_i\psi)(\partial_m\psi)\partial_j\partial_l\psi \Big) + \hat{\theta}\sum_{i,j=0}^{n} \widetilde{a}_{ij}(\partial_i\psi)\partial_j\psi
\end{aligned}
$$

$$
\begin{aligned}
&= \sum_{i,j=0}^{n} \widetilde{b}_{ij,\theta}(\partial_i\psi)\partial_j\psi + 2\hat{\theta}\sum_{i,j=0}^{n} \widetilde{a}_{ij}(\partial_i\psi)\partial_j\psi - \sum_{i,j,m=0}^{n}(\theta_i\widetilde{a}_{mj}+\theta_j\widetilde{a}_{mi})(\partial_m\psi)(\partial_i\psi)\partial_j\psi \\
&= \sum_{i,j=0}^{n} \widetilde{b}_{ij,\theta}(\partial_i\psi)\partial_j\psi + 2Y\left(\hat{\theta} - \sum_{i=0}^{n}\theta_i(\partial_i\psi)\right),
\end{aligned}
$$

we obtain

$$
\begin{aligned}
Prin2b &= \int_Q 2s^3\lambda^3\varphi^3\left(\sum_{i,j=0}^{n}\widetilde{b}_{ij,\theta}(\partial_i\psi)\partial_j\psi + 2Y\left(\hat{\theta} - \sum_{i=0}^{n}\theta_i(\partial_i\psi)\right)\right)|w|^2dxdt \\
&\geq \int_Q 2s^3\lambda^3\varphi^3\left(\sum_{i,j=0}^{n}\widetilde{b}_{ij,\theta}(\partial_i\psi)\partial_j\psi\right)|w|^2dxdt \\
&\quad - \int_Q 4s^3\lambda^3\varphi^3|Y|\left|\hat{\theta} - \sum_{i=0}^{n}\theta_i(\partial_i\psi)\right||w|^2dxdt \\
&\geq \int_Q 2s^3\lambda^3\varphi^3\left(\sum_{i,j=0}^{n}\widetilde{b}_{ij,\theta}(\partial_i\psi)\partial_j\psi\right)|w|^2dxdt - \int_Q 2s^3\lambda^4\varphi^3|Y|^2|w|^2dxdt \\
&\quad - \int_Q 2s^3\lambda^2\varphi^3\left|\hat{\theta} - \sum_{i=0}^{n}\theta_i(\partial_i\psi)\right|^2|w|^2dxdt.
\end{aligned}
$$

Here in the last inequality, we applied the Cauchy–Schwarz inequality. Therefore, we derive

$$
\begin{aligned}
Prin2 \geq& \int_Q 2s^3\lambda^3\varphi^3\left(\sum_{i,j=0}^{n}\widetilde{b}_{ij,\theta}(\partial_i\psi)\partial_j\psi\right)|w|^2dxdt + \int_Q 2s^3\lambda^4\varphi^3|Y|^2|w|^2dxdt \\
&- \int_Q 2s^3\lambda^2\varphi^3\left|\hat{\theta} - \sum_{i=0}^{n}\theta_i(\partial_i\psi)\right|^2|w|^2dxdt.
\end{aligned} \tag{7.10}
$$

Recall that $\widetilde{B}_\theta := (\widetilde{b}_{ij,\theta})_{i,j=0,\ldots,n}$ is assumed to be strictly positive definite. Then (7.8)–(7.10) indicate

$$
\begin{aligned}
&2(P_1w, P_2w)_{L^2(Q)} \\
&\geq c_0\int_Q s^3\lambda^3\varphi^3|w|^2dxdt + c_0\int_Q s\lambda\varphi|\nabla_{x,t}w|^2dxdt - C\int_Q s\lambda^2\varphi(s^2\varphi^2+\lambda^2)|w|^2dxdt
\end{aligned}
$$

for some positive constant $c_0 > 0$. We fix $\lambda \geq 1$ large (e.g., $\lambda \geq \lambda_0 := \frac{2C}{c_0}$) and then take $s \geq s_0(\lambda_0) \geq 1$ so that the third term on the RHS can be absorbed by the first term on the RHS. By noting

$$\|e^{s\varphi}Lu\|^2_{L^2(Q)} = \left\| P_1 w + P_2 w - s(L\varphi)w - s\gamma w - \sum_{k=1}^{n} \gamma_k(\partial_k w) - \gamma_0(\partial_t w)\right\|^2_{L^2(Q)}$$

$$\geq \sum_{k=1}^{2} \| P_k(e^{s\varphi}u)\|^2_{L^2(Q)} + 2(P_1 w, P_2 w)_{L^2(Q)} - C(\|(s\varphi)^{\frac{1}{2}}w\|^2_{L^2(Q)} + \|\nabla_{x,t}w\|^2_{L^2(Q)}),$$

we change w back to u and finally obtain

$$\sum_{k=1}^{2} \| P_k(e^{s\varphi}u)\|^2_{L^2(Q)} + \|(s\varphi)^{\frac{1}{2}}e^{s\varphi}\nabla_{x,t}u\|^2_{L^2(Q)} + \|(s\varphi)^{\frac{3}{2}}e^{s\varphi}u\|^2_{L^2(Q)}$$
$$\leq C\|e^{s\varphi}Lu\|^2_{L^2(Q)}$$

for $s \geq s_0$. This proves Theorem 7.1 with

Next we give the proof to Theroem 7.2. Actually we calculate $\widetilde{b}_{ij,\theta}$ in details as follows:

$$\widetilde{b}_{00,\theta} = 2\partial_0^2\psi - 2\theta_0\partial_0\psi + \hat{\theta},$$

$$\widetilde{b}_{0j,\theta} = \widetilde{b}_{j0,\theta} = \sum_{m=1}^{n}(-2a_{mj}\partial_m\partial_0\psi - \partial_0(a_{mj})\partial_m\psi + \theta_0 a_{mj}\partial_m\psi) - \theta_j(\partial_0\psi),$$
$$j = 1, ..., n,$$

$$\widetilde{b}_{ij,\theta} = \sum_{m=1}^{n}\left(\sum_{l=1}^{n}(a_{il}\partial_l(a_{mj}) + a_{jl}\partial_l(a_{mi}) - a_{ml}\partial_l(a_{ij})) + \theta_i a_{mj} + \theta_j a_{mi}\right)\partial_m\psi$$
$$+ \sum_{m,l=1}^{n} 2a_{il}a_{mj}\partial_m\partial_l\psi + \partial_0(a_{ij})\partial_0\psi - a_{ij}\hat{\theta}, \quad i, j = 1, ..., n.$$

In particular, if a_{ij} is independent of t: $\partial_0(a_{ij}) = 0$ in Ω for any $i, j = 1, ..., n$, then we choose $\psi(x,t) = d(x) - \beta t^2$ for $(x,t) \in \overline{Q}$. Moreover, we take $\theta_0 = 0$. Thus, $\widetilde{b}_{ij,\theta}$ read

$$\widetilde{b}_{00,\theta} = \hat{\theta} - 4\beta, \quad \widetilde{b}_{0j,\theta} = \widetilde{b}_{j0,\theta} = 2\beta t\theta_j, \quad j = 1, ..., n,$$

$$\widetilde{b}_{ij,\theta} = \sum_{m=1}^{n}\left(\sum_{l=1}^{n}(a_{il}\partial_l(a_{mj}) + a_{jl}\partial_l(a_{mi}) - a_{ml}\partial_l(a_{ij})) + \theta_i a_{mj} + \theta_j a_{mi}\right)\partial_m d$$
$$+ \sum_{m,l=1}^{n} 2a_{il}a_{mj}\partial_m\partial_l d - a_{ij}\hat{\theta}, \quad i, j = 1, ..., n.$$

By recalling the symmetric matrix $B_\theta = (b_{ij,\theta})_{i,j=1,...,n}$ defined in Theorem 7.2, we obtain

$$\widetilde{b}_{00,\theta} = \hat{\theta} - 4\beta, \quad \widetilde{b}_{0j,\theta} = \widetilde{b}_{j0,\theta} = 2\beta t\theta_j, \quad j = 1, ..., n,$$
$$\widetilde{b}_{ij,\theta} = b_{ij,\theta} - a_{ij}\hat{\theta}, \quad i, j = 1, ..., n.$$

According to the positivity assumption on B_θ, we have

$$\sum_{i,j=1}^{n} \widetilde{b}_{ij,\theta}\xi_i\xi_j = \sum_{i,j=1}^{n} b_{ij,\theta}\xi_i\xi_j - \hat{\theta}\sum_{i,j=1}^{n} a_{ij}\xi_i\xi_j \geq (r_0 - \sigma_m\hat{\theta})|\xi|^2, \quad \xi \in R^n$$

for some constant $\sigma_m > 0$ such that

$$\sum_{i,j=1}^{n} a_{ij}(x)\xi_i\xi_j \leq \sigma_m|\xi|^2, \quad x \in \overline{\Omega},\ \xi \in \mathbb{R}^n.$$

Hence we take $\hat{\theta} = \frac{r_0}{2\sigma_m}$ such that

$$\sum_{i,j=1}^{n} \widetilde{b}_{ij,\theta}\xi_i\xi_j \geq \frac{r_0}{2}|\xi|^2.$$

Therefore

$$\sum_{i,j=0}^{n} \widetilde{b}_{ij,\theta}\xi_i\xi_j \geq \frac{r_0}{2}|\xi|^2 + (\frac{r_0}{2\sigma_m} - 4\beta)\xi_0^2 + 4\beta t\xi_0\sum_{j=1}^{n}\theta_j\xi_j. \tag{7.11}$$

By noting $2(\xi_0\theta_j)\xi_j \geq -(\xi_0\theta_j)^2 - \xi_j^2$, the above inequality yields

$$\begin{aligned}\sum_{i,j=0}^{n} \widetilde{b}_{ij,\theta}\xi_i\xi_j &\geq \frac{r_0}{2}|\xi|^2 + (\frac{r_0}{2\sigma_m} - 4\beta)\xi_0^2 - 2\beta t\xi_0^2\sum_{j=1}^{n}\theta_j^2 - 2\beta t|\xi|^2\\ &\geq (\frac{r_0}{2} - 2T\beta)|\xi|^2 + \left(\frac{r_0}{2\sigma_m} - 2(2 + T\sum_{j=1}^{n}\theta_j^2)\beta\right)\xi_0^2.\end{aligned}$$

Then $\beta \leq \beta_0 := \min\{\frac{r_0}{8T}, \frac{r_0}{8\sigma_m(2+T\|\theta_j\|^2_{L^\infty(Q)})}, j = 1, ..., n\}$ implies

$$\sum_{i,j=0}^{n} \widetilde{b}_{ij,\theta}(x,t)\xi_i\xi_j \geq \min\{\frac{r_0}{4}, \frac{r_0}{4\sigma_m}\}|\xi|^2 \quad \text{for any } (x,t) \in \overline{Q},$$

which indicates $\widetilde{B}_\theta$ is strictly positively defined in Q. Consequently, by applying Theorem 7.1, we proved Theorem 7.2.

Proposition 7.1 follows immediately after Theorem 7.2. Here we give a comment on $\beta_0 > 0$. In the case that we fix $\theta_k = 0$, $k = 1, ..., n$. We find from (7.11) that

$$\sum_{i,j=0}^{n} \widetilde{b}_{ij,\theta}\xi_i\xi_j \geq \frac{r_0}{2}|\xi|^2 + (\frac{r_0}{2\sigma_m} - 4\beta)\xi_0^2.$$

Then we can choose $\beta_0 = \frac{r_0}{16\sigma_m}$ which is also independent of T.

7.3 Realizations of the Weight Functions

In this section, we give several simple examples on the choices of weight function in some specified cases of the coefficients $a_{ij}(x,t), i, j = 1, ..., n$. It is sufficient to check the positivity of the matrix B or B_θ or $\widetilde{B}_\theta$.

Example 1 $a_{ij}(x,t) = a_{ij}$ for $i, j = 1, ..., n$ are constants.
By choosing $\psi(x,t) = |x - x_0|^2 - \beta t^2$ with $\beta > 0$ small enough, immediately we have

$$B = 4AA^T > 0,$$

which satisfies the conditions of Proposition 7.1.

Example 2 $a_{ij}(x,t) = \delta_{ij}a(x)$ for $i, j = 1, ..., n$ satisfy $I - \frac{|x-x_0|^2}{2a}\nabla^2 a > 0$.
In this case, we fix

$$d(x) = f(a)\widetilde{d}(x), \quad \widetilde{d}(x) = \frac{1}{2}|x - x_0|^2, \quad x \in \overline{\Omega}$$

with $x_0 \notin \overline{\Omega}$. Then for $i, j = 1, ..., n$, we have

$$\begin{aligned}
\partial_i d &= f'(a)(\partial_i a)\widetilde{d} + f(a)\partial_i \widetilde{d}, \\
\partial_j \partial_i d &= f''(a)(\partial_i a)(\partial_j a)\widetilde{d} + f'(a)(\partial_j \partial_i a)\widetilde{d} + f'(a)[(\partial_i a)(\partial_j \widetilde{d}) + (\partial_j a)(\partial_i \widetilde{d})] \\
&\quad + \delta_{ij} f(a).
\end{aligned}$$

Thus,

$$\begin{aligned}
b_{ij} =& 2a^2\partial_j\partial_i d + a(\partial_i a)\partial_j d + a(\partial_j a)\partial_i d - \delta_{ij}a(\nabla a \cdot \nabla d) \\
=& 2a^2(\partial_j\partial_i a)f'(a)\widetilde{d} + 2a(\partial_i a)(\partial_j a)(af''(a) + f'(a))\widetilde{d} \\
&+ a(2af'(a) + f(a))[(\partial_i a)(\partial_j \widetilde{d}) + (\partial_j a)(\partial_i \widetilde{d})] \\
&+ \delta_{ij}a(2af(a) - f'(a)|\nabla a|^2\widetilde{d} - f(a)\nabla a \cdot \nabla \widetilde{d}).
\end{aligned}$$

By taking $f(a) = a^{-\frac{1}{2}}$, we obtain

$$\begin{aligned}
b_{ij} &= -a^{\frac{1}{2}}(\partial_j\partial_i a)\widetilde{d} + \frac{1}{2}a^{-\frac{1}{2}}(\partial_i a)(\partial_j a)\widetilde{d} + \delta_{ij}(2a^{\frac{3}{2}} + \frac{1}{2}a^{-\frac{1}{2}}|\nabla a|^2\widetilde{d} - a^{\frac{1}{2}}\nabla a \cdot \nabla \widetilde{d}) \\
&= \delta_{ij}\Big(a^{\frac{3}{2}} + \frac{1}{4}a^{-\frac{1}{2}}(|\nabla a|^2|x - x_0|^2 - |\nabla a \cdot (x - x_0)|^2) \\
&\quad + \frac{1}{4}a^{-\frac{1}{2}}|\nabla a \cdot (x - x_0) - 2a|^2\Big) - a^{\frac{1}{2}}(\partial_j\partial_i a)\widetilde{d} + \frac{1}{2}a^{-\frac{1}{2}}(\partial_i a)(\partial_j a)\widetilde{d}.
\end{aligned}$$

Hence

$$B \geq a^{\frac{3}{2}}I - \frac{1}{2}a^{\frac{1}{2}}|x - x_0|^2(\nabla^2 a) > 0.$$

Example 3 $a_{ij}(x,t) = a_{ij}(x)$ for $i, j = 1, ..., n$ satisfy that P generated by

$$p_{ij} = \sum_{k=1}^{n} a_{nk}\partial_k(a_{ij}) - \sum_{k=1}^{n-1}(a_{ik}\partial_k(a_{jn}) + a_{jk}\partial_k(a_{in})), \quad i, j = 1, ..., n$$

is strictly positive definite. That is, there exists a positive constant $r_0 > 0$ such that

$$\sum_{i,j=1}^{n} p_{ij}(x)\xi_i\xi_j \geq r_0|\xi|^2, \quad \text{for any } x \in \overline{\Omega} \text{ and } \xi \in R^n.$$

The assumptions in this example come from [5, Lemma 2.2]. Here we show our main theorems simply cover this specific example.

We choose $d(x) = (x_n - \gamma)^2 + \sum_{j=1}^{n-1}|x_j|^2$ for $x \in \Omega$ where $\gamma > 0$ is a large number. This is the same choice as [5, Lemma 2.2]. Then we calculate B_θ as follows:

$$\begin{aligned} b_{ij,\theta} &= 2\sum_{m=1}^{n}\Big(\sum_{l=1}^{n}(a_{il}\partial_l(a_{mj}) + a_{jl}\partial_l(a_{mi}) - a_{ml}\partial_l(a_{ij})) + \theta_i a_{mj} + \theta_j a_{mi}\Big)x_m \\ &+ 4\sum_{k=1}^{n} a_{ik}a_{kj} + 2\gamma\Big(\sum_{k=1}^{n}(a_{nk}\partial_k(a_{ij}) - a_{ik}\partial_k(a_{nj}) - a_{jk}\partial_k(a_{ni})) + \theta_i a_{nj} + \theta_j a_{ni}\Big). \end{aligned}$$

Owing to the fact that γ is large, we only need to check the last term which includes the highest power of γ. By setting

$$\begin{aligned} q_{ij,\theta} &:= \sum_{k=1}^{n}(a_{nk}\partial_k(a_{ij}) - a_{ik}\partial_k(a_{nj}) - a_{jk}\partial_k(a_{ni})) + \theta_i a_{nj} + \theta_j a_{ni} \\ &= p_{ij} + a_{ni}(\theta_j - \partial_n(a_{nj})) + a_{nj}(\theta_i - \partial_n(a_{ni})), \end{aligned}$$

we find the choice of θ_k: $\theta_k = \partial_n(a_{nk})$, $k = 1, ..., n$. Thus, we have

$$q_{ij,\theta} = p_{ij}$$

which is positively defined in virtue of our assumption imposed on $\{a_{ij}\}_{i,j=1,...,n}$. Finally, we take $\gamma \geq \gamma_0 = \gamma_0(\max_{i,j=1,...,n}\|a_{ij}\|_{W^{1,\infty}(\Omega)}, \sup_{x\in\overline{\Omega}}|x|_{l^\infty}, r_0)$, so that $B_\theta = (b_{ij,\theta})_{i,j=1,...,n}$ is positively defined.

Example 4 $a_{ij}(x,t) = a_{ij}(x)$ for $i, j = 1, 2$ satisfy

$$a_{11} = 1 + |x_1|^2, \quad a_{12} = a_{21} = x_1, \quad a_{22} = 1.$$

Under a common choice $d(x) = |x_1|^2 + |x_2 - p|^2$ with $(0, p) \notin \overline{\Omega}$, we have

$$
\begin{aligned}
b_{11} &= 4a_{11}^2 + 4a_{12}^2 + \big(2a_{11}(\partial_1 a_{m1}) - a_{m1}(\partial_1 a_{11})\big)\partial_m d \\
&= 4(1+|x_1|^2)^2 + 4|x_1|^2 + 4|x_1|^2(1+|x_1|^2) + 4(x_2 - p) \\
&= 4\big(2x_1^4 + 4x_1^2 + 1 + (x_2 - p)\big) \\
b_{12} = b_{21} &= 4a_{11}a_{12} + 4a_{12}a_{22} + \big(a_{11}(\partial_1 a_{m2}) + a_{21}(\partial_1 a_{m1}) - a_{m1}(\partial_1 a_{12})\big)\partial_m d \\
&= 4x_1(1+|x_1|^2) + 4x_1 + 4x_1^3 \\
&= 8x_1(x_1^2+1) \\
b_{22} &= 4a_{12}^2 + 4a_{22}^2 + \big(2a_{21}(\partial_1 a_{m2}) - a_{m1}(\partial_1 a_{22})\big)\partial_m d \\
&= 4|x_1|^2 + 4 + 4|x_1|^2 \\
&= 4(2x_1^2+1)
\end{aligned}
$$

Direct calculation yields that B is strictly positive defined if and only if

$$
\left(2x_1^2+1\right)\left(x_1^2 + x_2 + \frac{1}{2} - p\right) + \frac{1}{2} > 0.
$$

Therefore, we pick up p small enough (e.g., $p \le p_0 := \min_{x\in\overline{\Omega}}(x_2 + x_1^2 + \frac{1}{2})$) so that B is strictly positive definite.

Acknowledgements The author is supported by Grant-in-Aid for Scientific Research(S) 15H05740 of Japan Society for the Promotion of Science (JSPS). This work is also supported by A3 Foresight Program "Modeling and Computation of Applied Inverse Problems" of JSPS.

References

1. A. Amirov, M. Yamamoto, A timelike Cauchy problem and an inverse problem for general hyperbolic equations. Appl. Math. Lett. **21**, 885–891 (2008)
2. M. Bellassoued, M. Yamamoto, *Carleman Estimates and Applications to Inverse Problems for Hyperbolic Systems* (Springer-Japan, Tokyo, 2017)
3. O.Yu. Imanuvilov, On Carleman estimates for hyperbolic equations. Asymptot. Anal. **32**, 185–220 (2002)
4. V. Isakov, *Inverse Problems for Partial Differential Equations* (Springer, Berlin, 2006)
5. V. Isakov, N. Kim, Weak Carleman estimates with two large parameters for second order operators and applications to elasticity with residual stress. Appl. Math. **27**(2), 799–825 (2010)
6. O.Yu. Imanuvilov, M. Yamamoto, Global uniqueness and stability in determining coefficients of wave equations. Comm. Part. Differ. Equ. **26**, 1409–1425 (2001)
7. M.V. Klibanov, A. Timonov, *Carleman Estimates for Coefficient Inverse Problems and Numerical Applications* (VSP, Utrecht, 2004)
8. I. Lasiecka, R. Triggiani, P. Yao, Inverse/observability estimates for second-order hyperbolic equations with variable coefficients. J. Math. Anal. Appl. **235**, 13–57 (1999)
9. R. Triggiani, P. Yao, Carleman estimates with no lower-order terms for general riemann wave equations. global uniqueness and observability in one shot. Appl. Math. Optim. **46**, 331–375 (2002)
10. M. Yamamoto, Carleman estimates for parabolic equations and applications. Inverse Probl. **25**(12), 123013 (2009)

Part II
Regularization Theory of Inverse Problems

Chapter 8
A Priori Parameter Choice in Tikhonov Regularization with Oversmoothing Penalty for Non-linear Ill-Posed Problems

Bernd Hofmann and Peter Mathé

Abstract We study Tikhonov regularization for certain classes of non-linear ill-posed operator equations in Hilbert spaces. Emphasis is on the case where the solution smoothness fails to have a finite penalty value, as in the preceding study *Tikhonov regularization with oversmoothing penalty for non-linear ill-posed problems in Hilbert scales*. Inverse Problems 34(1), 2018, by the same authors. Optimal order convergence rates are established for the specific a priori parameter choice, as used for the corresponding linear equations.

Keywords Tikhonov regularization · Oversmoothing penalty · A priori parameter choice · Non-linear ill-posed problems · Hilbert scales

MSC 2010 65J22 · 47J06 · 65J20

8.1 Introduction

The present paper is closely related to the recent work [4] of the authors published in the journal *Inverse Problems* devoted to the Tikhonov regularization for non-linear operator equations with oversmoothing penalties in Hilbert scales. Here we adopt the model and terminology. Since ibidem convergence rates of the optimal order were only proven for the discrepancy principle as an a posteriori choice of the regularization parameter, we try here to close a gap in regularization theory by extending the same rate results to the case of appropriate a priori choices. This is in good coinci-

B. Hofmann (✉)
Faculty of Mathematics, Chemnitz University of Technology, 09107 Chemnitz, Germany
e-mail: bernd.hofmann@tu-chemnitz.de
URL: https://www.tu-chemnitz.de/mathematik/ip/

P. Mathé
Weierstraß Institute for Applied Analysis and Stochastics,
Mohrenstraße 39, 10117 Berlin, Germany
e-mail: peter.mathe@wias-berlin.de

J. Cheng et al. (eds.), *Inverse Problems and Related Topics*,
Springer Proceedings in Mathematics & Statistics 310,
https://doi.org/10.1007/978-981-15-1592-7_8

dence with the corresponding results for linear operator equations presented in the seminal paper [8], where the same a priori parameter choice successfully allowed for order optimal convergence rates also in the case of oversmoothing penalties.

We consider the approximate solution of an operator equation

$$F(x) = y, \tag{8.1}$$

which models an inverse problem with an (in general) non-linear forward operator $F : \mathcal{D}(F) \subseteq X \to Y$ mapping between the infinite dimensional Hilbert spaces X and Y, with domain $\mathcal{D}(F)$. By $x^\dagger$ we denote a solution to (8.1) for given right-hand side y. As a consequence of the 'smoothing' property of F, which is typical for inverse problems, the non-linear equation (8.1) is *locally ill-posed* at the solution point $x^\dagger \in \mathcal{D}(F)$ (cf. [5, Def. 2]), which in particular means that stability estimates of the form

$$\|x - x^\dagger\|_X \le \varphi(\|F(x) - F(x^\dagger)\|_Y) \tag{8.2}$$

cannot hold for all $x \in \mathcal{D}(F)$ in an arbitrarily small ball around $x^\dagger$ and for strictly increasing continuous functions $\varphi : [0, \infty) \to [0, \infty)$ with $\varphi(0) = 0$. However, inequalities similar to (8.2), called *conditional stability estimates*, can hold on the one hand if the admissible range of $x \in \mathcal{D}(F)$ is restricted to densely defined subspaces of X. In this context, we refer to the seminal paper [1] as well as to [2, 6, 7] and references therein. On the other hand, they can hold for all $x \in \mathcal{D}(F)$ if the term $\|x - x^\dagger\|_X$ on the left-hand side of the inequality (8.2) is replaced with a weaker distance measure, for example a weaker norm (cf. [3] and references therein). In this paper, we follow a combination of both approaches in a Hilbert scale setting.

Based on noisy data $y^\delta \in Y$, obeying the deterministic noise model

$$\|y - y^\delta\|_Y \le \delta \tag{8.3}$$

with *noise level* $\delta > 0$, we use within the domain

$$\mathcal{D} := \mathcal{D}(F) \cap \mathcal{D}(B) \neq \emptyset$$

minimizers $x_\alpha^\delta \in \mathcal{D}$ of the Tikhonov functional T_α^δ solving the extremal problem

$$T_\alpha^\delta(x) := \|F(x) - y^\delta\|_Y^2 + \alpha\|B(x - \bar{x})\|_X^2 \to \min, \quad \text{subject to} \quad x \in \mathcal{D}, \tag{8.4}$$

as stable approximate solutions (regularized solutions) to $x^\dagger$. Above, the element $\bar{x} \in \mathcal{D}$ is a given smooth reference element, and $B : \mathcal{D}(B) \subset X \to X$ is a densely defined, unbounded, linear self-adjoint operator, which is strictly positive such that we have for some $m > 0$

$$\|Bx\|_X \ge m\|x\|_X, \quad \text{for all} \quad x \in \mathcal{D}(B). \tag{8.5}$$

This operator B generates a *Hilbert scale* $\{X_\tau\}_{\tau\in\mathbb{R}}$ with $X_0 = X$, $X_\tau = \mathcal{D}(B^\tau)$, and with corresponding norms $\|x\|_\tau := \|B^\tau x\|_X$.

Here we shall focus on the case of an oversmoothing penalty, which means that $x^\dagger \notin \mathcal{D}(B)$ such that $T_\alpha^\delta(x^\dagger) = \infty$. In this case, the regularizing property $T_\alpha^\delta(x_\alpha^\delta) \le T_\alpha^\delta(x^\dagger)$ does not provide additional value here. Throughout this paper, we assume the operator F to be sequentially weakly continuous and its domain $\mathcal{D}(F)$ to be a convex and closed subset of X, which makes the general results of [9, Sect. 4.1.1] on existence and stability of the regularized solutions x_α^δ applicable.

The paper is organized as follows: In Sect. 8.2 we formulate non-linearity and smoothness assumptions, which are required for obtaining a convergence rate result for Tikhonov regularization in Hilbert scales in the case of oversmoothing penalties under an appropriate a priori choice of the regularization parameter. Also in Sect. 8.2 we formulate the main theorem. Its proof will then follow from two propositions which are stated. Section 8.3 is devoted to proving both propositions. Section 8.4 completes the paper with some concluding discussions.

8.2 Assumptions and Main Result

In accordance with the previous study [4] we make the following additional assumption on the structure of non-linearity for the forward operator F with respect to the Hilbert scale generated by the operator B. Sufficient conditions and examples for this non-linearity assumption can be found in the appendix of [4].

Assumption 8.1 (*Non-linearity structure*) There is a number $a > 0$, and there are constants $0 < c_a \le C_a < \infty$ such that

$$c_a\|x - x^\dagger\|_{-a} \le \|F(x) - F(x^\dagger)\|_Y \le C_a\|x - x^\dagger\|_{-a} \quad \text{for all } x \in \mathcal{D}. \tag{8.6}$$

The left-hand inequality of condition (8.6) implies that, for the right-hand side y, there is no solution to (8.1) which belongs to $\mathcal{D}$. Moreover note that the parameter $a > 0$ in Assumption 8.1 can be interpreted as *degree of ill-posedness* of the mapping F at $x^\dagger$.

The solution smoothness is measured with respect to the generator B of the Hilbert scale as follows. We fix the reference element $\bar{x} \in \mathcal{D}$, occurring in the penalty functional of T_α^δ.

Assumption 8.2 (*Solution smoothness*) There are $0 < p < 1$ and $E < \infty$ such that $x^\dagger \in \mathcal{D}(B^p)$ and

$$x^\dagger - \bar{x} \in M_{p,E} := \left\{x \in X_p, \quad \|x\|_p := \|B^p x\|_X \le E\right\}. \tag{8.7}$$

Moreover, we assume that $x^\dagger$ is an interior point of $\mathcal{D}(F)$, but $x^\dagger \notin \mathcal{D}(B)$.

We shall analyze the error behavior of the minimizer x_α^δ to the Tikhonov functional T_α^δ for the following specific a priori parameter choice.

Assumption 8.3 (*A priori parameter choice*) Given the noise level $\delta > 0$, an ill-posedness degree $a > 0$ (cf. Assumption 8.1) and the solution smoothness $p \in (0, 1)$ (cf. Assumption 8.2), let

$$\alpha_* = \alpha_*(\delta) := \delta^{\frac{2(a+1)}{a+p}}. \tag{8.8}$$

We shall occasionally use the identity $\frac{\delta}{\sqrt{\alpha_*}} = \delta^{\frac{p-1}{a+p}}$, and we highlight that for this parameter choice we have $\frac{\delta}{\sqrt{\alpha_*(\delta)}} \to \infty$ as $\delta \to 0$, for $0 < p < 1$.

The main result is as follows.

Theorem 8.1 *Under the assumptions stated above let $x_{\alpha_*}^\delta$ be the minimizer of the Tikhonov functional $T_{\alpha_*}^\delta$ for the a priori choice α_* from* (8.8). *Then we have the convergence rate*

$$\|x_{\alpha_*}^\delta - x^\dagger\|_X = \mathcal{O}\left(\delta^{\frac{p}{a+p}}\right) \quad \text{as} \quad \delta \to 0. \tag{8.9}$$

This asymptotics is an immediate consequence of the following two propositions, the proofs of which will be given in the next section.

Proposition 8.1 *Under the a priori choice α_* from* (8.8) *and for sufficiently small $\delta > 0$ we have that*

$$\|x_{\alpha_*}^\delta - x^\dagger\|_{-a} \le K\delta \tag{8.10}$$

holds for some positive constant K.

Proposition 8.2 *Under the a priori choice α_* from* (8.8) *and for sufficiently small $\delta > 0$ we have that*

$$\|x_{\alpha_*}^\delta - x^\dagger\|_p \le \tilde{E} \tag{8.11}$$

holds for some positive constant $\tilde{E}$.

Proof of Theorem 8.1. Taking into account the assertions of Propositions 8.1 and 8.2, the convergence rate (8.9) follows directly from the interpolation inequality of the Hilbert scale $\{X_\tau\}_{\tau \in \mathbb{R}}$, applied here in the form

$$\|x_{\alpha_*}^\delta - x^\dagger\|_X \le \|x_{\alpha_*}^\delta - x^\dagger\|_{-a}^{\frac{p}{a+p}} \|x_{\alpha_*}^\delta - x^\dagger\|_p^{\frac{a}{a+p}}. \tag{8.12}$$

Thus, the proof of the theorem is complete.

8.3 Proofs of Propositions 8.1 and 8.2

Propositions 8.1 and 8.2 yield bounds in the (weak) $(-a)$-norm and in the (strong) (p)-norm, respectively. For the proofs we shall use *auxiliary elements* x_{α_*}, constructed as follows. Precisely, in conjunction with the Tikhonov functional T_α^δ from (8.4) we consider the artificial Tikhonov functional

$$T_{-a,\alpha}(x) := \|x - x^\dagger\|_{-a}^2 + \alpha\|B(x - \bar{x})\|_X^2, \tag{8.13}$$

which is well-defined for all $x \in X$. Let x_α be the minimizers of $T_{-a,\alpha}$ over all $x \in X$. These are, for all $\alpha > 0$, independent of the noise level $\delta > 0$. Recall now the parameter choice (8.8) from Assumption 8.3. For this choice of the regularization parameter the estimates from [4, Prop. 2] yield immediately the following assertions.

Proposition 8.3 *Suppose that $x^\dagger \in M_{p,E}$ for some $0 < p < 1$, and let x_α be the minimizer of $T_{-a,\alpha}$. Given $\alpha_* > 0$ as in Assumption 8.3 the resulting element x_{α_*} obeys the bounds*

$$\|x_{\alpha_*} - x^\dagger\|_X \le E\delta^{p/(a+p)}, \tag{8.14}$$

$$\|B^{-a}(x_{\alpha_*} - x^\dagger)\|_X \le E\delta, \tag{8.15}$$

$$\|B(x_{\alpha_*} - \bar{x})\|_X \le E\delta^{(p-1)/(a+p)} = E\frac{\delta}{\sqrt{\alpha_*}}, \tag{8.16}$$

and

$$T_{-a,\alpha}(x_{\alpha_*}) \le 2E^2\delta. \tag{8.17}$$

Moreover, we have that

$$\|x_{\alpha_*} - \bar{x}\|_p \le E, \text{ and } \|x_{\alpha_*} - x^\dagger\|_p \le E. \tag{8.18}$$

Notice, that in contrast to the solution element $x^\dagger$ the auxiliary element x_{α_*} belongs to $\mathcal{D}$, provided that δ is small enough, and hence we can use the minimizing property

$$T_{\alpha_*}^\delta(x_{\alpha_*}^\delta) \le T_{\alpha_*}^\delta(x_{\alpha_*}). \tag{8.19}$$

We derive the following consequence of Proposition 8.3 formulated in Proposition 8.4.

Proposition 8.4 *Let $x_{\alpha_*}^\delta$ be the minimizer of $T_{\alpha_*}^\delta$ for the Tikhonov functional T_α^δ from (8.4) with the choice α_*, as in Assumption 8.3, of the regularization parameter $\alpha > 0$. Then we have for sufficiently small $\delta > 0$ that*

$$\|F(x_{\alpha_*}^\delta) - y^\delta\|_Y \le C\delta, \tag{8.20}$$

and

$$\|B(x_{\alpha_*}^\delta - \bar{x})\|_X \le C\frac{\delta}{\sqrt{\alpha_*}}, \tag{8.21}$$

where $C := \left((C_a E + 1)^2 + E^2\right)^{1/2}$.

Proof Using (8.19) it is enough to bound $T_{\alpha_*}^\delta(x_{\alpha_*})$ by $C^2\delta^2$. We first argue that x_{α_*}, the minimizer of the auxiliary functional (8.13) for $\alpha = \alpha_*$, belongs to the set

$\mathcal{D}(F) \cap \mathcal{D}(B)$. Indeed, the bound (8.16) shows that $x_{\alpha_*} \in \mathcal{D}(B)$, and the bound (8.14) indicates that $x_{\alpha_*} \in \mathcal{D}(F)$ for sufficiently small $\delta > 0$, where we use that $x^\dagger$ is an interior point of $\mathcal{D}(F)$. Now we find from (8.15) and (8.16), and from the right-hand inequality of (8.6), that

$$\begin{aligned} T^\delta_{\alpha_*}(x_{\alpha_*}) &\leq \left(\|F(x_{\alpha_*}) - F(x^\dagger)\|_Y + \|F(x^\dagger) - y^\delta\|_Y\right)^2 + \alpha_* \|B(x_{\alpha_*} - \bar{x})\|_X^2 \\ &\leq \left(C_a \|x_{\alpha_*} - x^\dagger\|_{-a} + \delta\right)^2 + E^2 \alpha_* \delta^{2(p-1)/(a+p)} \\ &\leq (C_a E\delta + \delta)^2 + E^2\delta^2 \\ &= \left((C_a E + 1)^2 + E^2\right)\delta^2. \end{aligned}$$

This completes the proof of Proposition 8.4. □

Now we turn to the proofs of Propositions 8.1 and 8.2.

Proof of Proposition 8.1. Here we use the left-hand side in the non-linearity condition from Assumption 8.1 and find for sufficiently small $\delta > 0$

$$\begin{aligned} \|x^\delta_{\alpha_*} - x^\dagger\|_{-a} &\leq \frac{1}{c_a} \|F(x^\delta_{\alpha_*}) - F(x^\dagger)\|_Y \\ &\leq \frac{1}{c_a}\left(\|F(x^\delta_{\alpha_*}) - y^\delta\|_Y + \|F(x^\dagger) - y^\delta\|_Y\right) \\ &\leq \frac{1}{c_a}(C\delta + \delta) = \frac{1}{c_a}(C+1)\,\delta = K\delta, \end{aligned}$$

where C is the constant from Proposition 8.4 and $K := \frac{1}{c_a}(C+1)$. The proof is complete.

In order to establish the bound occurring in Proposition 8.2 we start with the following estimate.

Lemma 8.1 *Let α_* be as in Assumption 8.3. Then there is a constant $\tilde{C}$ such that we have for sufficiently small $\delta > 0$ that*

$$\|B(x^\delta_{\alpha_*} - x_{\alpha_*})\|_X \leq \tilde{C}\frac{\delta}{\sqrt{\alpha_*}}.$$

Proof By using the triangle inequality we find that

$$\|B(x^\delta_{\alpha_*} - x_{\alpha_*})\|_X \leq \|B(x_{\alpha_*} - \bar{x})\|_X + \|B(x^\delta_{\alpha_*} - \bar{x})\|_X$$

The first summand on the right was bounded in (8.16), and the second was bounded in Proposition 8.4 for sufficiently small $\delta > 0$. This yields the assertion with $\tilde{C} := C + E$, where C is the constant from Proposition 8.4. □

This allows for the following result.

Proposition 8.5 *There is a constant* $\bar{E}$ *such that we have for the parameter* α_* *from Assumption 8.3 and for sufficiently small* $\delta > 0$ *that*

$$\|x_{\alpha_*}^{\delta} - x_{\alpha_*}\|_p \leq \bar{E}.$$

Proof Again, we use the interpolation inequality, now in the form of

$$\|x_{\alpha_*}^{\delta} - x_{\alpha_*}\|_p \leq \|x_{\alpha_*}^{\delta} - x_{\alpha_*}\|_1^{\frac{a+p}{a+1}} \|x_{\alpha_*}^{\delta} - x_{\alpha_*}\|_{-a}^{\frac{1-p}{a+1}}.$$

The norm in the first factor was bounded in Lemma 8.1. The norm in the second factor can be bounded from

$$\|x_{\alpha_*}^{\delta} - x_{\alpha_*}\|_{-a} \leq \|x_{\alpha_*}^{\delta} - x^{\dagger}\|_{-a} + \|x_{\alpha_*} - x^{\dagger}\|_{-a}. \tag{8.22}$$

Now we discuss both summands in the right-hand side of the inequality (8.22). The first summand was bounded in Proposition 8.1 by a multiple of δ, and the same holds true for the second summand by virtue of (8.15). Therefore, there is a constant $\bar{E}$ such that

$$\|x_{\alpha_*}^{\delta} - x_{\alpha_*}\|_p \leq \bar{E} \left(\frac{\delta}{\sqrt{\alpha_*}} \right)^{\frac{a+p}{a+1}} \delta^{\frac{1-p}{a+1}} = \bar{E}\, \delta\, \alpha_*^{-\frac{a+p}{2(a+1)}} = \bar{E}.$$

This completes the proof of Proposition 8.5. □

We have gathered all auxiliary estimates in order to turn to the final proof.
Proof of Proposition 8.2. The estimate (8.11) is an immediate consequence of Proposition 8.5 and of the second bound from (8.18) yielding the constant $\tilde{E} := \bar{E} + E$, overall. This completes the proof.

8.4 Conclusions

We have shown that under the non-linearity Assumption 8.1 and the solution smoothness given as in Assumption 8.2 the a priori regularization parameter choice

$$\alpha_* = \alpha_*(\delta) = \delta^{\frac{2(a+1)}{a+p}} = \delta^{2 - \frac{2(p-1)}{a+p}}$$

allows for the order optimal convergence rate (8.9) for all $0 < p < 1$. The obtained rate from Theorem 8.1 is valid for $0 < p \leq a + 1$ when using the same a priori choice of the regularization parameter from Assumption 8.3. In all cases, we have that $\alpha_*(\delta) \to 0$ as $\delta \to 0$. In the regular case with $1 < p \leq a + 1$ we also find the usual convergence $\frac{\delta^2}{\alpha(\delta)} \to 0$ as $\delta \to 0$. For the borderline case $p = 1$ the quotient $\frac{\delta^2}{\alpha_*(\delta)}$ is constant. However, in the oversmoothing case $0 < p < 1$, as considered here, we find that $\frac{\delta^2}{\alpha_*(\delta)} \to \infty$ as $\delta \to 0$.

We stress another observation. In the regular case with $p > 1$ we have the *convergence property*

$$\lim_{\delta \to 0} \|x_{\alpha_*}^{\delta} - x^{\dagger}\|_X = 0\,, \tag{8.23}$$

which is a consequence of the sequential weak compactness of a ball (in the Hilbert space X), and of the Kadec–Klee property (in the Hilbert space X). This cannot be shown for $0 < p < 1$ without using additional smoothness of the solution $x^{\dagger}$. Hence, the convergence property (8.23) as an implication of Theorem 8.1 is essentially based on the existence of a positive value p in (8.7) expressing solution smoothness.

Acknowledgements The research was financially supported by Deutsche Forschungsgemeinschaft (DFG-grant HO 1454/12-1) and by the Weierstrass Institute for Applied Analysis and Stochastics, Berlin.

References

1. J. Cheng, M. Yamamoto, One new strategy for a priori choice of regularizing parameters in Tikhonov's regularization. Inverse Probl. **16**(4), L31–L38 (2000)
2. J. Cheng, B. Hofmann, S. Lu, The index function and Tikhonov regularization for ill-posed problems. J. Comput. Appl. Math. **265**, 110–119 (2014)
3. H. Egger, B. Hofmann, Tikhonov regularization in Hilbert scales under conditional stability assumptions. Inverse Probl. **34**(11), 115015 (17 pp) (2018)
4. B. Hofmann, P. Mathé, Tikhonov regularization with oversmoothing penalty for non-linear ill-posed problems in Hilbert scales. Inverse Probl. **34**(1), 015007 (14 pp) (2018)
5. B. Hofmann, O. Scherzer, Factors influencing the ill-posedness of nonlinear problems. Inverse Probl. **10**(6), 1277–1297 (1994)
6. B. Hofmann, M. Yamamoto, On the interplay of source conditions and variational inequalities for nonlinear ill-posed problems. Appl. Anal. **89**(11), 1705–1727 (2010)
7. T. Hohage, F. Weidling, Variational source conditions and stability estimates for inverse electromagnetic medium scattering problems. Inverse Probl. Imaging **11**(1), 203–220 (2017)
8. F. Natterer, Error bounds for Tikhonov regularization in Hilbert scales. Appl. Anal. **18**(1–2), 29–37 (1984)
9. T. Schuster, B. Kaltenbacher, B. Hofmann, K.S. Kazimierski, *Regularization Methods in Banach Spaces, Radon Series on Computational and Applied Mathematics*, vol. 10 (Walter de Gruyter, Berlin, 2012)

Chapter 9
Case Studies and a Pitfall for Nonlinear Variational Regularization Under Conditional Stability

Daniel Gerth, Bernd Hofmann and Christopher Hofmann

Abstract Conditional stability estimates are a popular tool for the regularization of ill-posed problems. A drawback in particular under nonlinear operators is that additional regularization is needed for obtaining stable approximate solutions if the validity area of such estimates is not completely known. In this paper we consider Tikhonov regularization under conditional stability estimates for nonlinear ill-posed operator equations in Hilbert scales. We summarize assertions on convergence and convergence rate in three cases describing the relative smoothness of the penalty in the Tikhonov functional and of the exact solution. For oversmoothing penalties, for which the true solution no longer attains a finite value, we present a result with modified assumptions for a priori choices of the regularization parameter yielding convergence rates of optimal order for noisy data. We strongly highlight the local character of the conditional stability estimate and demonstrate that pitfalls may occur through incorrect stability estimates. Then convergence can completely fail and the stabilizing effect of conditional stability may be lost. Comprehensive numerical case studies for some nonlinear examples illustrate such effects.

Keywords Nonlinear inverse problems · Conditional stability · Tikhonov regularization · Oversmoothing penalties · Hilbert scales · Convergence rates

MSC 2010 47J06 · 65J20 · 47A52

D. Gerth · B. Hofmann (✉) · C. Hofmann
Faculty of Mathematics, Chemnitz University of Technology, 09107 Chemnitz, Germany
e-mail: bernd.hofmann@mathematik.tu-chemnitz.de

D. Gerth
e-mail: daniel.gerth@mathematik.tu-chemnitz.de

C. Hofmann
e-mail: christopher.hofmann@mathematik.tu-chemnitz.de

J. Cheng et al. (eds.), *Inverse Problems and Related Topics*,
Springer Proceedings in Mathematics & Statistics 310,
https://doi.org/10.1007/978-981-15-1592-7_9

9.1 Introduction

Regularization theory for nonlinear ill-posed inverse problems is always a challenging endeavor. In contrast to linear inverse problems, where the theory is rather coherent and well-developed (see, for example, the monographs [5, 19]), the nonlinear theory is harder to grasp. Numerous assumptions exist in the literature that restrict the nonlinear behavior of the forward operator in such a way that stable approximate solutions exist which converge to the exact solution in the limit of vanishing data noise. It is important to keep in mind that the nonlinearity conditions only hold locally. A main goal of this paper is to show that this can be a pitfall, as incorrect localization leads to the loss of the stabilizing property. A second objective of the paper is to verify theoretical convergence results in numerical examples, as well as pointing out some open questions. To this end, we focus here on regularization in Hilbert scales. This technique is a common approach for the regularization of ill-posed problems and we refer, for example, to [20, 25, 26]. Going into detail of our paper, we consider in this paper the stable approximate solution of the nonlinear operator equation

$$F(x) = y \tag{9.1}$$

by variational (Tikhonov-type) regularization. Equation (9.1) serves as a model for an inverse problem where the nonlinear forward operator $F : \mathcal{D}(F) \subseteq X \to Y$ maps between the infinite dimensional real Hilbert spaces X and Y with domain $\mathcal{D}(F)$. The symbols $\|\cdot\|_X$, $\|\cdot\|_Y$ and $\langle\cdot,\cdot\rangle_X$, $\langle\cdot,\cdot\rangle_Y$ designate the norms and inner products of the spaces X and Y, respectively. Instead of the exact right-hand side $y = F(x^\dagger)$, with the uniquely determined preimage $x^\dagger \in \mathcal{D}(F)$, we assume to know a noisy element $y^\delta \in Y$ satisfying the noise model

$$\|y - y^\delta\|_Y \le \delta \tag{9.2}$$

with some noise level $\delta > 0$. Based on this data element $y^\delta \in Y$ we use as approximations to $x^\dagger$ global minimizers $x_\alpha^\delta \in \mathcal{D}(F)$ of the extremal problem

$$T_\alpha^\delta(x) := \|F(x) - y^\delta\|_Y^2 + \alpha\|Bx\|_X^2 \to \min, \quad \text{subject to } x \in \mathcal{D}(F). \tag{9.3}$$

Here, $B\colon \mathcal{D}(B) \subset X \to X$ is a densely defined, unbounded, linear, and self-adjoint operator which is strictly positive such that $\|Bx\|_X \ge c_B\|x\|_X$ holds for all $x \in \mathcal{D}(B)$. Such operators B generate a Hilbert scale $\{X_\nu\}_{\nu\in\mathbb{R}}$, where $X_\nu = \mathcal{D}(B^\nu)$ coincides with the range $\mathcal{R}(B^{-\nu})$ of the operator $B^{-\nu}$. In particular $X_0 = X$, and we set $\|x\|_\nu := \|B^\nu x\|_X$ for the norm of the Hilbert scale element $x \in X_\nu$. With this, the specific Tikhonov functional $T_\alpha^\delta : X \to [0, \infty]$ in (9.3) is the weighted sum of the quadratic misfit functional $\|F(\cdot) - y^\delta\|_Y^2$ and the Hilbert-scale penalty functional $\|B\,\cdot\|_X^2 = \|\cdot\|_1^2$, where the regularization parameter $\alpha > 0$ acts as weight factor. Note that no generality is lost by considering only the penalty in the 1-norm $\|\cdot\|_1$, since one can always rescale the operator B to obtain $Bx\|_X = \|(B^{1/p})^p x\|_X = \|\tilde{B}^p\|_X$ for $p > 0$, i.e, one obtains a penalty of arbitrary index p in the Hilbert scale generated by the operator $\tilde{B} := B^{\frac{1}{p}}$. Finally, we mention that for $x \in \mathcal{D}(F)$ we set $T_\alpha^\delta(x) := +\infty$ if $x \notin \mathcal{D}(B)$, and that the Tikhonov functional attains a well-defined value $0 \le T_\alpha^\delta(x) < +\infty$ if $x \in \mathcal{D} := \mathcal{D}(F) \cap \mathcal{D}(B) \neq \emptyset$.

A typical phenomenon of the nonlinear equation (9.1) as a model for an inverse problem is *local ill-posedness* at the solution point $x^\dagger \in \mathcal{D}(F)$ (cf. [16, Def. 2] or [15, Def. 3]), which means that inequalities of the form

$$\|x - x^\dagger\|_X \le K\,\varphi(\|F(x) - F(x^\dagger)\|_Y) \quad \text{for all} \quad x \in \mathcal{B}_r^X(x^\dagger) \cap \mathcal{D}(F) \tag{9.4}$$

cannot hold for any positive constants K, r and any index function φ.[1] However, the inverse problem literature offers numerous examples, where the left-hand term $\|x - x^\dagger\|_X$ in (9.4) is replaced with a weaker norm $\|x - x^\dagger\|_{-a}$ $(a > 0)$ and a corresponding conditional stability estimate takes place. In the sequel, we restrict our considerations to the concave index functions $\varphi(t) = t^\gamma$ of Hölder-type with exponents $0 < \gamma \le 1$ and hence to *conditional stability estimates* of the form

$$\|x - x^\dagger\|_{-a} \le K\,\|F(x) - F(x^\dagger)\|_Y^\gamma \quad \text{for all} \quad x \in Q \cap \mathcal{D}(F) \tag{9.5}$$

with some index $a > 0$, which can be interpreted as *degree of ill-posedness* of F at $x^\dagger$, a suitable subset Q in X which acts as the aforementioned localization of the nonlinearity condition, and a constant $K > 0$ that may depend on Q.

Let us consider the situation that $x^\dagger \in Q$ and Q is known. Then one may employ a least squares iteration process of minimizing the norm square

$$\|F(x) - y^\delta\|_Y^2 \to \min, \quad \text{subject to} \quad x \in Q \cap \mathcal{D}(F). \tag{9.6}$$

The minimizers x_{ls} of (9.6) satisfy $\|F(x_{ls}) - y^\delta\| \le \delta$ by definition and due to $x^\dagger \in Q$. Hence we have convergence $\|x_{ls}^\delta - x^\dagger\|_{-a} \to 0$ as $\delta \to 0$ of these least squares-type solutions to $x^\dagger$ in the norm of the space X_{-a} which is weaker than the one in X.

To achieve convergence and even convergence rates in the norm of X, additional smoothness $x^\dagger \in X_p$ for some $p > 0$ is needed. If the approximate solutions $x_{ls}^\delta \in Q \cap \mathcal{D}(F)$ also possess such smoothness with $\|x_{ls}^\delta\|_p$ uniformly bounded for all $0 < \delta \le \bar{\delta}$, then, with $-a < t \le p$ the interpolation inequality in Hilbert scales (see [18]) applies in the form

$$\|x\|_t \le \|x\|_{-a}^{\frac{p-t}{p+a}} \|x\|_p^{\frac{t+a}{p+a}} \tag{9.7}$$

for all $x \in X_p$. Hence we derive from (9.5) and (9.7) with $t = 0$ and by the triangle inequality that

$$\|x_{ls}^\delta - x^\dagger\|_X \le \bar{K}\,\delta^{\frac{\gamma p}{p+a}}$$

[1] Throughout, $\mathcal{B}_r^H(\bar{x})$ denotes a closed ball in the Hilbert space H around $\bar{x} \in H$ with radius $r > 0$. Furthermore, we call a function $\varphi\colon [0, \infty) \to [0, \infty)$ index function if it is continuous, strictly increasing and satisfies the boundary condition $\varphi(0) = 0$.

for sufficiently small $\delta > 0$ and some constant $\bar{K}$. A way to ensure the property that the approximate solutions belong to $X_p \cap Q \cap \mathcal{D}(F)$ is to use regularized solutions which minimize the Tikhonov functional $\|F(x) - y^\delta\|_Y^2 + \alpha \|B^s x\|_X^2$, subject to $x \in Q \cap \mathcal{D}(F)$, where $s \ge p$ is required. Hence, Tikhonov-type regularization is here an auxiliary tool which complements the conditional stability estimate (9.5) in order to obtain stable approximate solutions measured in the norm of X.

On the other hand, we have to take into account the frequently occurring situation that the set Q in (9.5) is not or not completely known and a minimization process according to (9.6) is impossible, because of a not completely known set of constraints for the optimization problem. Nevertheless, a combination of the conditional stability estimate (9.5) with variational regularization of the form (9.3) can be successful. For a systematic treatment of convergence results in the context of regularization theory we will distinguish the following cases relating the smoothness of the solution $x^\dagger$ and of the approximate solutions x_α^δ implied by the functional (9.3):

Case distinction

(a) Classical regularization: $x^\dagger \in X_p$ for $p > 1$, which means that $\|Bx^\dagger\|_X^2 < +\infty$ and there is some source element $w \in X_\varepsilon$ ($\varepsilon > 0$) such that $x^\dagger = B^{-1}w$;
(b) Matching smoothness: $x^\dagger \in X_1$, i.e. $\|Bx^\dagger\|_X < \infty$, but $x^\dagger \notin X_{1+\varepsilon}$ for all $\varepsilon > 0$.
(c) Oversmoothing regularization: $x^\dagger \in X_p$ for some $0 < p < 1$, but $x^\dagger \notin X_1$, i.e. $\|Bx^\dagger\|_X = +\infty$.

The goal of this paper is to discuss the different opportunities and limitations for convergence and rates of regularized solutions x_α^δ in the situations (a), (b), and (c), respectively. It is organized as follows: Sect. 9.2 recalls assertions on convergence of regularized solutions in cases (a) and (b). Moreover, usual technical assumptions on forward operator, its domain and the exact solution are listed. In Sect. 9.3, Hölder rate results under conditional stability estimates are summarized for the cases of classical regularization and matching smoothness. The rate result of Proposition 9.4 for the oversmoothing case (c) is of specific interest. It requires two-sided inequalities as conditional stability estimates, whereas in cases (a) and (b) only one-sided inequalities are needed. Three inverse model problems of ill-posed nonlinear equations covering all cases (a), (b), and (c) are outlined in Sect. 9.4, for which numerical case studies are presented in Sect. 9.5. The proof of Proposition 9.4 is given in the appendix.

9.2 Convergence

In this section we collect properties of the regularized solutions x_α^δ obtained as solutions of the optimization problem (9.3) for the cases (a), (b), and (c) in different ways. Throughout this paper we suppose that the following assumption concerning the nonlinear forward operator F and the solvability of the operator equation (9.1) holds true.

Assumption 9.1 The operator $F : D(F) \subseteq X \to Y$ is weak-to-weak sequentially continuous and its domain $D(F)$ is a convex and closed subset of X. For the right-hand side $y = F(x^\dagger) \in Y$ under consideration let $x^\dagger \in \mathcal{D}(F)$ be the uniquely determined solution to the operator equation (9.1).

Under the setting introduced in Sect. 9.1, the penalty $\|Bx\|_X^2$ as part of the Tikhonov functional T_α^δ in (9.3) is a non-negative, convex, and sequentially lower semi-continuous functional. Moreover, this functional is *stabilizing* in the sense that all its sublevel sets are weakly sequently compact in X. Taking also into account Assumption 9.1, the Assumptions 3.11 and 3.22 of [24] are satisfied and the assertions from [24, Sect. 4.1.1] apply, which ensure *existence* and *stability* of the regularized solutions x_α^δ in our present Hilbert scale setting, consistent for all three cases (a), (b), and (c).

We emphasize at this point that we always have $x_\alpha^\delta \in X_1$ by definition of the minimizers in (9.3), but only in the cases (a) and (b) one can take profit of the inequality

$$T_\alpha^\delta(x_\alpha^\delta) \le T_\alpha^\delta(x^\dagger), \tag{9.8}$$

which implies for all $\alpha > 0$ that

$$\|x_\alpha^\delta\|_1 \le \sqrt{\|x^\dagger\|_1^2 + \frac{\delta^2}{\alpha}}. \tag{9.9}$$

In the case (c), however, due to $x^\dagger \notin X_1$ and hence $\|x^\dagger\|_1 = +\infty$ we have no such uniform bounds of $\|x_\alpha^\delta\|_1$ from above. On the contrary, in [7] it was shown that $\|x_\alpha^\delta\|_1 \to \infty$ as $\delta \to 0$ is necessary even for weak convergence of the regularizers x_α^δ to $x^\dagger$.

In order to obtain *convergence* of the regularized solutions x_α^δ to $x^\dagger$ as $\delta \to 0$, the interplay of the noise level and the choice of the regularization parameter $\alpha > 0$, which we choose either a priori $\alpha = \alpha(\delta)$ or a posteriori $\alpha = \alpha(\delta, y^\delta)$, must be appropriate. In the literature, this interplay is typically controlled by the limit conditions

$$\alpha \to 0 \quad \text{and} \quad \frac{\delta^2}{\alpha} \to 0 \quad \text{as} \quad \delta \to 0. \tag{9.10}$$

In our case (a) this is a sufficient description.

Proposition 9.1 *Let the regularization parameter* $\alpha > 0$ *fulfill the conditions* (9.10). *Then we have under Assumption 9.1 and for case (a), i.e. for* $1 < p < \infty$, *by setting* $\alpha_n = \alpha(\delta_n)$ *or* $\alpha_n = \alpha(\delta_n, y^{\delta_n})$, $x_n = x_{\alpha_n}^{\delta_n}$, *that for* $\delta_n \to 0$ *as* $n \to \infty$

$$\lim_{n\to\infty} \|x_n\|_1 = \|x^\dagger\|_1,$$

and

$$\lim_{n\to\infty} \|x_n - x^\dagger\|_\nu = 0 \quad \textit{for all} \quad 0 \le \nu \le 1.$$

Proof The proof follows along the lines of Theorem 4.3 and Corollary 4.6 from [24]. □

As we will see in Proposition 9.2 in the next section, the optimal parameter choice fulfills the conditions (9.10) in case (a). In case (b), where the smoothness of $x^\dagger$ coincides with the smoothness of the regularization, i.e., $p = 1$, the matter becomes unclear. On one hand, it is easily seen that Proposition 9.1 holds in the exact same way for case (b), which is a consequence of (9.9) holding in both cases. Hence, we have the following corollary:

Corollary 9.1 *Under the assumptions of Proposition 9.1, in particular for the cases (a) and (b) and for a regularization parameter choice satisfying* (9.10), *we have that the regularized solutions x_α^δ belong to the ball $\mathcal{B}_r^{X_\nu}(x^\dagger)$ for prescribed values $r > 0$ and $0 \le \nu \le 1$ whenever $\delta > 0$ is sufficiently small.*

The surprising difference between the cases (a) and (b) on the other hand, is that the optimal choice of the regularization parameter for (b) (we show in Proposition 9.3 below that $\alpha \sim \delta^2$ yields the optimal convergence rate) violates the second condition in (9.10). Since obviously a convergence rate implies norm convergence, this means that the condition $\delta^2/\alpha \to 0$ in (9.10) is not necessary but sufficient for convergence, at least in case (b).

In case (c) with oversmoothing penalty, the inequality (9.8) and consequently (9.9) are missing. Results of Proposition 9.1 and Corollary 9.1 in general do not apply in that case. One cannot even show weak convergence $x_n \rightharpoonup x^\dagger$ in X, and regularized solutions x_α^δ need not belong to a ball $\mathcal{B}_r^X(x^\dagger)$ with small radius $r > 0$ if $\delta > 0$ is sufficiently small. As will be shown in Proposition 9.4 of Sect. 9.3 (see also [13, 14]), convergence rates can be proven under stronger conditions also for (c), where we have some $0 < p < 1$ such that $x^\dagger \in X_p$. The key to these results was the appropriate choice of α either by an a priori or a posteriori parameter choice. In particular, $\frac{\delta^2}{\alpha} \to \infty$ as $\delta \to 0$, which violates (9.10), is typical there. The interplay of α and δ will be in the focus of our numerical case studies in Sect. 9.5 below.

9.3 Convergence Rate Results

In this section, we are going to discuss convergence rate results for cases (a) and (b) on one hand, but also (c) on the other hand. In addition to Assumption 9.1 some versions of conditional stability estimates have to be imposed which, in combination with the smoothness assumptions $x^\dagger \in X_p$, are essentially hidden forms of source conditions for the solution $x^\dagger$.

In Assumption 9.2 we first consider the situation for the setting $Q := \mathcal{B}_\rho^{X_1}(0)$. This model setting was comprehensively discussed and illustrated by examples of associated nonlinear inverse problems in the papers [3, 4, 13, 17]. Here we have evidently $x^\dagger \in Q$ for the cases (a) and (b) whenever $\|x^\dagger\|_1 \le \rho$.

Assumption 9.2 Let for fixed $a > 0$ and $0 < \gamma \le 1$ the conditional stability estimates

$$\|x - x^\dagger\|_{-a} \le K(\rho)\, \|F(x) - F(x^\dagger)\|_Y^\gamma \quad \text{for all} \;\; x \in \mathcal{B}_\rho^{X_1}(0) \cap \mathcal{D}(F) \tag{9.11}$$

hold, where constants $K(\rho) > 0$ are supposed to exist for all radii $\rho > 0$.

Then the following proposition, which is a direct consequence of [4, Theorem 2.1] when adapting the corresponding proof, yields an order optimal convergence rate in case (a).

Proposition 9.2 *Under Assumptions 9.1 and 9.2 and for* $x^\dagger \in X_p$ *with* $1 < p \le a + 2$ *we have the rate of convergence of regularized solutions* $x_\alpha^\delta \in \mathcal{D}(F) \cap \mathcal{D}(B)$ *to the solution* $x^\dagger \in \mathcal{D}(F) \cap \mathcal{D}(B)$ *as*

$$\|x_\alpha^\delta - x^\dagger\|_X = \mathcal{O}\left(\delta^{\frac{\gamma p}{p+a}}\right) \quad \text{as} \;\; \delta \to 0, \tag{9.12}$$

provided that the regularization parameter $\alpha = \alpha(\delta)$ *is chosen a priori as*

$$\alpha(\delta) \sim \delta^{2-2\gamma\frac{p-1}{p+a}}. \tag{9.13}$$

We easily see that the convergence results of Proposition 9.1 apply here for $p > 1$ and that in particular (9.13) implies (9.10). The additional smoothness of $x^\dagger$, which is always required to obtain convergence rates in regularization of ill-posed problems appears in Hilbert scales in form $x^\dagger = B^{-p}v$ with some source element $v \in X$.

Remark 9.1 We mention that along the lines of [4, Theorem 2.2] the rate (9.12) can also be shown under the assumptions of Proposition 9.2 when the regularization parameter $\alpha = \alpha(\delta, y^\delta)$ is chosen a posteriori by a sequential discrepancy principle.

The modified version of the rate result for case (b) is as follows:

Proposition 9.3 *Under the Assumptions 9.1 and 9.2 and for* $x^\dagger \in X_1$ *we have the rate of convergence of regularized solutions* $x_\alpha^\delta \in \mathcal{D}(F) \cap \mathcal{D}(B)$ *to the solution* $x^\dagger \in \mathcal{D}(F) \cap \mathcal{D}(B)$ *as*

$$\|x_\alpha^\delta - x^\dagger\|_X = \mathcal{O}\left(\delta^{\frac{\gamma}{1+a}}\right) \quad \text{as} \;\; \delta \to 0, \tag{9.14}$$

if the regularization parameter $\alpha = \alpha(\delta)$ *is chosen a priori as*

$$\alpha(\delta) \sim \delta^2. \tag{9.15}$$

Proof By the standard technique of variational regularization under conditional stability estimates (cf. [4, Proof of Theorem 1.1] or [24, Sect. 4.2.5]) we obtain for the choice (9.15) of the regularization parameter and by using the conditional stability estimate (9.11) the inequality

$$\|x_\alpha^\delta - x^\dagger\|_{-a} \le C\delta^\gamma, \tag{9.16}$$

where the constant $C > 0$ via ρ and $K(\rho)$ depends on $\|x^\dagger\|_1$ and on upper and lower bounds of δ^2/α. Combining this with the interpolation inequality (9.7), taking $t = 0$ and $s = 1$, and applying the triangle inequality provides us with the estimate

$$\|x_\alpha^\delta - x^\dagger\|_X \le C(\|x_\alpha^\delta\|_1 + \|x^\dagger\|_1)^{\frac{a}{1+a}}\,\delta^{\frac{\gamma}{1+a}}.$$

Due to (9.9) the norm $\|x_\alpha^\delta\|_1$ is uniformly bounded by a finite constant for $\alpha(\delta)$ from (9.15). This yields the rate (9.14) and completes the proof. Finally, we should note that the inequality (9.16) can only be established, because constants $K(\rho) > 0$ in (9.11) exist for arbitrarily large $\rho > 0$. □

In the borderline case (b) we have also a borderline a priori choice of the regularization parameter which contradicts the second limit condition in (9.10) such that the quotient $\frac{\delta^2}{\alpha}$ is uniformly bounded below by a positive constant and above by a finite constant.

In Assumption 9.3 we consider alternatively the situation that $Q := \mathcal{B}_r^X(x^\dagger)$. This model, which is illustrated by Example 9.1 in Sect. 9.4 below, is typical for conditional stability estimates that arise from nonlinearity conditions imposed on the forward operator F in a neighbourhood of the solution $x^\dagger$. In this context, the radius $r > 0$ which restricts the validity area of stability estimates can be rather small. In all cases of the Case distinction we have here $x^\dagger \in Q \cap \mathcal{D}(F)$, but only for (a) and (b) also $x^\dagger \in \mathcal{D}(F) \cap \mathcal{D}(B)$.

Assumption 9.3 Let for fixed $a > 0$ and $0 < \gamma \le 1$ the conditional stability estimate

$$\|x - x^\dagger\|_{-a} \le K(r)\,\|F(x) - F(x^\dagger)\|_Y^\gamma \quad \text{for all } \; x \in \mathcal{B}_r^X(x^\dagger) \cap \mathcal{D}(F) \tag{9.17}$$

hold, where the constant $K(r) > 0$ depends on the largest admissible radius $r > 0$.

Corollary 9.2 *The assertion of Proposition 9.2 remains true if Assumption 9.2 is replaced with Assumption 9.3.*

Proof To see the validity of Proposition 9.2 under Assumption 9.3 in case (a) of the Case distinction, where the regularization parameter choice satisfies (9.10), it is enough to take the assertion of Corollary 9.1 into account. This assertion implies that for sufficiently small $\delta > 0$ the regularized solutions x_α^δ belong to the ball $\mathcal{B}_r^X(x^\dagger)$ for prescribed $r > 0$. Then the conditional stability estimate (9.17) applies and yields the convergence rate (9.12) along the lines of the proof of [4, Theorem 2.1]. □

In case (b), however, for the choice (9.15) of Proposition 9.3 the condition (9.10) fails and even if $\delta > 0$ is sufficiently small, it cannot be shown that $x_\alpha^\delta \in \mathcal{B}_r^X(x^\dagger)$ for prescribed $r > 0$. Consequently, the conditional stability estimate (9.17) need not hold for the regularized solutions $x = x_\alpha^\delta$ and the rate assertion (9.14) of Proposition 9.3 is only valid under Assumption 9.3 if constants $K(r) > 0$ in (9.17) exist for arbitrarily large $r > 0$. This is, however, the case in the exponential growth model of Example 9.1 below.

Now we turn to the cases with oversmoothing penalty, where $x^\dagger \notin X_1$ and restrict ourselves to $\gamma = 1$ in the conditional stability estimates. As is well-known since the paper by Natterer [21], convergence rates in this case require lower and upper estimates of $\|F(x) - F(x^\dagger)\|_Y$ by multiples of the term $\|x - x^\dagger\|_{-a}$. We start with a corresponding analytical result. The goal of the case studies in Sect. 9.5 below is to gain further insight into the behavior of regularized solutions in case (c) for a priori and a posteriori choices of the regularization parameter.

Assumption 9.4 Let $a > 0$. Moreover, let $x^\dagger$ be an interior point of $\mathcal{D}(F)$ such that for the radius $r > 0$ we have $\mathcal{B}_r^X(x^\dagger) \subset \mathcal{D}(F)$ and the two estimates

$$\underline{K}\,\|x - x^\dagger\|_{-a} \le \|F(x) - F(x^\dagger)\|_Y \quad \text{for all} \quad x \in \mathcal{D}(F) \cap \mathcal{D}(B) = \mathcal{D}(F) \cap X_1 \tag{9.18}$$

and

$$\|F(x) - F(x^\dagger)\|_Y \le \overline{K}\,\|x - x^\dagger\|_{-a} \quad \text{for all} \quad x \in \mathcal{B}_r^X(x^\dagger) \cap X_1 \tag{9.19}$$

hold true, where $0 < \underline{K} \le \overline{K} < \infty$ are constants.

Proposition 9.4 *Let $x^\dagger \in X_p$ for some $0 < p < 1$, but $x^\dagger \notin X_1$. Under the Assumptions 9.1 and 9.4 we then have the rate of convergence of regularized solutions to the exact solution as*

$$\|x_{\alpha_*}^\delta - x^\dagger\|_X = \mathcal{O}\left(\delta^{\frac{p}{p+a}}\right) \qquad as \quad \delta \to 0, \tag{9.20}$$

if the regularization parameter is chosen a priori as

$$\alpha_* = \alpha(\delta) = \delta^{2-\frac{2(p-1)}{p+a}}. \tag{9.21}$$

The proof of Proposition 9.4 is given in the appendix along the lines of [14, Theorem 1], where we set for simplicity $\bar{x} = 0$. Note that Theorem 1 in [14] refers to a simplified version of the pair of estimates (9.18) and (9.19), which are ibid both assumed to hold for all $x \in \mathcal{D}(F)$. As the proof in the appendix shows, the upper estimate (9.19) is only exploited by auxiliary elements x_α, which belong to $\mathcal{B}_r^X(x^\dagger) \cap X_1$ for sufficiently small $\alpha > 0$. On the other hand, there are no arguments for restricting the noisy regularized solutions x_α^δ to small balls. Consequently, the lower estimate (9.18) needs to hold for all elements in $\mathcal{D}(F) \cap X_1$. This is an essential drawback for the application of Proposition 9.4 to practical problems. An analogue of Proposition 9.4 for the discrepancy principle as parameter choice rule can be formulated and proven along the lines of the paper [13].

As already mentioned in Sect. 9.2, we stress again that, despite the assertion of Proposition 9.4, norm convergence of regularized solutions cannot be shown in general for case (c), not even weak convergence in X can be established. Evidently the parameter choice (9.21) violates (9.10) since we have

$$\alpha(\delta) \to 0 \quad \text{and} \quad \frac{\delta^2}{\alpha(\delta)} = \delta^{\frac{2(p-1)}{p+a}} \to \infty \quad \text{as} \quad \delta \to 0.$$

It appears that

$$\alpha(\delta) \to 0 \quad \text{and} \quad \frac{\delta^2}{\alpha(\delta)} \to \infty \quad \text{as} \quad \delta \to 0$$

tends to be the typical situation in the oversmoothing case (c), at least for regularization parameters yielding optimal convergence rates. Numerical case studies below support this conjecture. A similar behavior of the regularization parameters was noted for oversmoothing ℓ^1-regularization [7].

To conclude and summarize this section, we stress that in all cases of the Case distinction, we have under the appropriate conditional stability assumption (to show the similarities between the cases, we fix $\gamma = 1$ for (a), (b), and (c) for the next assertion) and for $x^\dagger \in X_p$ for some $p > 0$ the convergence rate

$$\|x_\alpha^\delta - x^\dagger\|_X = \mathcal{O}\left(\delta^{\frac{p}{p+a}}\right) \quad \text{as} \quad \delta \to 0 \tag{9.22}$$

under both the discrepancy principle and the a priori parameter choice

$$\alpha(\delta) = \delta^{2-2\frac{p-1}{p+a}} = \delta^{\frac{2(a+1)}{a+p}}. \tag{9.23}$$

Hence, we obtain the same parameter choice and the same convergence rate as in the case of a linear operator. Namely in [21], (9.22) and (9.23) were obtained for a linear operator $A : X \to Y$ under a two-sided inequality

$$\underline{K}\|x\|_{-a} \le \|Ax\|_Y \le \overline{K}\|x\|_{-a} \quad \text{for all} \quad x \in X,$$

in analogy to the estimates from Assumption 9.4.

9.4 Examples

In the following, we introduce two nonlinear inverse problems of type (9.1), for which we will investigate the analytic results from the previous section numerically. Before doing so, we will introduce two similar, but different Hilbert scales used as penalty in the minimization problem (9.3) and as measure of the solution smoothness. On one hand, we consider the standard Sobolev-scale $H^p[0, 1]$. For integer values of $p \ge 0$, these function spaces consist of functions whose p-th derivative is still in $L^2(0, 1)$. For real parameters of $p > 0$, the spaces can be defined by an interpolation argument [1]. Using Fourier-analysis, one can define a norm in $H^p[0, 1]$ via

$$\|x\|^2_{H^p[0,1]} := \int_{\mathbb{R}} (1+|\xi|^2)^p \, |\hat{x}(\xi)|^2 \, d\xi, \tag{9.24}$$

where $\hat{x}$ is the Fourier-transform of x. Then $x \in H^p[0,1]$ iff $\|x\|_{H^p[0,1]} < \infty$. The Sobolev scale for $p \ge 0$ does not constitute a Hilbert scale in the strict sense, but for each $0 < p^* < \infty$ there is an operator $B : L^2(0,1) \to L^2(0,1)$ such that $\{X_p\}_{0 \le p \le p^*}$ is a Hilbert scale [22]. This is not an issue in numerical experiments. Note that the norm (9.24) is easy to implement, in particular it allows a precise gauging of the solution smoothness.

The reason why the Sobolev scale does not form a Hilbert scale for arbitrary values of p lies in the boundary values. In order to generate a full Hilbert scale $\{X_\tau\}_{\tau \in \mathbb{R}}$, we exploit the simple integration operator

$$[Jh](t) := \int_0^t h(\tau)d\tau \quad (0 \le t \le 1) \tag{9.25}$$

of Volterra-type mapping in $X = Y = L^2(0,1)$ and set

$$B := (J^*J)^{-1/2}. \tag{9.26}$$

By considering the Riemann–Liouville fractional integral operator J^p and its adjoint $(J^*)^p = (J^p)^*$ for $0 < p \le 1$ we have that

$$X_p = \mathcal{D}(B^p) = \mathcal{R}((J^*J)^{p/2}) = \mathcal{R}((J^*)^p),$$

cf. [9, 10, 23], and hence by [10, Lemma 8]

$$X_p = \begin{cases} H^p[0,1] & \text{for } 0 < p < \frac{1}{2} \\ \{x \in H^{\frac{1}{2}}[0,1] : \int\limits_0^1 \frac{|x(t)|^2}{1-t} dt < \infty\} & \text{for } \quad p = \frac{1}{2} \\ \{x \in H^p[0,1] : x(1) = 0\} & \text{for } \frac{1}{2} < p \le 1 \end{cases}, \tag{9.27}$$

where the fractional Sobolev spaces $H^p[0,1]$ occur. One can also show that

$$X_p = \{x \in H^p[0,1] : x(1) = 0\} \quad \text{for} \quad 1 < p < \frac{3}{2}. \tag{9.28}$$

On the other hand, it is well-known that $X_2 \subset X_p \subset X_1$ for $1 < p < 2$ and that X_2 is characterized by

$$X_2 = \{x \in H^2[0,1] : x'(0) = 0, \quad x(1) = 0\} \tag{9.29}$$

in an explicit manner, see for example [19, Example 2.1.5]. We omit discussing higher smoothness spaces since we will not consider those in our examples. In the following we present examples that show and illustrate the occurrence of the cases (a), (b) and

(c) of the Case distinction in Sect. 9.1. Note that for $0 < p < \frac{1}{2}$ the Sobolev-scale $H^p[0,1]$ and the Hilbert scale $\{X_p\}_{p>0}$ induced by J coincide.

Model problem 9.1 (*Exponential growth model*) The following exponential growth model has been previously discussed in the literature, and we refer for more details and properties to [11, Sect. 3.1] and [12]. To identify the time dependent growth rate $x(t)$ $(0 \le t \le T)$ of a population we use observations $y(t)$ $(0 \le t \le T)$ of the time-dependent size of the population with initial size $y(0) = y_0 > 0$, where the O.D.E. initial value problem

$$y'(t) = x(t)\, y(t) \quad (0 < t \le T), \qquad y(0) = y_0,$$

is assumed to hold. For simplicity let in the sequel $T := 1$ and consider the space setting $X = Y := L^2(0,1)$. Then we simply derive the nonlinear forward operator $F : x \mapsto y$ mapping in the real Hilbert space $L^2(0,1)$ as

$$[F(x)](t) = y_0 \exp\left(\int_0^t x(\tau)d\tau\right) \qquad (0 \le t \le 1), \tag{9.30}$$

with full domain $\mathcal{D}(F) = L^2(0,1)$ and with the Fréchet derivative

$$[F'(x)h](t) = [F(x)](t) \int_0^t h(\tau)d\tau \quad (0 \le t \le 1, \quad h \in X).$$

It can be shown that there is some constant $\hat{K} > 0$ such that for all $x \in X$ the inequality

$$\|F(x) - F(x^\dagger) - F'(x^\dagger)(x - x^\dagger)\|_Y \le \hat{K}\, \|F(x) - F(x^\dagger)\|_Y\, \|x - x^\dagger\|_X \tag{9.31}$$

is valid. By applying the triangle inequality to (9.31) we obtain the estimate

$$\|F'(x^\dagger)(x - x^\dagger)\|_Y \le (\hat{K}\, \|x - x^\dagger\|_X + 1)\, \|F(x) - F(x^\dagger)\|_Y \le \check{K}(r)\, \|F(x) - F(x^\dagger)\|_Y \tag{9.32}$$

for all $x \in \mathcal{B}_r^X(x^\dagger)$, where the constant $\check{K}(r) > 0$ attains the form $\check{K}(r) := r\hat{K} + 1$ for arbitrary $r > 0$.

Using the Hilbert scale generated by the operator J from (9.25), taking into account that $\|Jh\|_Y = \|(J^*J)^{1/2}h\|_X = \|B^{-1}h\|_X = \|h\|_{-1}$ for all $h \in X$, and that there is some $0 < \underline{c} < \infty$ such that $\underline{c} \le [F(x^\dagger)](t)$ $(0 \le t \le 1)$ for the multiplier function in $F'(x^\dagger)$, there is a constant $0 < c_0 < \infty$ satisfying

$$c_0\, \|x - x^\dagger\|_{-1} = c_0\, \|J(x - x^\dagger)\|_Y \le \|F'(x^\dagger)(x - x^\dagger)\|_Y \qquad \text{for all} \quad x \in X.$$

This implies by formula (9.32) the estimate

$$\|x - x^\dagger\|_{-1} \le K(r)\, \|F(x) - F(x^\dagger)\|_Y \qquad \text{for all} \quad x \in \mathcal{B}_r^X(x^\dagger). \tag{9.33}$$

This estimate is of the form (9.17) with $a := 1$ and $K(r) := \frac{r\hat{K}+1}{c_0}$. But it is specific for this example that there exist constants $K(r) > 0$ for arbitrarily large radii $r > 0$ such that (9.33) is valid.

The case (a) of the Case distinction in Sect. 9.1 occurs due to formula (9.28) if the solution is sufficiently smooth, i.e. $x^\dagger \in H^p[0,1]$ for some $p > 1$ and, for the Hilbert scale induced by J, it fulfills the necessary boundary conditions. Case (b) will be the subject of Model problem 9.3 below. The oversmoothing case (c) of the Case distinction occurs either if the solution is insufficiently smooth, i.e. $x^\dagger \in H^p[0,1]$ for $0 < p < 1$, or in case of the Hilbert scale induced by J, one might have $x^\dagger \in H^p[0,1]$ for $p \ge 1$ but the boundary condition $x^\dagger(1) = 0$ fails. Due to formula (9.27) we then have $x^\dagger \in X_p$ for all $p < 1/2$, but $x^\dagger \notin X_p$ for all $p > 1/2$ and consequently also $x^\dagger \notin X_1$. This is, for example, the case for the constant function $x^\dagger(t)=1$ $(0 \le t \le 1)$. We complete this example with the remark that due to formula (9.29) a function $x^\dagger \in H^2[0,1]$, like the function $x^\dagger(t) = -(t-0.5)^2 + 0.25$ used in the case studies below and satisfying $x^\dagger(1) = 0$, does not belong to X_2 whenever its first derivative at $t = 0$ does not vanish.

Model problem 9.2 (*Autoconvolution*) As a second problem, we consider under the same space setting $X = Y := L^2(0,1)$ the autoconvolution operator on the unit interval defined as

$$[F(x)](s) = \int_0^s x(s-t)x(t)dt \quad (0 \le s \le 1), \tag{9.34}$$

with full domain $\mathcal{D}(F) = L^2(0,1)$. This operator and the associated nonlinear operator equation (9.1) with applications in statistics and physics have been discussed early in the literature of inverse problems (cf. [8]). Due to extensions in laser optics, the deautoconvolution problem was comprehensively revisited recently (see, e.g., [2] and [6]). Even though F from (9.34) is a non-compact operator, we have for all $x \in X$ a compact Fréchet derivative

$$[F'(x)h](s) = 2\int_0^s x(s-t)\,h(t)dt \quad (0 \le s \le 1, \quad h \in X).$$

Taking the Hilbert scale $\{X_\tau\}_{\tau\in\mathbb{R}}$ based on the operator B from (9.26) and the integral operator J from (9.25), we see for the specific solution

$$x^\dagger(t) = 1 \quad (0 \le t \le 1) \tag{9.35}$$

that

$$\|F'(x^\dagger)h\|_Y = 2\|Jh\|_Y = 2\|B^{-1}h\|_X = \|h\|_{-1} \qquad \text{for all} \quad h \in X.$$

Unfortunately no estimate of the form (9.32) is available, because such estimates with F-differences on the right-hand side are not known for the autoconvolution operator. However, as a condition characterizing the nonlinearity of F the inequality

$$\|F(x) - F(x^\dagger) - F'(x^\dagger)(x - x^\dagger)\|_Y = \|F(x - x^\dagger)\|_Y \le \|x - x^\dagger\|_X^2 \qquad \text{for all} \quad x \in X$$

is valid. Thus we have for all $x \in X$ and $x^\dagger$ from (9.35), by using the triangle inequality,

$$\|x - x^\dagger\|_{-1} \le \frac{1}{2}\|F(x) - F(x^\dagger) - F'(x^\dagger)(x - x^\dagger)\|_Y + \frac{1}{2}\|F(x) - F(x^\dagger)\|_Y$$
$$\le \frac{1}{2}\|F(x) - F(x^\dagger)\|_Y + \frac{1}{2}\|x - x^\dagger\|_X^2.$$

Using the interpolation inequality (9.7) in the form

$$\|h\|_X^2 \le \|h\|_{-1}\|h\|_1 \qquad \text{for all} \quad h \in X_1$$

we derive for $x - x^\dagger \in X_1$ the inequality

$$\|x - x^\dagger\|_{-1} \le \frac{1}{2}\|F(x) - F(x^\dagger)\|_Y + \frac{1}{2}\|x - x^\dagger\|_1\|x - x^\dagger\|_{-1}$$

and, if moreover $\|x - x^\dagger\|_1 \le \kappa < 2$, even the conditional stability estimate

$$\|x - x^\dagger\|_{-1} \le \frac{1}{2-\kappa}\|F(x) - F(x^\dagger)\|_Y \qquad \text{for all} \quad x - x^\dagger \in \mathcal{B}_\kappa^{X_1}(0). \tag{9.36}$$

The estimate (9.36) can only unfold a stabilizing effect if approximate solutions x are such that $x - x^\dagger \in \mathcal{B}_\kappa^{X_1}(0)$ for some $\kappa < 2$. For $x^\dagger$ from (9.35) with $x^\dagger(1) = 1 \neq 0$ we have $x^\dagger \notin X_1$, but regularized solutions $x = x_\alpha^\delta$ solving the extremal problem (9.3) have by definition the property $x_\alpha^\delta \in X_1$, which implies that $x_\alpha^\delta - x^\dagger \notin X_1$. This is a pitfall, because convergence assertions for x_α^δ as $\delta \to 0$ are missing in case (c) and thus the behaviour of x_α^δ remains completely unclear.

Model problem 9.3 (Situation of $x^\dagger$ meeting case (b)) It is not straight forward to construct an example for case (b) of Case distinction. We base our construction on the observation that the series $\sum_{n=2}^{\infty} \frac{1}{n(\log n)^2}$ is convergent, i.e. it characterizes a finite value, whereas the series $\sum_{n=2}^{\infty} \frac{n^\varepsilon}{n(\log n)^2}$ is divergent for all $\varepsilon > 0$, i.e. we have $\sum_{n=2}^{\infty} \frac{n^\varepsilon}{n(\log n)^2} = \infty$. In order to be able to use the model operators and the Hilbert scale introduced before, we use the following integral formulation.

Lemma 9.1 *The improper integral $\int_2^\infty \frac{1}{x^\eta \log^2(x)}\, dx$ converges for $\eta \ge 1$ and diverges for $\eta < 1$.*

Proof It is

$$\int \frac{1}{x^\eta \log^2(x)}\, dx = (1-\eta)E((1-\eta)\log x) - \frac{x^{1-\eta}}{\log x} + C,$$

$C \in \mathbb{R}$, where $E(z) := \int_z^\infty \frac{e^{-t}}{t}\, dt$. The claim follows since

$$\lim_{x\to\infty} (1-\eta)E((1-\eta)\log x) - \frac{x^{1-\eta}}{\log x} = \begin{cases} 0 & \eta \geq 1 \\ \infty & \eta < 1 \end{cases}.$$

Hence, we construct exact solutions $x^\dagger$ via their Fourier transform

$$\hat{x}^\dagger(\xi) := \begin{cases} 0 & |\xi| < 2 \\ \left(\frac{1}{|\xi|(\log|\xi|)^2(1+|\xi|^2)^p}\right)^{\frac{1}{2}} & |\xi| \geq 2 \end{cases} \tag{9.37}$$

to obtain, after an inverse Fourier transform, solutions $x^\dagger \in H^p[0,1]$, but for arbitrarily small $\epsilon > 0$ $x \notin H^{p+\epsilon}[0,1]$. Namely, for this $x^\dagger$, the H^p-norm in Fourier-domain (9.24) reads

$$\|x_s^\dagger\|_{H^p}^2 := \int_{\mathbb{R}} \frac{1}{|\xi|(\log|\xi|)^2}\, d\xi \leq \infty.$$

Since we have little control over the boundary values through this approach, we will only consider the Hilbert scale induced by (9.26) for $0 < p < \frac{1}{2}$.

9.5 Case Studies

In this section we provide numerical evidence for the behavior of regularized solutions x_α^δ with respect to the Case distinction from Sect. 9.1 and the Model problems from Sect. 9.4.

9.5.1 Numerical Studies for Model Problem 9.1

We consider the forward operator F from (9.30) in the setting $X = Y = L^2(0,1)$, $\mathcal{D}(F) = X$. As was shown, a conditional stability estimate of the form (9.17) is valid there with $a = 1$ and $\gamma = 1$ (cf. formula (9.33)). It must be emphasized that Assumption 9.3 applies even in an extended manner, which means that there are finite constants $K(r) > 0$ for arbitrarily large radii $r > 0$ such that (9.33) is valid.

In our first set of experiments, we will investigate the interplay between the value $p \in (0,1)$, α-rates of the regularization parameter and error rates of regularized solutions x_α^δ using several test cases. To this end, we consider five reference solutions as shown in Fig. 9.1. Of these examples, only RS5 fulfills the boundary condition

$x(1) = 0$, hence RS1–RS4 can only be an element of X_p for $0 < p < \frac{1}{2}$. Since for RS5 $x'(0) \neq 0$, we have in this case $x^\dagger \in H^p[0, 1]$ for $p \leq \frac{3}{2}$.

To confirm our theoretical findings of Sect. 9.3, we solve (9.3) using, after discretization via the trapezoidal rule for the integral, the MATLAB®-function `fmincon`. Typically, we use a discretization level $N = 200$. To the simulated data $y = F(x^\dagger)$ we add random noise for which we prescribe the *relative error* $\bar{\delta}$ such that $\|y - y^\delta\| = \bar{\delta}\|y\|$, i.e., we have (9.2) with $\delta = \bar{\delta}\|y\|$. To obtain the X_1 norm in the penalty, we set $\|\cdot\|_1 = \|\cdot\|_{H^1[0,1]}$ and additionally force the boundary condition $x(1) = 0$. The regularization parameter α is chosen as $\alpha_{DP} = \alpha(\delta, y^\delta)$ using the discrepancy principle, i.e.,

$$\delta \leq \|F(x^\delta_{\alpha_{DP}}) - y^\delta\|_Y \leq C\delta, \tag{9.38}$$

with some prescribed multiplier $C > 1$. Unless otherwise noted $C = 1.1$ was used. From Sect. 9.3, we know that this should yield a α-rate similar to the a-priori choice

$$\alpha(\delta) \sim \delta^{\frac{4}{p+1}},$$

cf. formula (9.23), that has already been used by Natterer in [21] for linear problems in the case of oversmoothing penalties. We should also be able to observe the order optimal convergence rate

$$\|x^\delta_{\alpha(\delta)} - x^\dagger\|_X = \mathcal{O}(\delta^{\frac{p}{p+1}}) \qquad \text{as} \qquad \delta \to 0.$$

Since $x^\dagger$ is known, we can compute the regularization errors $\|x^\delta_\alpha - x^\dagger\|_X$. We interpret this as a function of δ and make a regression for the model function

$$\|x^\delta_\alpha - x^\dagger\|_X \leq c_x \delta^{\kappa_x}, \tag{9.39}$$

and similarly we estimate the function behind the regularization parameter through the ansatz

$$\alpha \sim c_\alpha \delta^{\kappa_\alpha}. \tag{9.40}$$

Comparing (9.39) and the predicted rate (9.22), we have $\kappa_x = \frac{p}{a+p}$, hence we can estimate the smoothness of the solution as $p = \frac{a\kappa_x}{1-\kappa_x}$. Recall that $a = 1$ in this example. Results on regularized solutions $x^\delta_{\alpha_{DP}}$ with the discrepancy principle for all five reference solutions are summarized in Table 9.1. From this estimated p, we can calculate the a-priori parameter choice (9.23) and compare it to the measured one. Results on regularized solutions $x^\delta_{\alpha_{DP}}$ with the discrepancy principle for all five reference solutions are summarized in Table 9.1.

As discussed before, the reference solutions RS1, RS2, RS3 and RS4 all belong to case (c) of the Case distinction whereas RS5 belongs to case (a). Our computed results fit to this narrative. For RS1–RS4 we obtain an estimated $p < 1$ and $\kappa_\alpha > 2$, i.e., $\delta^2/\alpha \to \infty$ as $\delta \to 0$. As expected we have $0 < \kappa_\alpha < 2$ for RS5 ($p > 1$) together

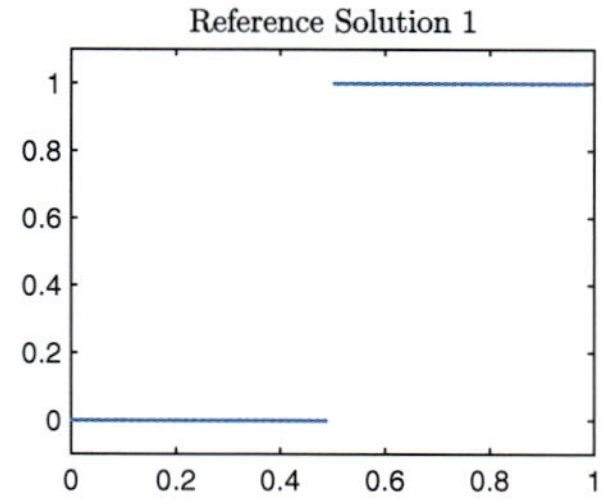

RS1: $$x(t) = \begin{cases} 0 & (0 \leq t \leq 0.5) \\ 1 & (0.5 < t \leq 1) \end{cases}$$

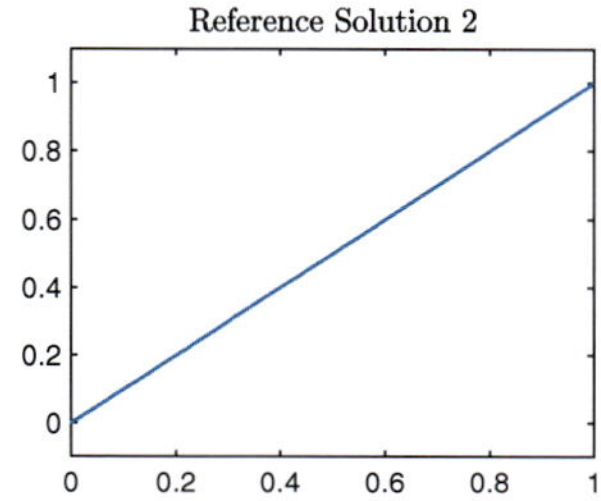

RS2: $$x^\dagger(t) = t \quad (0 \leq t \leq 1)$$

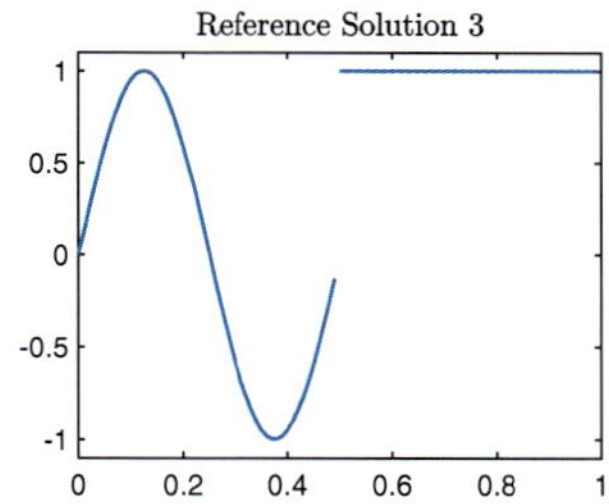

RS3: $$x^\dagger(t) = \begin{cases} \sin(4\pi t) & (0 \leq t \leq 0.5) \\ 1 & (0.5 < t \leq 1) \end{cases}$$

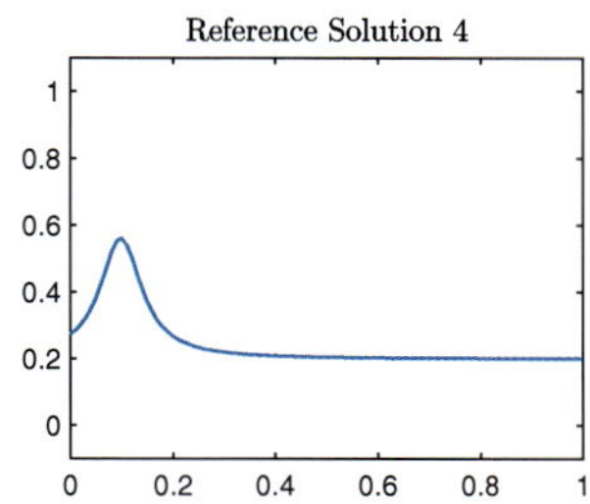

RS4: $x^\dagger(t) = 0.2 + \frac{0.36}{1+100(2.05t-0.2)^2} \; (0 \leq t \leq 1)$

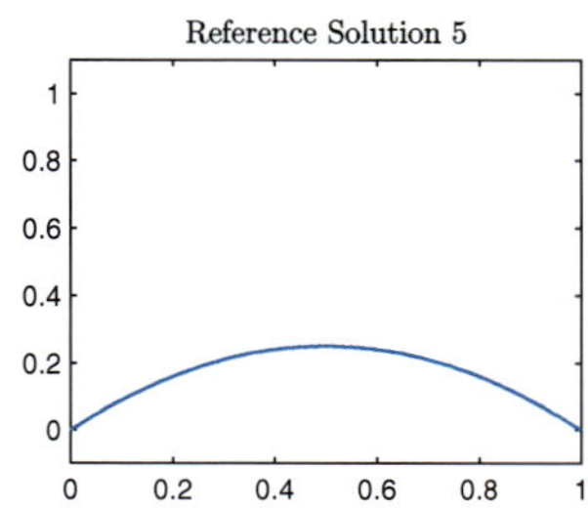

RS5: $x^\dagger(t) = -(t-0.5)^2 + 0.25 \quad (0 \leq t \leq 1)$

Fig. 9.1 Reference solutions used in the first series of experiments. Due to the failure of the respective boundary conditions, we have $x^\dagger \in X_p, 0 < p < \frac{1}{2}$ for RS1–RS4, and for RS5 we find $x^\dagger \in X_p$ with $p \leq \frac{3}{2}$

Table 9.1 Model problem 9.1: Numerically computed convergence rates (9.39) and α-rates (9.40) for the five test cases with estimated values p from the index κ_x, characterizing approximately the smoothness of the exact solution in these test cases

RS	c_x	κ_x	c_α	κ_α	est. p	$\frac{4}{p+1}$
1	0.9578	0.3276	5.8483	2.7950	0.4871	2.6898
2	0.9017	0.3426	13.5714	2.8609	0.5212	2.6290
3	1.6102	0.4110	0.2782	2.3221	0.6978	2.3560
4	0.2571	0.2582	462.1747	2.7974	0.3481	2.9671
5	0.8868	0.6135	25.8986	1.9546	1.5875	1.5459

with $\frac{\delta^2}{\alpha_{DP}} \to 0$ as $\delta \to 0$. For RS1 we know that p is bounded above by 0.5 and the estimated value 0.487 fits well. In particular in the oversmoothing cases, we have an excellent fit between the α-rates from the discrepancy principle and the a-priori choice based on our estimate of p.

As a second scenario for this model problem, we use the Sobolev scale $H^p[0,1]$, $0 < p < \frac{1}{2}$ to investigate a particular case of (c) in our Case distinction. Using the Fourier transform, we construct our solutions $x^\dagger$ such that $\hat{x}^\dagger(\xi) := (1+|\xi|)^2)^{-\frac{p}{2}-\frac{1}{4}}$ which yields solutions $x^\dagger \in H^{p-\epsilon}[0,1]$ for all $\epsilon > 0$, but $x^\dagger \notin H^p[0,1]$. This follows, because $\int_{\mathbb{R}}(1+|\xi|^2)^\nu d\xi$ converges for $\nu < -1/2$ (but diverges for $\nu \geq -1/2$) and the definition of the H^p-norm (9.24) in Fourier domain. We take $p = \frac{1}{3}$ and in principal repeat the previous experiments with the new solutions. The main difference is that we now minimize a Tikhonov functional with variable penalty smoothness,

$$T^\delta_{\alpha,s}(x) := \|F(x) - y^\delta\|^2_Y + \alpha\|x\|^2_{H^s[0,1]} \to \min, \quad \text{subject to } x \in \mathcal{D}(F). \quad (9.41)$$

From Sect. 9.2 we would expect $\delta^2/\alpha \to 0$ for $s < p = \frac{1}{3}$, $\delta^2/\alpha \approx \text{const}$ for $s = p = \frac{1}{3}$, and $\delta^2/\alpha \to \infty$ for $s > p = \frac{1}{3}$. The numerical results confirm this behavior, see Fig. 9.2 for a plot and Table 9.2 for the regression results along (9.39) and (9.40). Note that in particular the exponent in the convergence rate κ_x remains approximately constant as predicted by the theory.

Note that the "bumpy" structure in Fig. 9.2 and related plots below are due to the discrepancy principle as exemplified in Fig. 9.3 for $s = 0.9$.

9.5.2 *Numerical Studies for Model Problem 9.2*

We now turn to the autoconvolution operator F from (9.34). In this context, we consider only the specific solution $x^\dagger(t) = 1$ $(0 \leq t \leq 1)$, where $x^\dagger \notin X_1$, and the minimization problem (9.3). It must be emphasized that here, in contrast to model problem 9.1, Assumption 9.3 does not apply for any radius $r > 0$. It is therefore completely unclear which behavior the regularized solutions x^δ_α show when the noise

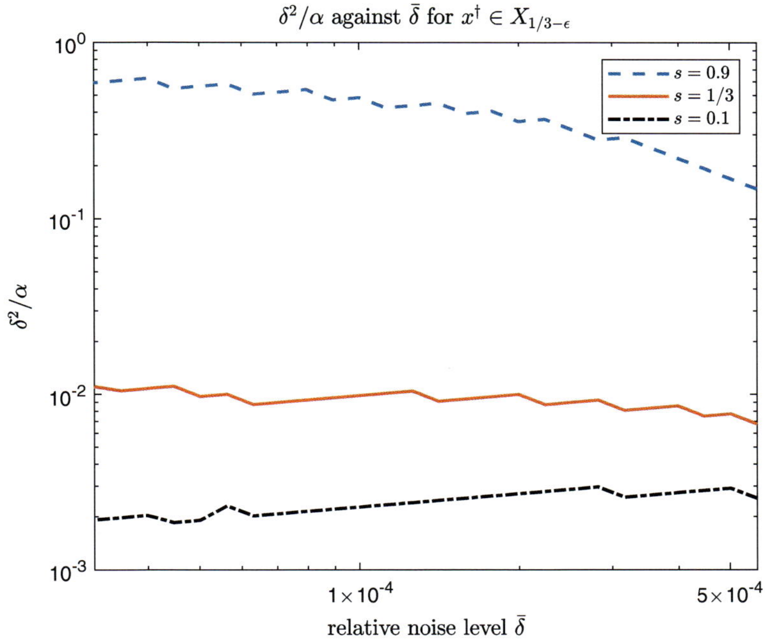

Fig. 9.2 Model problem 9.1: $\frac{\delta^2}{\alpha}$ for decreasing noise level and various s in (9.41); $x^\dagger \in X_{1/3-\epsilon}$. In subcase **a** (black, dash-dotted) we have $\frac{\delta^2}{\alpha} \to 0$ as $\delta \to 0$, in subcase **b** (red, solid) the quotient stays constant, and in subcase **c** (blue, dashed) we observe the predicted blow up $\frac{\delta^2}{\alpha} \to \infty$ as $\delta \to 0$

level δ tends to zero. It is a pitfall for exploiting Tikhonov regularization to get stable approximations for $x^\dagger$ when the regularized solutions x_α^δ from case (c), i.e., $0 < p < 1$, do not meet the validity area of the conditional stability estimate even if $\delta > 0$ is sufficiently small. Then due to $x_\alpha^\delta \notin Q$ estimates of type (9.5) are useless, convergence $x_\alpha^\delta \to x^\dagger$ as $\delta \to 0$ cannot be ensured, and the behaviour of the regularized solutions remains unclear. Such a situation occurs, as shown before, in Model problem 9.2 with $Q = \mathcal{B}_\kappa^{X_1}(0)$ and $x^\dagger \notin Q$ for $x^\dagger \equiv 1$. The following numerical case studies for that situation enlighten the properties of x_α^δ. For the test computations, again the discrepancy principle $\alpha_{DP} = \alpha(\delta, y^\delta)$ according to (9.38) has been used with $C = 1.3$ and a discretization of $N = 200$ grid points over the interval [0, 1]. In particular, we demonstrate that $\|x_{\alpha_{DP}}^\delta - x^\dagger\|_X$ does not tend to zero for $\delta \to 0$.

Figure 9.4 shows the regularization error $\|x_{\alpha_{DP}}^\delta - x^\dagger\|_X$ depending on the relative noise level $\bar{\delta}$. It can be seen, that $\|x_{\alpha_{DP}}^\delta - x^\dagger\|_X$ decreases for decreasing noise levels $\bar{\delta}$ whenever $\bar{\delta} \geq 2.9 \cdot 10^{-4}$. If $\bar{\delta}$ falls below the value $2.9 \cdot 10^{-4}$, then the monotonicity turns around and $\|x^\dagger - x_\alpha^\delta\|_X$ begins to grow. As illustrated in the overview of Fig. 9.5, the regularized solutions tend to oscillate for small $\delta > 0$, especially near the left and right boundaries of the interval [0, 1] in the sense of the Gibbs phenomenon. The Gibbs phenomenon at the right boundary $t = 1$ accompanies the required jump from one to zero between $x^\dagger \notin X_1$ and $x_\alpha^\delta \in X_1$. The oscillations blow up for small values

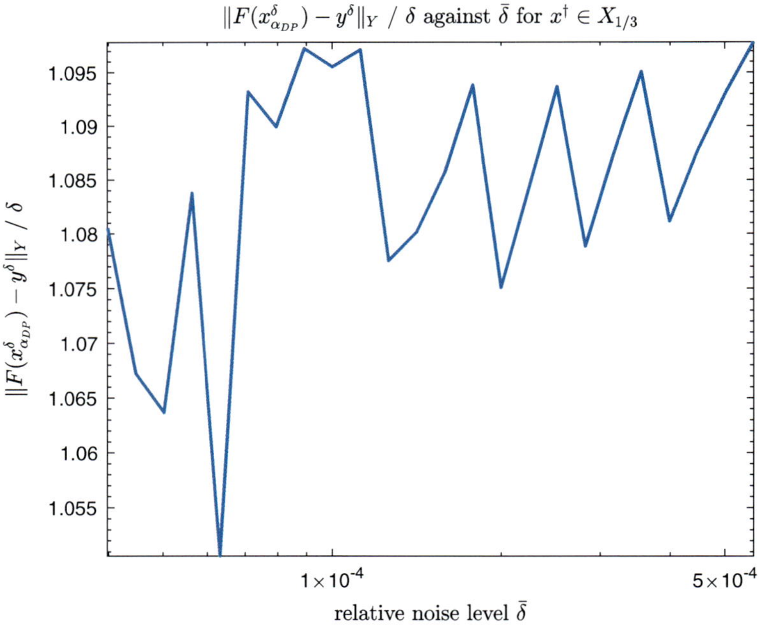

Fig. 9.3 Model problem 9.1: Discrepancy constant $C = \|F(x^\delta_{\alpha_{DP}}) - y^\delta\|_Y/\delta$ for decreasing noise level and s = 0.9 in (9.41); $x^\dagger \in X_{1/3-\epsilon}$

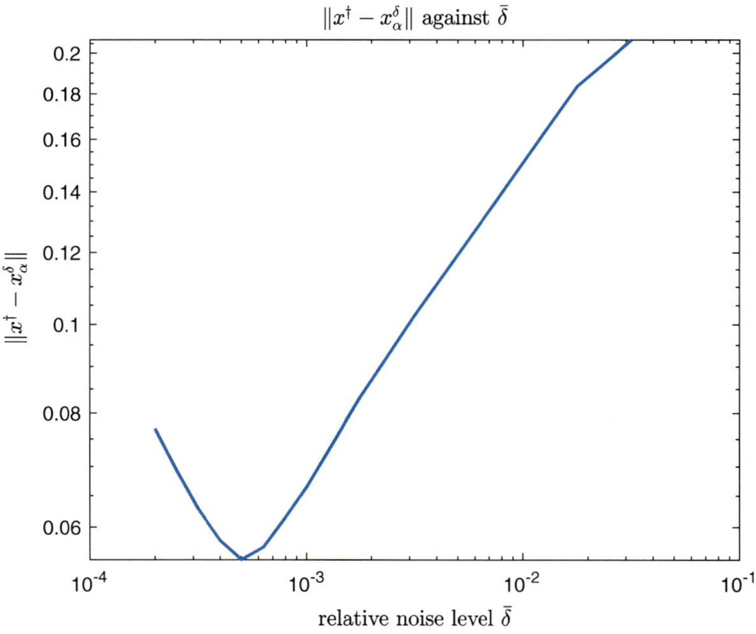

Fig. 9.4 Model problem 9.2: Plot of $\|x^\delta_{\alpha_{DP}} - x^\dagger\|_X$ against δ on a logarithmic scale for $x^\dagger \equiv 1$. For small values of $\bar{\delta}$ start to diverge, since $x^\dagger$ does not belong to the stability set $Q = \mathcal{B}^{X_1}_\kappa$

Table 9.2 Model problem 9.1: numerically computed convergence rates (9.39) and α-rates (9.40) for various s in (9.41) for given $x^\dagger \in X_{1/3-\epsilon}$

s	c_x	κ_x	c_α	κ_α
0.1	0.9460	0.2647	86.90	1.8168
0.33	1.1492	0.2828	337.42	2.1324
0.9	1.2633	0.2919	250.08	2.5319

Table 9.3 Model problem 9.3: Numerically computed convergence rates (9.39) and α-rates (9.40) in case (b) for various s in (9.41) and $x^\dagger \in X_{1/3}$

s	c_x	κ_x	c_α	κ_α
0.1	0.1275	0.2531	6.04e+02	1.6479
0.33	0.1228	0.2461	1.66e+03	1.8805
0.9	0.1229	0.2427	3.84e+04	2.5385

of δ (Fig. 9.5c–f) and indicate non-convergence of x_α^δ for $\delta \to 0$. Note that the Gibbs phenomenon starts to appear around the minimum of δ^2/α, compare to Fig. 9.4.

To confirm that this phenomenon is inherent to the oversmoothing situation, we consider again the Tikhonov functional (9.41) with H^s-penalty for $x^\dagger \equiv 1$, $s = 0.1$ and $s = 0.5$ respectively. As $x^\dagger \equiv 1 \in X_p$ for $0 < p < 1/2$ we expect similar asymptotic behavior of $\frac{\delta^2}{\alpha}$ for $\delta \to 0$ as at the end of Sect. 9.5.1. Figure 9.6 shows the result.

9.5.3 Numerical Studies for Model Problem 9.3

Based on the case destinction in Sect. 9.1 we now study the convergence rates and properties of $\frac{\delta^2}{\alpha}$ as δ decays to zero for the Model problem 9.3 in case (b) of Case distinction. Using the Sobolev-scale with norm (9.24) we define $x^\dagger \in X_p$, but $x^\dagger \notin X_{p+\epsilon}$ via (9.37). For given $x^\dagger \in X_p$ in the above sense, we then turn to the Tikhonov functional (9.41) with penalty in $H^s[0,1]$. Again we choose $p = \frac{1}{3}$ which means $x^\dagger \in X_{1/3}$ such that we can employ the theory from Model Problem 9.1. For $s > p$ we are in the classical setting and therefore expect $\frac{\delta^2}{\alpha} \to 0$ as $\delta \to 0$, for $s < p$ we are in a oversmoothing situation and expect that $\frac{\delta^2}{\alpha} \to \infty$. Letting $s = p$ yields precisely case (b) of the Case distinction, and $\frac{\delta^2}{\alpha}$ should remain approximately constant. The numerical results, see Fig. 9.7 and Table 9.3, confirm this. Note that, since $a = 1$ and $p = \frac{1}{3}$, we expect and obtain $\kappa_x = 0.25$. We also see that the $\kappa_\alpha < 2$ for $s = 0.1$, $\kappa > 2$ for $s = 0.9$, i.e. in the oversmoothing situation, and $k \approx 2$ and therefore $\frac{\delta^2}{\alpha}$ approximately constant for the situation where $x^\dagger$ and penalty term are of the same smoothness.

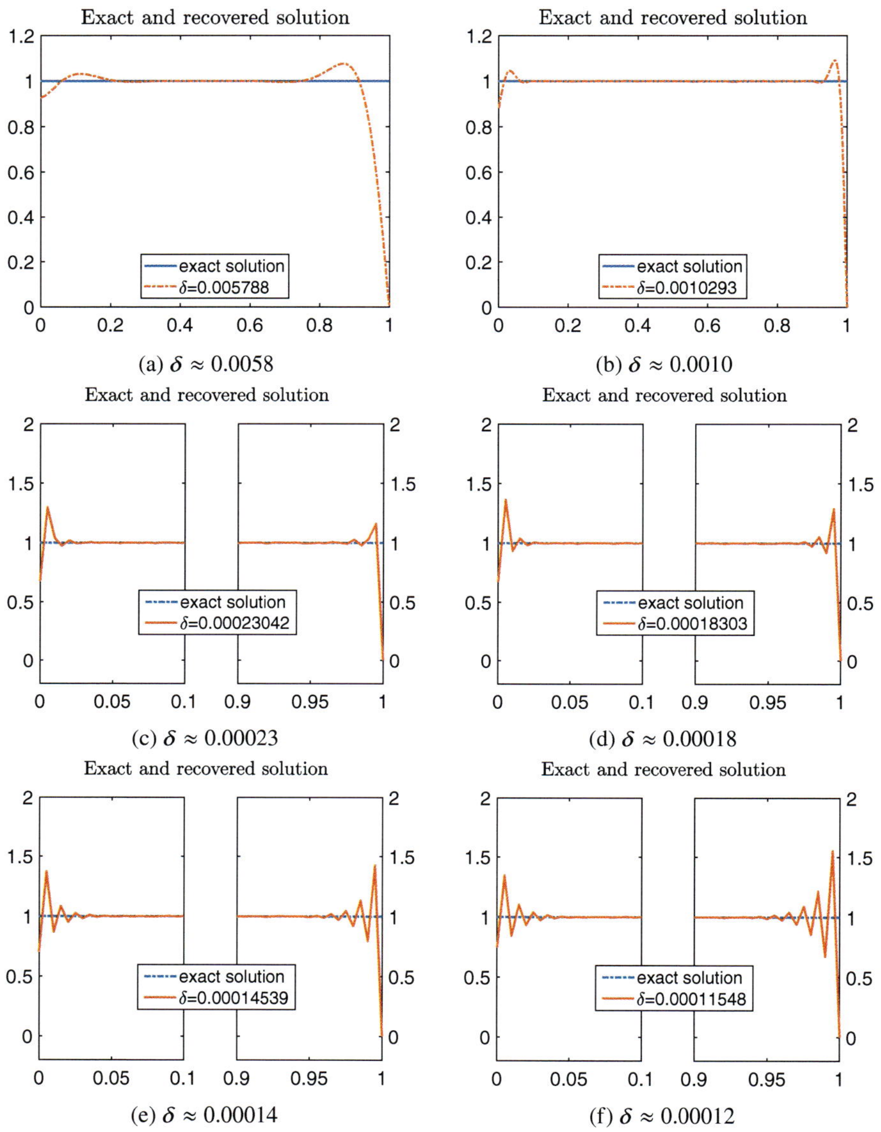

Fig. 9.5 Model problem 9.2: Regularized and exact solutions with various noise levels, $x^\dagger \equiv 1$. To improve the visibility of the blow-up at the boundaries, we omitted the middle part of the functions in the cases (**c**–**f**)

Acknowledgements We thank the colleagues Volker Michel and Robert Plato from the University of Siegen for a hint to the series that allowed us to formulate Model problem 9.3. The research was financially supported by Deutsche Forschungsgemeinschaft (DFG-grant HO 1454/12-1).

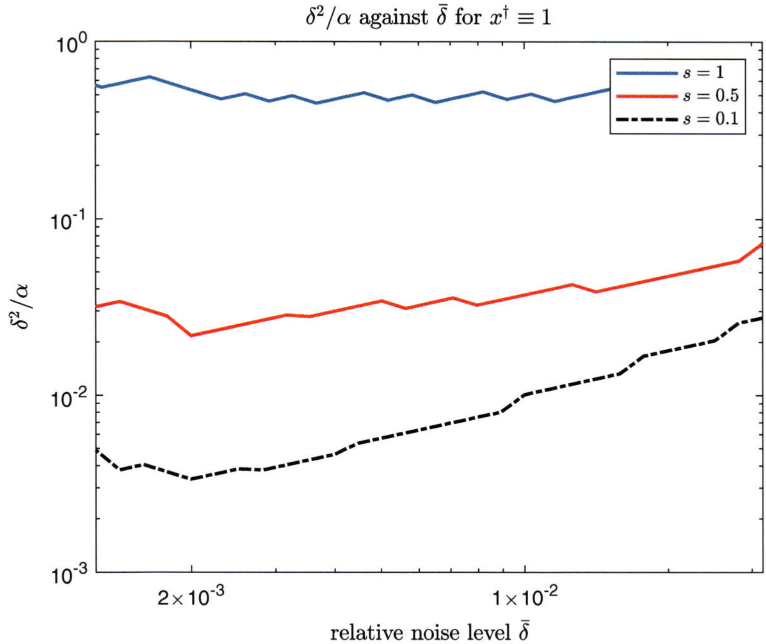

Fig. 9.6 Model problem 9.2: Realized values of $\frac{\delta^2}{\alpha}$ for decreasing noise level and various s with $x^\dagger \equiv 1$

Appendix: Proof of Proposition 9.4

In this proof we set $E := \|x^\dagger\|_p$. To prove the convergence rate result (9.20) under the a priori parameter choice (9.21) it is sufficient to show that for sufficiently small $\delta > 0$ there are two constants $K > 0$ and $\tilde{E} > 0$ such that the inequalities

$$\|x_{\alpha_*}^\delta - x^\dagger\|_{-a} \le K\delta \tag{9.42}$$

and

$$\|x_{\alpha_*}^\delta - x^\dagger\|_p \le \tilde{E} \tag{9.43}$$

hold. Namely, the convergence rate (9.20) follows directly from inequality chain

$$\|x_{\alpha_*}^\delta - x^\dagger\|_X \le \|x_{\alpha_*}^\delta - x^\dagger\|_{-a}^{\frac{p}{a+p}} \|x_{\alpha_*}^\delta - x^\dagger\|_p^{\frac{a}{a+p}} \le K^{\frac{p}{a+p}}\, \tilde{E}^{\frac{a}{a+p}}\, \delta^{\frac{p}{a+p}},$$

which is valid for sufficiently small $\delta > 0$ as a consequence of (9.42), (9.43) and of the interpolation inequality for the Hilbert scale $\{X_\tau\}_{\tau\in\mathbb{R}}$.

As an essential tool for the proof we use auxiliary elements x_α, which are for all $\alpha > 0$ the uniquely determined minimizers over all $x \in X$ of the artificial Tikhonov

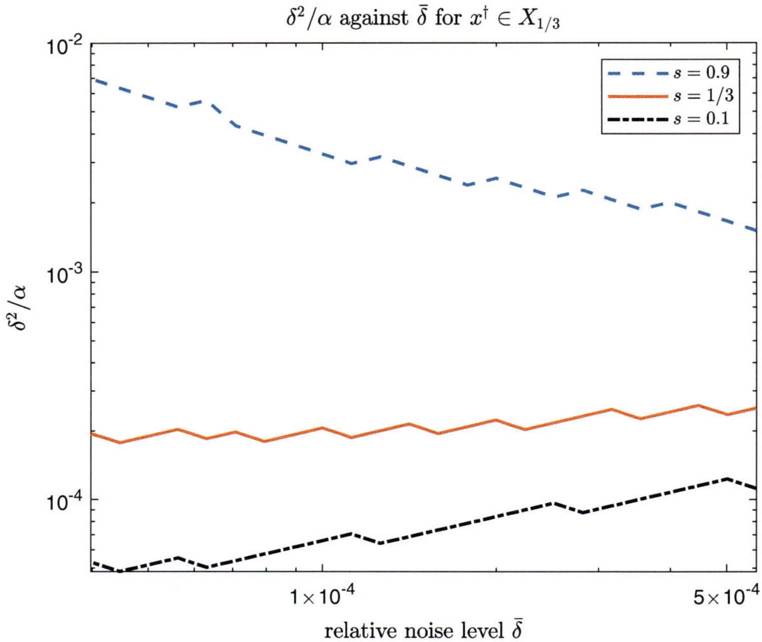

Fig. 9.7 Model problem 9.3: Realized values of $\frac{\delta^2}{\alpha}$ for decreasing noise level and various s in (9.41); $x^\dagger \in X_{1/3}$. In subcase **a** of the Case distinction (black, dash-dotted) we have $\frac{\delta^2}{\alpha} \to 0$ as $\delta \to 0$, in subcase **b** (red, solid) the quotient stays constant, and in subcase **c** (blue, dashed) we observe the predicted blow up $\frac{\delta^2}{\alpha} \to \infty$ as $\delta \to 0$

functional

$$T_{-a,\alpha}(x) := \|x - x^\dagger\|_{-a}^2 + \alpha \|Bx\|_X^2. \tag{9.44}$$

Note that the elements x_α are independent of the noise level $\delta > 0$ and belong by definition to X_1, which is in strong contrast to $x^\dagger \notin X_1$.

The following lemma is an immediate consequence of [13, Prop. 2], see also [14, Prop. 3].

Lemma 9.2 *Let $\|x^\dagger\|_p = E$ and x_α be the minimizer of the functional $T_{-a,\alpha}$ from* (9.44) *over all $x \in X$. Given $\alpha_* = \alpha_*(\delta) > 0$ as defined by formula* (9.21) *the resulting element x_{α_*} obeys the bounds*

$$\|x_{\alpha_*} - x^\dagger\|_X \le E\delta^{p/(a+p)}, \tag{9.45}$$

$$\|B^{-a}(x_{\alpha_*} - x^\dagger)\|_X \le E\delta, \tag{9.46}$$

$$\|Bx_{\alpha_*}\|_X \le E\delta^{(p-1)/(a+p)} = E\frac{\delta}{\sqrt{\alpha_*}} \tag{9.47}$$

and

$$\|x_{\alpha_*} - x^\dagger\|_p \le E.$$

Due to (9.45) we have $\|x_{\alpha_*} - x^\dagger\|_X \to 0$ as $\delta \to 0$. Hence by Assumption 9.4, in particular because $x^\dagger$ is an interior point of $\mathcal{D}(F)$, for sufficiently small $\delta > 0$ the element x_{α_*} belongs to $\mathcal{B}_r^X(x^\dagger) \subset \mathcal{D}(F)$ and moreover with $x_{\alpha_*} \in X_1$ the inequality (9.19) applies for $x = x_{\alpha_*}$ and such small δ.

Instead of the inequality (9.8), which is missing in case of oversmoothing penalties, we can use here the inequality

$$T_{\alpha_*}^\delta(x_{\alpha_*}^\delta) \le T_{\alpha_*}^\delta(x_{\alpha_*}). \tag{9.48}$$

as minimizing property for the Tikhonov functional. Using (9.48) it is enough to bound $T_{\alpha_*}^\delta(x_{\alpha_*})$ by $\overline{C}^2\delta^2$ with

$$\overline{C} := \left((\overline{K}E + 1)^2 + E^2\right)^{1/2} \tag{9.49}$$

in order to obtain the estimates

$$\|F(x_{\alpha_*}^\delta) - y^\delta\|_Y \le \overline{C}\delta \tag{9.50}$$

and

$$\|Bx_{\alpha_*}^\delta\|_X \le \overline{C}\frac{\delta}{\sqrt{\alpha_*}}. \tag{9.51}$$

Since the inequality (9.19) applies for $x = x_{\alpha_*}$ and sufficiently small $\delta > 0$, we can estimate for such δ as follows:

$$\begin{aligned}
T_{\alpha_*}^\delta(x_{\alpha_*}) &\le \left(\|F(x_{\alpha_*}) - F(x^\dagger)\|_Y + \|F(x^\dagger) - y^\delta\|_Y\right)^2 + \alpha_*\|Bx_{\alpha_*}\|_X^2 \\
&\le \left(\overline{K}\|x_{\alpha_*} - x^\dagger\|_{-a} + \delta\right)^2 + E^2\alpha_*\delta^{2(p-1)/(a+p)} \\
&\le \left(\overline{K}E\delta + \delta\right)^2 + E^2\delta^2 \\
&= \left((\overline{K}E + 1)^2 + E^2\right)\delta^2.
\end{aligned}$$

This ensures the estimates (9.50) and (9.51). Based on this we are going now to show that an inequality (9.42) is valid for some $K > 0$. Here, we use the inequality (9.18) of Assumption 9.4, which applies for $x = x_{\alpha_*}^\delta$, and we find

$$\begin{aligned}\|x_{\alpha_*}^\delta - x^\dagger\|_{-a} &\le \frac{1}{\underline{K}}\|F(x_{\alpha_*}^\delta) - F(x^\dagger)\|_Y \\ &\le \frac{1}{\underline{K}}\left(\|F(x_{\alpha_*}^\delta) - y^\delta\|_Y + \|F(x^\dagger) - y^\delta\|_Y\right) \\ &\le \frac{1}{\underline{K}}\left(\overline{C}\delta + \delta\right) = \frac{1}{\underline{K}}\left(\overline{C} + 1\right)\delta = K\delta,\end{aligned}$$

where $\overline{C}$ is the constant from (9.49) and we derive $K := \frac{1}{\underline{K}}\left(\overline{C} + 1\right)$.

Secondly, we still have to show the existence of a constant $\tilde{E} > 0$ such that the inequality (9.43) holds. By using the triangle inequality in combination with (9.51) and (9.47) we find that

$$\|B(x_{\alpha_*}^\delta - x_{\alpha_*})\|_X \le \|Bx_{\alpha_*}^\delta\|_X + \|Bx_{\alpha_*}\|_X \le (\overline{C} + E)\frac{\delta}{\sqrt{\alpha_*}}.$$

Again, we use the interpolation inequality and can estimate further as

$$\|x_{\alpha_*}^\delta - x_{\alpha_*}\|_p \le \|x_{\alpha_*}^\delta - x_{\alpha_*}\|_1^{\frac{a+p}{a+1}} \|x_{\alpha_*}^\delta - x_{\alpha_*}\|_{-a}^{\frac{1-p}{a+1}}$$

$$\le \left((\overline{C} + E)\frac{\delta}{\sqrt{\alpha_*}}\right)^{\frac{a+p}{a+1}} \left(\|x_{\alpha_*}^\delta - x^\dagger\|_{-a} + \|x^\dagger - x_{\alpha_*}\|_{-a}\right)^{\frac{1-p}{a+1}}$$

$$\le \left((\overline{C} + E)\frac{\delta}{\sqrt{\alpha_*}}\right)^{\frac{a+p}{a+1}} ((K + E)\delta)^{\frac{1-p}{a+1}}$$

$$\left((\overline{C} + E)\delta^{(p-1)/(a+p)}\right)^{\frac{a+p}{a+1}} ((K + E)\delta)^{\frac{1-p}{a+1}} =: \bar{E}.$$

Finally, we have now

$$\|x_{\alpha_*}^\delta - x^\dagger\|_p \le \|x_{\alpha_*}^\delta - x_{\alpha_*}\|_p + \|x_{\alpha_*} - x^\dagger\|_p \le \bar{E} + E =: \tilde{E}.$$

This shows (9.43) and thus completes the proof of Proposition 9.4. □

References

1. R.A. Adams, J.F.J. Fournier, *Sobolev Spaces* (Elsevier/Academic, Amsterdam, 2003)
2. S. Bürger, B. Hofmann, About a deficit in low order convergence rates on the example of autoconvolution. Appl. Anal. **94**, 477–493 (2015)
3. J. Cheng, M. Yamamoto, On new strategy for a priori choice of regularizing parameters in Tikhonov's regularization. Inverse Probl. **16**, L31–L38 (2000)
4. H. Egger, B. Hofmann, Tikhonov regularization in Hilbert scales under conditional stability assumptions. Inverse Probl. **34**, 115015 (17 pp) (2018)

5. H.W. Engl, M. Hanke, A. Neubauer, *Regularization of Inverse Problems*, Mathematics and its Applications, vol. 375 (Kluwer Academic Publishers Group, Dordrecht, 1996)
6. J. Flemming, *Variational Source Conditions, Quadratic Inverse Problems, Sparsity Promoting Regularization – New Results in Modern Theory of Inverse Problems and an Application in Laser Optics* (Birkhäuser, Basel, 2018)
7. D. Gerth, B. Hofmann, Oversmoothing regularization with ℓ^1-penalty term. AIMS Math. **4**(4), 1223–1247 (2019)
8. R. Gorenflo, B. Hofmann, On autoconvolution and regularization. Inverse Probl. **10**, 353–373 (1994)
9. R. Gorenflo, Y. Luchko, M. Yamamoto, Time-fractional diffusion equation in the fractional Sobolev spaces. Fract. Calc. Appl. Anal. **18**(3), 799–820 (2015)
10. R. Gorenflo, M. Yamamoto, Operator-theoretic treatment of linear Abel integral equations of first kind. Jpn. J. Indust. Appl. Math. **16**(1), 137–161 (1999)
11. C.W. Groetsch, *Inverse Problems in the Mathematical Sciences*, Vieweg Mathematics for Scientists and Engineers (Vieweg, Braunschweig, 1993)
12. B. Hofmann, A local stability analysis of nonlinear inverse problems, in *Inverse Problems in Engineering - Theory and Practice*,ed. by D. Delaunay et al. (The American Society of Mechanical Engineers, New York 1998), pp.313–320
13. B. Hofmann, P. Mathé, Tikhonov regularization with oversmoothing penalty for non-linear ill-posed problems in Hilbert scales. Inverse Probl. **34**, 015007 (14 pp) (2018)
14. B. Hofmann, P. Mathé, A priori parameter choice in Tikhonov regularization with oversmoothing penalty for non-linear ill-posed problems. To appear in this volume (Chapter 8). https://doi.org/10.1007/978-981-15-1592-7_8
15. B. Hofmann, R. Plato, On ill-posedness concepts, stable solvability and saturation. J. Inverse Ill-Posed Probl. **26**, 287–297 (2018)
16. B. Hofmann, O. Scherzer, Factors influencing the ill-posedness of nonlinear problems. Inverse Probl. **10**, 1277–1297 (1994)
17. B. Hofmann, M. Yamamoto, On the interplay of source conditions and variational inequalities for nonlinear ill-posed problems. Appl. Anal. **89**, 1705–1727 (2010)
18. S. Krein, Y. Petunin, Scales of Banach spaces. Russ. Math. Surv. **21**, 85–159 (1966)
19. A.K. Louis *Inverse und Schlecht Gestellte Probleme* (Teubner, Stuttgart, 1989)
20. M.T. Nair, S. Pereverzev, U. Tautenhahn, Regularization in Hilbert scales under general smoothing conditions. Inverse Probl. **21**(6), 1851–1869 (2005)
21. F. Natterer, Error bounds for Tikhonov regularization in Hilbert scales. Appl. Anal. **18**(1–2), 29–37 (1984)
22. A. Neubauer, When do Sobolev spaces form a Hilbert scale? Proc. Am. Math. Soc. **103**, 557–562 (1988)
23. R. Plato, B. Hofmann, P. Mathé, Optimal rates for Lavrentiev regularization with adjoint source conditions. Math. Comp. **87**(310), 785–801 (2018)
24. T. Schuster, B. Kaltenbacher, B. Hofmann, K.S. Kazimierski, Regularization methods in Banach spaces, in Radon Series on Computational and Applied Mathematics, vol. 10 (Walter de Gruyter, Berlin/Boston, 2012)
25. U. Tautenhahn, Error estimates for regularized solutions of nonlinear ill-posed problems. Inverse Probl. **10**, 485–500 (1994)
26. U. Tautenhahn, On a general regularization scheme for nonlinear ill-posed problems II: regularization in Hilbert scales. Inverse Probl. **14**, 1607–1616 (1998)

Chapter 10
Regularized Reconstruction of the Order in Semilinear Subdiffusion with Memory

Mykola Krasnoschok, Sergei Pereverzyev, Sergii V. Siryk and Nataliya Vasylyeva

Abstract For $\nu \in (0, 1)$, we analyze the semilinear integro-differential equation on the multidimensional space domain $\Omega \subset \mathbb{R}^n$ in the unknown $u = u(x, t)$:

$$\mathbf{D}_t^\nu u - \mathcal{L}_1 u - \int_0^t \mathcal{K}(t-s)\mathcal{L}_2 u(\cdot, s)ds = f(x, t, u) + g(x, t)$$

where $\mathbf{D}_t^\nu$ is the Caputo fractional derivative and $\mathcal{L}_1$ and $\mathcal{L}_2$ are uniform elliptic operators of the second order with time-dependent smooth coefficients. We obtain the explicit formula reconstructing the order of the fractional derivative ν for small time state measurements. The formula gives rise to a regularization algorithm for calculating ν from possibly noisy measurements. We present several numerical tests illustrating the algorithm when it is equipped with quasi-optimality criteria for choosing the regularization parameters.

Keywords Materials with memory · Subdiffusion quasilinear equations · Caputo derivative · Inverse problem · Regularization method · Quasioptimality approach

M. Krasnoschok · N. Vasylyeva (✉)
Institute of Applied Mathematics and Mechanics of NASU, G. Batyuka st. 19, 84100 Sloviansk, Ukraine
e-mail: nataliy_v@yahoo.com

M. Krasnoschok
e-mail: iamm012@ukr.net

S. Pereverzyev
Johann Radon Institute, Altenbergstrasse 69, 4040 Linz, Austria
e-mail: sergei.pereverzyev@oeaw.ac.at

S. V. Siryk
Igor Sikorsky Kyiv Polytechnic Institute, Prospect Peremohy 37, 03056 Kyiv, Ukraine
e-mail: accandar@gmail.com

J. Cheng et al. (eds.), *Inverse Problems and Related Topics*, Springer Proceedings in Mathematics & Statistics 310,
https://doi.org/10.1007/978-981-15-1592-7_10

10.1 Introduction

Fractional partial differential equations have applications in many fields, including mathematical modeling [24], electromagnetism [9], polymer science [3], viscoelasticity [28], hydrology [4, 12], geophysics [34], biophysics [8], finance [25], and prediction of extreme events like earthquake [5]. Over the last two decades the standard diffusive transport has been found to be inadequate to explain a wide-range phenomena that are observed in experiments [10, 12, 13, 29]. The feature of these anomalies in diffusion/transport processes is that the mean square displacement of the diffusing species $\langle(\Delta\mathbf{x})^2\rangle$ scales as a nonlinear power law in time, i.e. $\langle(\Delta\mathbf{x})^2\rangle \sim t^\nu$, $\nu > 0$ [24]. For a subdiffusive process, the value of ν is such that $0 < \nu < 1$, while for normal diffusion $\nu = 1$, and for a superdiffusive process, we have $\nu > 1$. However, sometime a value of the subdiffusion order is not given a priori.

In this paper, we discuss an approach to the reconstruction of a subdiffusion order ν from small time state measurements. To this end, we analyze an inverse problem of recovering the order ν.

Let Ω be a bounded domain in $\mathbb{R}^n$ with a sufficiently smooth boundary $\partial\Omega$, $\partial\Omega \in \mathbf{C}^{k+\alpha}$, $k \geq 2$. For an arbitrary given time $T > 0$ we denote

$$\Omega_T = \Omega \times (0, T) \quad \text{and} \quad \partial\Omega_T = \partial\Omega \times [0, T].$$

For $\nu \in (0, 1)$, we consider the nonlinear equation with the unknown function $u = u(x, t) : \Omega_T \to \mathbb{R}$,

$$\mathbf{D}_t^\nu u - \mathcal{L}_1 u - \mathcal{K} \star \mathcal{L}_2 u = f(x, t, u) + g(x, t), \ (x, t) \in \Omega_T, \tag{10.1}$$

supplemented with the initial condition

$$u(x, 0) = u_0(x) \text{ in } \bar{\Omega}, \tag{10.2}$$

subject either to the Dirichlet boundary condition (**DBC**)

$$u(x, t) = 0 \text{ on } \partial\Omega_T, \tag{10.3}$$

or to the condition of the third kind (**III BC**)

$$\mathcal{M}_1 u + \mathcal{K}_1 \star \mathcal{M}_2 u + \sigma u = 0 \quad \text{on } \partial\Omega_T, \tag{10.4}$$

where a positive number σ, the functions u_0, g, f, the operators $\mathcal{L}_1$, $\mathcal{L}_2$, $\mathcal{M}_1$, $\mathcal{M}_2$, and the memory kernels $\mathcal{K}$, $\mathcal{K}_1$ are assumed to be given. Here, the *star* denotes the usual time-convolution product on $(0, t)$, namely

$$(h_1 \star h_2)(t) := \int_0^t h_1(t - s)h_2(s)ds, \ t > 0,$$

while the symbol $\mathbf{D}_t^\nu$ stands the Caputo fractional derivative of order ν with respect to time t (see e.g. (2.4.1) in [18]), defined as

$$\mathbf{D}_t^\nu u(x,t) = \frac{1}{\Gamma(1-\nu)}\frac{\partial}{\partial t}\int\limits_0^t \frac{[u(x,\tau)-u(x,0)]}{(t-\tau)^\nu}d\tau,\ \ \nu \in (0,1)$$

with Γ being the Euler Gamma-function. An equivalent definition in the case of absolutely continuous functions reads

$$\mathbf{D}_t^\nu u(x,t) = \frac{1}{\Gamma(1-\nu)}\int_0^t \frac{1}{(t-s)^\nu}\frac{\partial u}{\partial s}(x,s)ds.$$

Finally, coming to the involved operators $\mathcal{L}_i$ and $\mathcal{M}_i$, we consider two different cases. In the case of nonlinear sources (NS), i.e. $f(x,t,u) \neq 0$, and a multidimensional domain Ω ($n \geq 2$), $\mathcal{L}_i$ and $\mathcal{M}_i$ read as

$$\mathcal{L}_1 := \sum_{ij=1}^n \frac{\partial}{\partial x_i}\left(A_{ij}(x,t)\frac{\partial}{\partial x_j}\right) + \sum_{i=1}^n A_i(x,t)\frac{\partial}{\partial x_i} + A_0(x,t),$$

$$\mathcal{L}_2 := \sum_{ij=1}^n \frac{\partial}{\partial x_i}\left(A_{ij}(x,t)\frac{\partial}{\partial x_j}\right) + \sum_{i=1}^n B_i(x,t)\frac{\partial}{\partial x_i} + B_0(x,t), \tag{10.5}$$

$$\mathcal{M}_1 = \mathcal{M}_2 := \sum_{ij=1}^n A_{ij}(x,t)N_i(x)\frac{\partial}{\partial x_j},$$

where $N = \{N_1(x), ..., N_n(x)\}$ is the outward normal to $\partial\Omega$. In the case of a linear source (LS), i.e. $f \equiv 0$, or in the 1-dimensional case and NS, the operators have the form

$$\mathcal{L}_1 := \sum_{ij=1}^n a_{ij}(x,t)\frac{\partial^2}{\partial x_j \partial x_i} + \sum_{i=1}^n a_i(x,t)\frac{\partial}{\partial x_i} + a_0(x,t),$$

$$\mathcal{L}_2 := \sum_{ij=1}^n b_{ij}(x,t)\frac{\partial^2}{\partial x_j \partial x_i} + \sum_{i=1}^n b_i(x,t)\frac{\partial}{\partial x_i} + b_0(x,t), \tag{10.6}$$

$$\mathcal{M}_1 = \sum_{i=1}^n c_i(x,t)\frac{\partial}{\partial x_i} + c_0(x,t),$$

$$\mathcal{M}_2 = \sum_{i=1}^n d_i(x,t)\frac{\partial}{\partial x_i} + d_0(x,t).$$

To formulate the inverse problem, we introduce the observation data $\psi(t)$ for small time $t \in [0, t^\star)$, $t^\star < \min(1, T)$,

$$u(x_0, t) = \psi(t), \tag{10.7}$$

where the point $x_0 \in \Omega$ in the **DBC** case, and $x_0 \in \bar{\Omega}$ in the **III BC** case.

For the given functions f, g ψ and the given coefficients of the operators $\mathcal{L}_i$ and $\mathcal{M}_i$, $i = 1, 2$, the inverse problem (IP) is to find the order $\nu \in (0, 1)$, such that the solution of the direct problems (10.1)–(10.6) satisfies observation data (10.7) for small time t.

A motivation for the study of Eqs. (10.1) arises from theoretical and experimental investigations of materials with memory [3, 4, 6, 12, 28]. Indeed, in the mentioned papers, the authors stated that the passage of the fluid through the porous matrix may cause a local variation of the permeability. That implies the presence of "memory" in the matrix or in the fluid. In practice [6, 12], the flow may perturb the porous formation by causing particle migration resulting in pore clogging, or chemical reacting with the medium so to enlarge the pores or diminish their size. Moreover, in biological systems, the filtering properties of membranes may become saturated. To adequately represent the memory, the derivative of fractional order which weighs the 'past' of the solution is introduced in the constitutive equations. More discussion about physical phenomena that can be modeled by (10.1)–(10.6) with $\mathcal{K} \neq 0$ can be found in [3, 4, 6, 12, 19, 28].

In the literature on the inverse problems of reconstructing the order ν in (10.1) one can find two approaches mainly associated with the case $\mathcal{K} = 0$.

The first approach is connected with finding an explicit formula for ν in terms of small or large time measurements [15, 16], such as

$$\nu = \lim_{t \to 0} \frac{t u_t(x_0, t)}{u(x_0, t) - u_0(x_0)} \quad \text{and} \quad \nu = \lim_{t \to \infty} \frac{t u_t(x_0, t)}{u(x_0, t)},$$

$$\nu = \lim_{t \to 0} \frac{\ln |u(x_0, t) - u(x_0, 0)|}{\ln t}. \tag{10.8}$$

The second technique is started in [12] and related with the minimization of a certain functional depending on the solution of the corresponding direct problem and on given measurements either on the whole time interval $[0, T]$ or for the final time $t = T$ [17, 26, 31, 35].

Both of these approaches have certain advantages and disadvantages. Indeed, the second approach [17, 22, 26, 31, 35] needs not only the measurements but also all information on the coefficients and the right-hand sides in the direct problem, while a calculation by the explicit formula requires only the knowledge of the measurements. However, the first two explicit formulas in (10.8) have been proved only for a linear autonomous equation like (10.1) with $\mathcal{K} \equiv 0$, $f, g = 0$ and in the case of multidimensional domains of the special geometry. Moreover, these formulas involve the derivative, existence of which is not guaranteed in general by models (10.1)–(10.3).

The last formula in (10.8) has been obtained for a linear equation with time independent coefficients, with the memory term $\mathcal{K} \star u_{xx}$, and only in the one-dimensional domain Ω. Furthermore, problems considered in [15, 16] do not cover the third kind boundary conditions, such as (10.4).

All this narrows the scope of the application. Another important question that is understudied in the literature is what happens if known measurements are noisy and given for a finite number of time moments.

In this paper, we aim at resolving the above-mentioned issues. Namely, introducing non-vanishing memory kernels in (10.1)–(10.7) and analyzing this model in the fractional Hölder classes, we prove the explicit reconstruction formula for order ν:

$$\nu = \lim_{t \to 0} \frac{\ln |\psi(t) - u(x_0, 0)|}{\ln t}$$

where the only requirement for $x_0 \in \bar{\Omega}$ is

$$\mathcal{L}_1 u(x_0, t)|_{t=0} + f(x_0, 0, u(x_0, 0)) + g(x_0, 0) \neq 0.$$

As we mentioned above, this formula for ν was previously obtained in [16]. However, unlike [16], we establish this formula in the case of nonlinear nonautonomous Eq. (10.1) with more general representations of memory terms. Moreover, we study here not only Dirichlet boundary conditions but also **III BC** (10.4).

Then we show how to use this formula in the case where we have only noisy observations $\psi_\delta(t_k) \approx u(x_0, t_k)$ at a finite number N of time moments $t = t_k$, $k = 1, 2, ..., N$. To overcome this difficulty, we at first propose to reconstruct $\psi(t) = \psi_{\delta,\lambda}(t)$ by means of the regularized regression from the given noisy data ψ_δ, where the regularization is performed in the finite-dimensional space

$$\text{span}\{t^{\nu_i}, \mathcal{P}_j^{(0,-\gamma)}, \quad i = 1, 2, 3, \quad j = 1, 2, ..., m(N)\}$$

according to the Tikhonov scheme with the penalty term $\lambda \| \cdot \|_{L^2_{t^{-\gamma}}(0,t_N)}$. Here λ is the regularization parameter, ν_i are our initial guesses about ν (in particular, $\nu_i = 0$), $L^2_{t^{-\gamma}}(0, t_N)$ is a weighted space L^2 with the weight $t^{-\gamma}$, $\gamma \in (0, 1)$; $\mathcal{P}_j^{(0,-\gamma)}$ are Jacobi polynomials shifted to $[0, t_N]$.

Then, according to our formula, we consider the quantities

$$\nu(\lambda, t) = \frac{\ln |\psi_{\delta,\lambda}(t) - u(x_0, 0)|}{\ln t}$$

calculated for the sequences of (regularization) parameters: $\lambda \in \{\lambda_p\}$, $t \in \{\tilde{t}_q\}$.

Finally, the regularized reconstructor

$$\nu := \nu_{reg} = \nu(\tilde{\lambda}, \tilde{t})$$

is chosen from the set of approximate values $\{\nu(\lambda_p, \tilde{t}_q)\}$ by applying two-parameter quasi-optimality criterion [11] selecting $\tilde{\lambda} \in \{\lambda_p\}$, $\tilde{t} \in \{\tilde{t}_q\}$.

The paper is organized as follows. In the next section we introduce the function spaces and state some auxiliary results, which will play a key role in our analysis. The main theoretical results of the paper along with the general assumptions on the model are stated in Sect. 10.3. In Sect. 10.4, we discuss existence, uniqueness in (10.1)–(10.7), prove the explicit formula for ν and analyze the influence of noise. Finally, Sect. 10.5 is devoted to the description of the algorithm for regularized recovering of ν. The proposed method is illustrated by numerical examples.

10.2 Functional Spaces and Preliminaries

We carry out our analysis in the framework of the fractional Hölder spaces. For any Banach space $(\mathbf{X}, \|\cdot\|_{\mathbf{X}})$, we consider the usual spaces

$$\mathbf{C}([0,T],\mathbf{X}),\quad \mathbf{C}^{\beta}[0,T]\ \text{ and }\ W^{1,p}(\Omega),\quad L^{p}(\Omega),\quad \beta\in(0,1),\ p\in(1,+\infty).$$

Let

$$\langle u\rangle^{(\beta)}_{x,\Omega_T} = \sup\left\{\frac{|u(x_1,t)-u(x_2,t)|}{|x_1-x_2|^{\beta}} : x_1\neq x_2;\ x_1,x_2\in\bar{\Omega},\ t\in[0,T]\right\},$$

$$\langle u\rangle^{(\beta)}_{t,\Omega_T} = \sup\left\{\frac{|u(x,t_1)-u(x,t_2)|}{|t_1-t_2|^{\beta}} : x\in\bar{\Omega},\ t_1\neq t_2;\ t_1,t_2\in[0,T]\right\}.$$

We say that a function $u = u(x,t)$ belongs to the fractional Hölder spaces $\mathbf{C}^{l+\alpha,\frac{l+\alpha}{2}\beta}(\bar{\Omega}_T)$, $\alpha\in(0,1)$, $l=0,1,2$, if the function u together with its derivatives are continuous and the norms below are finite

$$\|u\|_{\mathbf{C}^{l+\alpha,\frac{l+\alpha}{2}\beta}(\bar{\Omega}_T)} = \|u\|_{\mathbf{C}([0,T],\mathbf{C}^{l+\alpha}(\bar{\Omega}))} + \sum_{|j|=0}^{l}\langle D_x^j u\rangle^{(\frac{l+\alpha-|j|}{2}\beta)}_{t,\Omega_T},\quad l=0,1,$$

$$\|u\|_{\mathbf{C}^{2+\alpha,\frac{2+\alpha}{2}\beta}(\bar{\Omega}_T)} = \|u\|_{\mathbf{C}([0,T],\mathbf{C}^{2+\alpha}(\bar{\Omega}))} + \|\mathbf{D}_t^{\beta}u\|_{\mathbf{C}^{\alpha,\frac{\alpha}{2}\beta}(\bar{\Omega}_T)} + \sum_{|j|=1}^{2}\langle D_x^j u\rangle^{(\frac{2+\alpha-|j|}{2}\beta)}_{t,\Omega_T}.$$

The properties of these spaces have been discussed in Sect. 2 [21]. In a similar way, for $l = 0, 1, 2$, we introduce the space $\mathbf{C}^{l+\alpha,\frac{l+\alpha}{2}\beta}(\partial\Omega_T)$.

In the paper, we also use the spaces $L^2_w(t_1,t_2)$ of real-valued functions that a square integrable with a positive weight $w(t)$. Note that $L^2_w(t_1,t_2)$ is a Hilbert space with the inner product

$$\langle f, g\rangle_w := \int_{t_1}^{t_2} w(t) f(t) g(t) dt$$

and the corresponding norm $\|\cdot\|_{L^2_w(t_1,t_2)}$.

We conclude this section with briefly review of the properties of Jacobi polynomials [1, 30]. Note that, these polynomials play a key role in our numerical schemes presented in Sect. 10.5.

Usual Jacobi polynomials $P_n^{a,b}(t)$ are defined on $(-1, 1)$ as

$$P_n^{a,b}(t) = \sum_{m=0}^{n} p_{n,m} (t-1)^{n-m} (t+1)^m,$$

where

$$p_{n,m} = \frac{1}{2^n} \binom{n+a}{m} \binom{n+b}{n-m}.$$

After shifting them to $(0, 1)$, we obtain the Jacobi polynomials defined on $(0, 1)$ as

$$\mathcal{P}_n^{(a,b)}(t) = P_n^{a,b}(2t-1).$$

For $a, b > -1$ these polynomials are orthogonal with respect to the weight function $\bar{w}(t) := (1-t)^a t^b$,

$$\int_0^1 \mathcal{P}_n^{(a,b)}(t) \mathcal{P}_k^{(a,b)}(t) \bar{w}(t) dt = \begin{cases} \frac{\Gamma(n+a+1)\Gamma(n+b+1)}{(2n+a+b+1)n!\Gamma(n+a+b+1)}, & n = k, \\ 0, & n \neq k. \end{cases}$$

Moreover, the straightforward calculations provide the identity

$$\int_0^T t^b \mathcal{P}_n^{(0,b)}(t/T) t^{\bar{\nu}} dt = \frac{T^{\bar{\nu}+b+1}}{\bar{\nu}+b+1} {}_2F_1(-n, b+n+1; 2+\bar{\nu}+b; 1),$$

where T is positive, $\bar{\nu} \in (0, 1)$ and ${}_2F_1(\cdot)$ is the hypergeometric function [1].

10.3 The Main Result

In the sequel we rely on the following assumptions concerning model (10.1)–(10.7).

h1 (Ellipticity conditions): There are positive constants μ_1, μ_2, $\mu_1 < \mu_2$, such that

$$\mu_1|\xi|^2 \le \sum_{ij=1}^{n} A_{ij}(x,t)\xi_i\xi_j \le \mu_2|\xi|^2,$$

$$\mu_1|\xi|^2 \le \sum_{ij=1}^{n} a_{ij}(x,t)\xi_i\xi_j \le \mu_2|\xi|^2,$$

for any $(x, t, \xi) \in \bar{\Omega}_T \times \mathbb{R}^n$. There exists a positive constant μ_3 such that

$$\left|\sum_{i=1}^{n} c_i(x,t)N_i(x)\right| \ge \mu_3.$$

h2 (Smoothness of the coefficients): For $\alpha \in (0,1)$ and $i, j = 1, \ldots, n$,

$$A_{ij}(x,t),\, A_i(x,t),\, B_i(x,t) \in \mathbf{C}^{1+\alpha,\frac{1+\alpha}{2}}(\bar{\Omega}_T),$$

$$A_0(x,t),\, B_0(x,t) \in \mathbf{C}^{\alpha,\frac{\alpha}{2}}(\bar{\Omega}_T); \tag{10.9}$$

$$a_{ij}(x,t),\, b_{ij}(x,t),\, a_i(x,t),\, b_i(x,t),\, a_0(x,t),\, b_0(x,t) \in \mathbf{C}^{\alpha,\frac{\alpha}{2}}(\bar{\Omega}_T),$$

$$c_i(x,t),\, d_i(x,t),\, c_0(x,t),\, d_0(x,t) \in \mathbf{C}^{1+\alpha,\frac{1+\alpha}{2}}(\bar{\Omega}_T). \tag{10.10}$$

Moreover, in the 1-dimensional case and $f \neq 0$, we additionally assume that

$$\frac{\partial a_{11}}{\partial x},\quad \frac{\partial b_{11}}{\partial x} \in \mathbf{C}(\bar{\Omega}_T). \tag{10.11}$$

h3 (Conditions on the kernel): Depending on the considered case, we assume the following:

$$\begin{cases} \mathcal{K}(t) \in L^1(0,T) & \text{in the LS case,} \\ \mathcal{K}(t) \in \mathbf{C}^1[0,T] & \text{in the NS case.} \end{cases} \tag{10.12}$$

h4 (Conditions on the given functions): The following inclusions hold with $\alpha, \beta \in (0,1)$

$$\mathcal{K}_1(t) \in L^1(0,T)$$

$$u_0(x) \in C^{2+\alpha}(\bar{\Omega}),$$

$$g(x,t) \in \mathbf{C}^{\alpha,\frac{\alpha}{2}}(\bar{\Omega}_T),$$

$$\psi(t) \in \mathbf{C}^{\beta}[0,t^\star],\ \mathbf{D}_t^{\beta}\psi(t) \in \mathbf{C}^{\frac{\alpha\beta}{2}}[0,t^\star].$$

h5 (Compatibility conditions): In the case of **DBC** (10.3), the following compatibility conditions hold for every $x \in \partial\Omega$ at the initial time $t = 0$

$$u_0(x) = 0 \quad \text{and} \quad 0 = \mathcal{L}_1 u_0(x)\Big|_{t=0} + f(x, 0, u_0) + g(x, 0),$$

while in the case of **III BC** (10.4) it is assumed that

$$\mathcal{M}_1 u_0(x)\Big|_{t=0} + \sigma u_0 = 0.$$

h6 (Conditions on the nonlinearity): For every $\varrho > 0$ and for any $(x_i, t_i, u_i) \in \bar{\Omega}_T \times [-\varrho, \varrho]$, there exists a constant $C_\varrho > 0$ such that

$$|f(x_1, t_1, u_1) - f(x_2, t_2, u_2)| \le C_\varrho\big(|x_1 - x_2| + |t_1 - t_2| + |u_1 - u_2|\big).$$

Moreover, there is a constant $L > 0$ such that the inequality

$$|f(x, t, u)| \le L[1 + |u|]$$

holds for any $(x, t, u) \in \bar{\Omega}_T \times \mathbb{R}$.

We are ready now to state our main results.

Theorem 10.1 *Let $T > 0$ be arbitrary fixed, $f \neq 0$, and let assumptions **h1**, **h3–h6** and (10.9) hold. We assume that*

$$\mathcal{L}_1 u_0(x_0)\Bigg|_{t=0} + f(x_0, 0, u_0(x_0)) + g(x_0, 0) \neq 0. \tag{10.13}$$

Then the pair $(\nu, u_\nu(x, t))$ solves (10.1)–(10.5), (10.7), where ν is defined as

$$\nu = \lim_{t \to 0} \frac{\ln|\psi(t) - u_0(x_0)|}{\ln t}, \tag{10.14}$$

and $u_\nu(x, t)$ is a unique solution of direct problem (10.1)–(10.5) satisfying regularity $u_\nu \in \mathbf{C}^{2+\alpha, \frac{\alpha+2}{2}\nu}(\bar{\Omega}_T)$.

Theorem 10.2 *Let $T > 0$ be arbitrary fixed, and $f \equiv 0$. Then under assumptions **h1**, **h3–h5**, (10.10), and (10.13), the relations (10.1)–(10.4), (10.6), (10.7) are satisfied by the pair $(\nu, u_\nu(x, t))$, where ν is defined by (10.14) and u_ν is a unique solution of the direct problem (10.1)–(10.4), (10.6), satisfying regularity $u_\nu \in \mathbf{C}^{2+\alpha, \frac{\alpha+2}{2}\nu}(\bar{\Omega}_T)$.*

Remark 10.1 In the case $\Omega \subset \mathbb{R}$, results of Theorems 10.1 and 10.2 hold if assumptions **h1**, **h4–h6**, (10.10), (10.11) and (10.13) are fulfilled, and the kernel $\mathcal{K}$ meets the requirement

$$|\mathcal{K}| \le C t^{-\theta} \quad \text{for any} \quad t \in [0, T] \quad \text{and} \quad \theta \in (0, 1).$$

Remark 10.2 Note that in NS case the assumptions on the kernel $\mathcal{K}$ in Theorem 10.1 and Remark 10.1 can be relaxed, provided that the nonlinearity f fulfills a stronger requirement. Namely, f has to satisfy global Lipschitz continuity, and $\mathcal{K} \in L^1(0, T)$.

Remark 10.3 Actually, a slight modification of the proof allows the same results to be obtained for (10.1)–(10.7) with inhomogeneous boundary conditions:

$$u(x,t) = \psi_1(x,t) \quad \text{on} \quad \partial\Omega_T,$$

$$\mathcal{M}_1 u + \mathcal{K}_1 \star \mathcal{M}_2 u + \sigma u = \psi_2(x,t) \quad \text{on} \quad \partial\Omega_T,$$

if $\psi_1 \in \mathbf{C}^{2+\alpha,\frac{2+\alpha}{2}}(\partial\Omega_T)$, $\psi_2 \in \mathbf{C}^{1+\alpha,\frac{1+\alpha}{2}}(\partial\Omega_T)$ and the corresponding compatibility conditions hold. The details are left to the interested reader.

10.4 Proof of the Main Results

10.4.1 Direct Problems: Solvability and Estimates

We begin this sections with consideration of the direct problems (10.1)–(10.6). The following theorem summarizes the results [19–21].

Theorem 10.3 *Let $T > 0$ be arbitrary fixed and $\beta \in (0,1)$. Then under the assumptions **h1**–**h6** the direct problem (10.1)–(10.6) with $\nu = \beta$ has a unique global classical solution $u \in \mathbf{C}^{2+\alpha,\frac{2+\alpha}{2}\beta}(\bar{\Omega}_T)$ admitting the estimate*

$$\|u\|_{\mathbf{C}^{2+\alpha,\frac{2+\alpha}{2}\beta}(\bar{\Omega}_T)} \leq C\Big[1 + \|g\|_{\mathbf{C}^{\alpha,\frac{\alpha}{2}\beta}(\bar{\Omega}_T)} + \|u_0\|_{\mathbf{C}^{2+\alpha}(\bar{\Omega})}\Big].$$

Moreover, if in (10.1) $f \equiv 0$ then

$$\|u\|_{\mathbf{C}^{2+\alpha,\frac{2+\alpha}{2}\beta}(\bar{\Omega}_T)} \leq C\Big[\|g\|_{\mathbf{C}^{\alpha,\frac{\alpha}{2}\beta}(\bar{\Omega}_T)} + \|u_0\|_{\mathbf{C}^{2+\alpha}(\bar{\Omega})}\Big].$$

Here the generic positive bounded quantity C depends only on the parameters in the model (10.1)–(10.6).

Note that Theorem 10.3 is also valid if the assumptions **h2** and **h4** are substituted for their weaker versions:

h2*:

$$A_{ij}(x,t),\, A_i(x,t),\, B_i(x,t) \in \mathbf{C}^{1+\alpha,\frac{1+\alpha}{2}\beta}(\bar{\Omega}_T),$$

$$A_0(x,t),\, B_0(x,t) \in \mathbf{C}^{\alpha,\frac{\alpha}{2}\beta}(\bar{\Omega}_T);$$

$$a_{ij}(x,t),\, b_{ij}(x,t),\, a_i(x,t),\, b_i(x,t),\, a_0(x,t),\, b_0(x,t) \in \mathbf{C}^{\alpha,\frac{\alpha}{2}\beta}(\bar{\Omega}_T),$$

$$c_i(x,t),\, d_i(x,t),\, c_0(x,t),\, d_0(x,t) \in \mathbf{C}^{1+\alpha,\frac{1+\alpha}{2}\beta}(\bar{\Omega}_T).$$

h4*:

$$g(x,t) \in \mathbf{C}^{\alpha,\frac{\alpha}{2}\beta}(\bar{\Omega}_T).$$

Theorem 10.3 guarantees continuous dependence of the solution of (10.1)–(10.6) on g and u_0. Now we analyze the dependence of the solution on the order ν and start with some auxiliary estimates which play a key role in our further discussion.

Proposition 10.1 *Let $0 < \nu_1 < \nu_2 < 1$ and the function $U(t)$ have the fractional derivative of the order ν_2 satisfying regularity condition $\mathbf{D}_t^{\nu_2} U \in \mathbf{C}^{\gamma}[0,T]$ with some fixed $\gamma \in (0,1)$. Then for any fixed T and $t \in [0,T]$ the following hold*

(i)

$$|1 - \Gamma(\nu_2 - \nu_1 + 1)| \leq C(\nu_2 - \nu_1),$$

(ii)

$$|\Gamma(1 - \nu_2) - \Gamma(1 - \nu_1)| \leq C(\nu_2 - \nu_1),$$

(iii)

$$\left| \int_0^t \Big[\frac{\tau^{-\nu_1}}{\Gamma(1-\nu_1)} - \frac{\tau^{-\nu_2}}{\Gamma(1-\nu_2)} \Big] d\tau \right| \leq C(1+T)(\nu_2 - \nu_1),$$

(iv)

$$|\mathbf{D}_t^{\nu_2} U - \mathbf{D}_t^{\nu_1} U| \leq C(\nu_2 - \nu_1)[1 + t + |\ln t| + t^{\nu_2 - \nu_1 + \gamma}]$$

$$\times (\|\mathbf{D}_t^{\nu_2} U\|_{\mathbf{C}[0,T]} + \langle \mathbf{D}_t^{\nu_2} U \rangle_{t,[0,T]}^{(\gamma)}),$$

(v)

$$\|\mathbf{D}_t^{\nu_2} U - \mathbf{D}_t^{\nu_1} U\|_{L^p(0,T)} \leq C(\nu_2 - \nu_1)(\|\mathbf{D}_t^{\nu_2} U\|_{\mathbf{C}[0,T]} + \langle \mathbf{D}_t^{\nu_2} U \rangle_{t,[0,T]}^{(\gamma)})$$

with some generic positive bounded quantity C.

The proof of Proposition 10.1 is technical and left to Appendix. Theorem 10.3 and Proposition 10.1 allow us to estimate the stability of the solution of (10.1)–(10.6) with respect to the order ν.

Lemma 10.1 *Let the assumptions of Theorem 10.3 hold and $0 < \beta_1 < \beta_2 < 1$. If u_{β_1} and u_{β_2} solve (10.1)–(10.6) with $\nu = \beta_1$ and $\nu = \beta_2$, correspondingly, then*

i: *for each $\varepsilon \in (0,T)$*

$$\|u_{\beta_1} - u_{\beta_2}\|_{\mathbf{C}^{2+\alpha,\frac{2+\alpha}{2}\beta_1}(\bar{\Omega}\times[\varepsilon,T])}$$

$$\leq C(\beta_2 - \beta_1)(1 + |\ln \varepsilon|)[\|\mathbf{D}_t^{\beta_2} u_{\beta_2}\|_{\mathbf{C}(\bar{\Omega}_T)} + \langle \mathbf{D}_t^{\beta_2} u_{\beta_2} \rangle_{t,\Omega_T}^{(\frac{\alpha\beta_2}{2})}]$$

$$\leq C(\beta_2 - \beta_1)(1 + |\ln \varepsilon|)\Big[1 + \|g\|_{\mathbf{C}^{\alpha,\frac{\alpha}{2}\beta_2}(\bar{\Omega}_T)} + \|u_0\|_{\mathbf{C}^{2+\alpha}(\bar{\Omega})}\Big];$$

ii: *in the one-dimensional case, i.e.* $\Omega \subset \mathbb{R}$,

$$\|u_{\beta_1} - u_{\beta_2}\|_{\mathbf{C}(\bar{\Omega}_T)} + \|u_{\beta_1} - u_{\beta_2}\|_{\mathbf{C}([0,T],W^{1,2}(\Omega))}$$

$$\leq C(\beta_2 - \beta_1)[\|\mathbf{D}_t^{\beta_2} u_{\beta_2}\|_{\mathbf{C}(\bar{\Omega}_T)} + \langle \mathbf{D}_t^{\beta_2} u_{\beta_2}\rangle_{t,\Omega_T}^{(\frac{\alpha\beta_2}{2})}]$$

$$\leq C(\beta_2 - \beta_1)\Big[1 + \|g\|_{\mathbf{C}^{\alpha,\frac{\alpha}{2}\beta_2}(\bar{\Omega}_T)} + \|u_0\|_{\mathbf{C}^{2+\alpha}(\bar{\Omega})}\Big];$$

iii: *in the multidimensional case, i.e.* $\Omega \subset \mathbb{R}^n$, $n \geq 2$, *and for* $p \geq 2$

$$\|u_{\beta_1} - u_{\beta_2}\|_{\mathbf{C}([0,T],L^p(\Omega))} + \|u_{\beta_1} - u_{\beta_2}\|_{\mathbf{C}([0,T],W^{1,2}(\Omega))}$$

$$\leq C(\beta_2 - \beta_1)[\|u_{\beta_2}\|_{\mathbf{C}(\bar{\Omega}_T)} + \|\mathbf{D}_t^{\beta_2} u_{\beta_2}\|_{\mathbf{C}(\bar{\Omega}_T)} + \langle \mathbf{D}_t^{\beta_2} u_{\beta_2}\rangle_{t,\Omega_T}^{(\frac{\alpha\beta_2}{2})}]$$

$$\leq C(\beta_2 - \beta_1)\Big[1 + \|g\|_{\mathbf{C}^{\alpha,\frac{\alpha}{2}\beta_2}(\bar{\Omega}_T)} + \|u_0\|_{\mathbf{C}^{2+\alpha}(\bar{\Omega})}\Big].$$

Moreover, if in addition $f \equiv 0$ *and* $\mathcal{K}, \mathcal{K}_1 \in \mathbf{C}^1[0,T]$, *then for* $p > n + \frac{1}{\beta_1}$

$$\|u_{\beta_1} - u_{\beta_2}\|_{\mathbf{C}^{\alpha,\frac{\alpha}{2}\beta_1}(\bar{\Omega}_T)} + \|u_{\beta_1} - u_{\beta_2}\|_{L^p((0,T),W^{2,p}(\Omega))}$$

$$\leq C(\beta_2 - \beta_1)\|\mathbf{D}_t^{\beta_2} u_{\beta_2} - \mathbf{D}_t^{\beta_1} u_{\beta_2}\|_{L^p(\Omega_T)}$$

$$\leq C(\beta_2 - \beta_1)\Big[1 + \|g\|_{\mathbf{C}^{\alpha,\frac{\alpha}{2}\beta_2}(\bar{\Omega}_T)} + \|u_0\|_{\mathbf{C}^{2+\alpha}(\bar{\Omega})}\Big].$$

Proof We will carry out the detailed proof of Lemma 10.1 in the **DBC** case. The **III BC** case can be analyzed in a similar way.

Note that the difference

$$V(x,t) = u_{\beta_2}(x,t) - u_{\beta_1}(x,t)$$

solves the problem

$$\begin{aligned} \mathbf{D}_t^{\beta_1} V - \mathcal{L}_1 V - \mathcal{K} \star \mathcal{L}_2 V &= \bar{g}(x,t) + \bar{f}(x,t,V) \quad \text{in} \quad \Omega_T, \\ V(x,0) &= 0 \quad \text{in} \quad \Omega, \\ V(x,t) &= 0 \quad \text{on} \quad \partial\Omega_T, \end{aligned} \tag{10.15}$$

where

$$\bar{f}(x,t,V) = f(x,t,u_{\beta_2}) - f(x,t,u_{\beta_2} - V) \quad \text{and} \quad \bar{g}(x,t) = \mathbf{D}_t^{\beta_2} u_{\beta_2} - \mathbf{D}_t^{\beta_1} u_{\beta_2}.$$

Then the estimates of Theorem 10.3 and Proposition 10.1 immediately give the inequality **(i)**.

Next, we focus on getting bounds in the one-dimensional case and begin to evaluate V in $\mathbf{C}([0,T], W^{1,2}(\Omega))$. To this end, we recast here step-by-step the arguments from the proof of Lemma 5.1 [19].

For $t \in [0,T]$ we consider the functions

$$\Theta_{\beta_1}(t) = \frac{t^{\beta_1 - 1}}{\Gamma(\beta_1)},$$

$$W_1(t) = \|V(\cdot,t)\|^2_{L^2(\Omega)} + \Theta_{\beta_1} \star \|V_x(\cdot,t)\|^2_{L^2(\Omega)},$$

$$W_2(t) = \|V_x(\cdot,t)\|^2_{L^2(\Omega)} + \Theta_{\beta_1} \star \|V_{xx}(\cdot,t)\|^2_{L^2(\Omega)}.$$

Then, as in [19], for each $t \in [0,T]$ we have the bounds

$$W_1(t) \leq C[\Theta_{\beta_1} \star \|\bar{g}\|^2_{L^2(\Omega)} + [\Theta_{\beta_1} + |\mathcal{K}|] \star W_1],$$

$$W_2(t) \leq C[\Theta_{\beta_1} \star \|\bar{g}\|^2_{L^2(\Omega)} + [\Theta_{\beta_1} + |\mathcal{K}|] \star W_2].$$

After that, the easy verified embedding $(\Theta_{\beta_1} + |\mathcal{K}|) \in L^1(0,T)$ and Gronwall-type inequality (4.3) from [21] entail the estimate

$$|W_1| + |W_2| \leq C \sup_{[0,T]} (\Theta_{\beta_1} \star \|\bar{g}\|^2_{L^2(\Omega)}),$$

that can be rewritten in terms of the difference $(u_{\beta_2} - u_{\beta_1})$ as

$$\|u_{\beta_2} - u_{\beta_1}\|^2_{\mathbf{C}([0,T], W^{1,2}(\Omega))} \leq C \sup_{[0,T]} (\Theta_{\beta_1} \star \|\bar{g}\|^2_{L^2(\Omega)}). \tag{10.16}$$

Estimating the right-hand side by means of Proposition 10.1 and making use the estimates from Theorem 10.3, we enhance inequality (10.16) to

$$\begin{aligned} \|u_{\beta_2} - u_{\beta_1}\|_{\mathbf{C}([0,T], W^{1,2}(\Omega))} &\leq C(\beta_2 - \beta_1)[\|\mathbf{D}_t^{\beta_2} u_{\beta_2}\|_{\mathbf{C}(\bar{\Omega}_T)} + \langle \mathbf{D}_t^{\beta_2} u_{\beta_2} \rangle^{(\frac{\alpha\beta_2}{2})}_{t,\Omega_T}] \\ &\leq C(\beta_2 - \beta_1)\Big[1 + \|g\|_{\mathbf{C}^{\alpha, \frac{\alpha}{2}\beta_2}(\bar{\Omega}_T)} + \|u_0\|_{\mathbf{C}^{2+\alpha}(\bar{\Omega})}\Big] \end{aligned} \tag{10.17}$$

with a generic positive bounded quantity C.

Finally, Sobolev embedding theorem (see, e.g. Sect. 5.4 [2]) and inequality (10.17) allow us to arrive at the estimate **(ii)** of this proposition. This completes our considerations in the one-dimensional case.

In the multidimensional case estimate (10.17) is obtained by the same arguments. Thus, in order to verify the statement **(iii)** of this proposition, we are left to estimate the function V in the space $\mathbf{C}([0,T], L^p(\Omega))$, $p \geq 2$.

Let $\bar{\mathcal{K}}$ be the conjugate kernel for $\mathcal{K}$ (the main properties of this kernel can be found in Proposition 4.4 of [21]) and let

$$\hat{g} = \bar{g} + \mathcal{K}(0)[\Theta_{1-\beta_1} - \Theta_{1-\beta_2}] \star u_{\beta_2} + \bar{\mathcal{K}}' \star [\Theta_{1-\beta_1} - \Theta_{1-\beta_2}] \star u_{\beta_2}.$$

Recasting the proof of Lemma 5.2 [21] in the case of problem (10.15) leads to the inequality

$$\int\limits_{\Omega} |V|^p dx \leq Cp\Theta_{\beta_1} \star \|\hat{g}\|^p_{L^p(\Omega)} + C(p-1)\Theta_{\beta_1} \star \|V\|^p_{L^p(\Omega)}$$

with $p \geq 2$. Then, the straightforward calculations with aid of Proposition 10.1 provide the estimate

$$\int\limits_{\Omega} |V|^p dx \leq C(\beta_2 - \beta_1)[\|\mathbf{D}_t^{\beta_2} u_{\beta_2}\|_{\mathbf{C}(\bar{\Omega}_T)} + \langle \mathbf{D}_t^{\beta_2} u_{\beta_2}\rangle^{(\frac{\alpha\beta_2}{2})}_{t,\Omega_T} + \|u_{\beta_2}\|_{\mathbf{C}(\bar{\Omega}_T)}]^p$$
$$+C(p-1)\Theta_{\beta_1} \star \|V\|^p_{L^p(\Omega)}.$$

Applying again Gronwall-type inequality and Theorem 10.3, we conclude

$$\|u_{\beta_2} - u_{\beta_1}\|_{\mathbf{C}([0,T],L^p(\Omega))} = \|V\|_{\mathbf{C}([0,T],L^p(\Omega))}$$
$$\leq C(\beta_2 - \beta_1)[\|\mathbf{D}_t^{\beta_2} u_{\beta_2}\|_{\mathbf{C}(\bar{\Omega}_T)} + \langle \mathbf{D}_t^{\beta_2} u_{\beta_2}\rangle^{(\frac{\alpha\beta_2}{2})}_{t,\Omega_T} + \|u_{\beta_2}\|_{\mathbf{C}(\bar{\Omega}_T)}]$$
$$\leq C(\beta_2 - \beta_1)\Big[1 + \|g\|_{\mathbf{C}^{\alpha,\frac{\alpha}{2}\beta_2}(\bar{\Omega}_T)} + \|u_0\|_{\mathbf{C}^{2+\alpha}(\bar{\Omega})}\Big].$$

Thus, to complete the proof, we are left to verify estimates in the linear case of (10.1). Recasting step-by-step the arguments from the proof of Proposition 5.4 [21] and applying the inequality above, we deduce

$$\|V\|_{\mathbf{C}^{\alpha,\frac{\alpha\beta_1}{2}}(\bar{\Omega}_T)} + \|V\|_{L^p((0,T),W^{2,p}(\Omega))}$$
$$\leq C[\|\mathbf{D}_t^{\beta_2} u_{\beta_2} - \mathbf{D}_t^{\beta_1} u_{\beta_2}\|_{L^p(\Omega_T)} + \|V\|_{\mathbf{C}([0,T],L^p(\Omega))}]$$
$$\leq C(\beta_2 - \beta_1)\Big[1 + \|g\|_{\mathbf{C}^{\alpha,\frac{\alpha}{2}\beta_2}(\bar{\Omega}_T)} + \|u_0\|_{\mathbf{C}^{2+\alpha}(\bar{\Omega})}\Big]$$
$$+C\|\mathbf{D}_t^{\beta_2} u_{\beta_2} - \mathbf{D}_t^{\beta_1} u_{\beta_2}\|_{L^p(\Omega_T)}$$

with $p > \beta_1^{-1} + n$. The proof is finished by applying Proposition 10.1, statement **(iii)** of this Lemma and then Theorem 10.3 to estimate the last term in the right-hand side of the inequality above. □

As we see, Lemma 10.1 can provide the continuous dependence of the solution u on the order $\nu = \beta$ in the case of the Sobolev spaces, but this result does not hold in the case of $\mathbf{C}^{2+\alpha, \frac{2+\alpha}{2}\beta}$. Most likely, in virtue of statement **(i)**, the stability could be expected in weighted fractional Hölder spaces. Analysis of this issue is left to a further work.

10.4.2 Solvability of IP, Explicit Formula to ν

The following result is a direct consequence of Theorem 10.3 and condition (10.7).

Proposition 10.2 *Let $\nu \in (0, 1)$, and let the assumptions of Theorems 10.1 and 10.2 hold. If the pair (ν, u_ν) satisfies (10.1)–(10.7), then*

$$\psi(t) = u_\nu(x_0, t) \in \mathbf{C}^\nu[0, t^\star],$$

$$\mathbf{D}_t^\nu \psi(t) = \mathbf{D}_t^\nu u_\nu(x_0, t) \in \mathbf{C}^{\frac{\alpha\nu}{2}}[0, t^\star],$$

$$\mathbf{D}_t^\nu \psi(0) = \mathcal{L}_1 u_0(x_0)|_{t=0} + f(x_0, 0, u_0(x_0)) + g(x_0, 0) \neq 0.$$

Now we begin to obtain the explicit formula for the order of the fractional derivative.

Lemma 10.2 *Let T be arbitrarily fixed and $\beta \in (0, 1)$. Let a function $G(t)$ and its fractional derivative $\mathbf{D}_t^\beta G(t)$ be continuous on $[0, T]$. Then for each $t \in [0, T]$ the function $G(t)$ allows the following representation*

$$G(t) = G(0) + t^\beta \frac{\mathbf{D}_t^\beta G(0)}{\Gamma(1+\beta)} + \frac{1}{\Gamma(\beta)} \int_0^t (t-\tau)^{\beta-1} [\mathbf{D}_\tau^\beta G(\tau) - \mathbf{D}_t^\beta G(0)] d\tau. \tag{10.18}$$

Assume in addition, that $\mathbf{D}_t^\beta G(0) \neq 0$ and

$$\omega(t) := \sup_{\tau \in [0,t]} |\mathbf{D}_\tau^\beta G(\tau) - \mathbf{D}_t^\beta G(0)| \to 0 \quad \text{as} \quad t \to 0,$$

then

$$\beta = \lim_{t \to 0} \frac{\ln |G(t) - G(0)|}{\ln t}. \tag{10.19}$$

Proof The direct calculations based on the smoothness of the function G give (10.18). Further, equality (10.18) provides the relation

$$\ln|G(t)-G(0)| = \beta \ln t + \ln\left|\frac{\mathbf{D}_t^\beta G(0)}{\Gamma(1+\beta)} + \frac{t^{-\beta}}{\Gamma(\beta)}\int_0^t (t-\tau)^{\beta-1}[\mathbf{D}_t^\beta G(\tau) - \mathbf{D}_t^\beta G(0)]d\tau\right|,$$

where

$$\left|\frac{t^{-\beta}}{\Gamma(\beta)}\int_0^t (t-\tau)^{\beta-1}[\mathbf{D}_t^\beta G(\tau) - \mathbf{D}_t^\beta G(0)]d\tau\right| \leq \frac{\omega(t)}{\Gamma(1+\beta)} \to 0 \quad \text{as} \quad t \to 0.$$

Thus, we achieve the representation

$$\frac{\ln|G(t)-G(0)|}{\ln t} = \beta + \frac{\ln\left|\frac{\mathbf{D}_t^\beta G(0)}{\Gamma(1+\beta)} + \frac{t^{-\beta}}{\Gamma(\beta)}\int_0^t (t-\tau)^{\beta-1}[\mathbf{D}_t^\beta G(\tau) - \mathbf{D}_t^\beta G(0)]d\tau\right|}{\ln t}$$

$$\to \beta \quad \text{as} \quad t \to 0,$$

that completes the proof of the lemma. □

Note that the right-hand side of (10.19) exists and is bounded under weaker conditions on the function $G(t)$.

Remark 10.4 It is apparent that if the function $G \in \mathbf{C}^\beta[0, T]$, $\beta \in (0, 1)$, then

$$0 < \lim_{t\to 0} \frac{\ln|G(t)-G(0)|}{\ln t} \leq \beta.$$

Returning to the observation $\psi(t)$ and taking into account condition **h4**, one can easily obtain

$$\sup_{\tau\in[0,t]} |\mathbf{D}_t^\beta \psi(\tau) - \mathbf{D}_t^\beta \psi(0)| \leq t^{\frac{\alpha\beta}{2}} \langle \mathbf{D}_t^\beta \psi \rangle_{t,[0,T]}^{(\frac{\alpha\beta}{2})} \to 0 \quad \text{as} \quad t \to 0. \tag{10.20}$$

Therefore, relation (10.20) together with Lemma 10.2, Theorem 10.3 and Proposition 10.2 provide the existence of the solution (ν, u_ν) of (10.1)–(10.7), where

$$\nu = \beta = \lim_{t\to 0} \frac{\ln|\psi(t) - u_0(x_0)|}{\ln t}.$$

10.4.3 Uniqueness in IP

Here we discuss the uniqueness of the pair (ν, u_ν) satisfying problem (10.1)–(10.7).

Theorem 10.4 *Under the assumptions **h1–h6** and restriction (10.13), the order* ν *can be identified uniquely by the observation data* $\psi(t)$ *using formula (10.14), and there is the unique pair* (ν, u_ν) *satisfying problem (10.1)–(10.7).*

Proof Assume contrary that there are two quantities ν_1 and ν_2 defined with (10.14) and corresponding to the same observation data (10.7) and the same right-hand sides and the coefficients in model (10.1)–(10.6). We denote by u_{ν_1} and u_{ν_2} the corresponding solutions of direct problem (10.1)–(10.6). To be specific, we assume that $0 < \nu_1 < \nu_2 < 1$. Proposition 10.2 and condition (10.13) provide

$$\psi(t) \in \mathbf{C}^{\nu_2}[0, t^*], \quad \mathbf{D}_t^{\nu_i}\psi(t) \in \mathbf{C}^{\alpha\nu_i/2}[0, t^*],$$

$$\psi(0) = u_0(x_0) \quad \text{and} \quad \mathbf{D}_t^{\nu_i}\psi(0) \neq 0. \tag{10.21}$$

Then Lemma 10.2 allows the representations

$$\psi(t) = u_0(x_0) + \frac{t^{\nu_i}}{\Gamma(1+\nu_i)}\mathbf{D}_t^{\nu_i}\psi(0) + \frac{1}{\Gamma(\nu_i)}\int_0^t \frac{[\mathbf{D}_\tau^{\nu_i}\psi(\tau) - \mathbf{D}_\tau^{\nu_i}\psi(0)]}{(t-\tau)^{1-\nu_i}}d\tau, \quad i = 1, 2,$$

telling us that for $t \in [0, t^\star]$ the equality holds

$$\frac{\mathbf{D}_t^{\nu_1}\psi(0)}{\Gamma(1+\nu_1)} + \frac{t^{-\nu_1}}{\Gamma(\nu_1)}\int_0^t \frac{[\mathbf{D}_\tau^{\nu_1}\psi(\tau) - \mathbf{D}_\tau^{\nu_1}\psi(0)]}{(t-\tau)^{1-\nu_1}}d\tau$$

$$= \frac{t^{\nu_2-\nu_1}}{\Gamma(1+\nu_2)}\mathbf{D}_t^{\nu_2}\psi(0) + \frac{t^{-\nu_1}}{\Gamma(\nu_2)}\int_0^t \frac{[\mathbf{D}_\tau^{\nu_2}\psi(\tau) - \mathbf{D}_\tau^{\nu_2}\psi(0)]}{(t-\tau)^{1-\nu_2}}d\tau. \tag{10.22}$$

It is apparent that for sufficiently small t we have

$$\frac{t^{-\nu_1}}{\Gamma(\nu_1)}\int_0^t \frac{[\mathbf{D}_\tau^{\nu_1}\psi(\tau) - \mathbf{D}_\tau^{\nu_1}\psi(0)]}{(t-\tau)^{1-\nu_1}}d\tau = O(t^{\alpha\nu_1/2}),$$

$$\frac{t^{-\nu_1}}{\Gamma(\nu_2)}\int_0^t \frac{[\mathbf{D}_\tau^{\nu_2}\psi(\tau) - \mathbf{D}_\tau^{\nu_2}\psi(0)]}{(t-\tau)^{1-\nu_2}}d\tau = O(t^{\nu_2-\nu_1+\alpha\nu_2/2}).$$

Then, passing to the limit in (10.22) as $t \to 0$, we obtain the equality

$$\frac{\mathbf{D}_t^{\nu_1}\psi(0)}{\Gamma(1+\nu_1)} = 0,$$

that contradicts conclusion (10.21) above.

This contradiction is resolved as soon as we admit $\nu_1 = \nu_2$. After that, Theorem 10.3 ensures the equality $u_{\nu_1} = u_{\nu_2}$, that completes the proof. □

10.4.4 Error Estimate for Noisy Data

Here we analyze formula (10.14) in the case, where $\psi(t)$ is replaced by the noisy data $\psi_\delta(t)$.

We assume that

$$|\psi(t) - \psi_\delta(t)| \leq C_1 \delta \Phi(t), \quad 0 \leq t \leq t^\star, \tag{10.23}$$

where $\Phi(t)$ is some nonnegative function, δ denotes the noise level and C_1 is some constant.

In the paper, we focus on the analysis of two types of the noise. The first-type noise (**FTN**) model corresponds to the case when

$$\Phi(0) = 0 \quad \text{and} \quad \Phi(t)t^{-\gamma} < C^\star \quad \text{for all} \quad \gamma \in (0, 1), \tag{10.24}$$

and the positive constant $C^\star$, while the second type noise (**STN**) model is characterized by

$$\Phi(0) = 0 \quad \text{and} \quad \Phi(t)t^{-\gamma} < |\ln t| \quad \text{for some fixed} \quad \gamma \in [\nu, 1). \tag{10.25}$$

Remark 10.5 It is easy to see that the function

$$\Phi(t) = \begin{cases} e^{-1/t}|\ln t| & \text{in } \mathbf{FTN} \text{ case,} \\ t^\gamma |\ln t| & \text{in } \mathbf{STN} \text{ case,} \end{cases}$$

satisfies conditions (10.24) and (10.25).

In the sequel we will deal with

$$\nu_\delta = \lim_{t \to 0} \frac{\ln|\psi_\delta(t) - u_0(x_0)|}{\ln t}.$$

Proposition 10.3 *Let the assumptions of Theorems 10.1 and 10.2 hold. If inequalities (10.23)–(10.25) are satisfied with* C_1 *and* $\delta \in (0, 1)$ *such that*

$$\frac{C_1}{||\mathcal{L}_1 u_0(x_0)|_{t=0} + f(x_0, 0, u_0(x_0)) + g(x_0, 0)| - C_1\delta|} < 1,$$

then

$$|\nu - \nu_\delta| = 0 \quad \text{in } \textbf{\textit{FTN}} \text{ case,}$$

and

$$|\nu - \nu_\delta| \le \frac{C_1\delta}{||\mathcal{L}_1 u_0(x_0)|_{t=0} + f(x_0, 0, u_0(x_0)) + g(x_0, 0)| - C_1\delta|} \quad \text{in } \textbf{\textit{STN}} \text{ case.}$$

Proof In view of condition (10.13), we can assume for simplicity that

$$C_0 := \mathcal{L}_1 u_0(x_0)|_{t=0} + f(x_0, 0, u_0(x_0)) + g(x_0, 0) > 0.$$

Consider the quantity

$$\Delta := \left| \frac{\psi(t) - \psi_\delta(t)}{\psi_\delta(t) - u_0(x_0)} \right|,$$

which due to (10.23) allows the bound

$$\Delta \le \frac{C_1\delta\Phi(t)}{||\psi(t) - u_0(x_0)| - |\psi_\delta(t) - \psi(t)||} \le \frac{C_1\delta\Phi(t)}{||\psi(t) - u_0(x_0)| - C_1\delta\Phi(t)|}. \tag{10.26}$$

Let us denote

$$C_2 = \langle \mathbf{D}_t^\nu \psi \rangle_{t,[0,t^*]}^{(\alpha\nu/2)}$$

and use representation (10.18) with $G(t) = \psi(t)$. Then

$$|\psi(t) - u_0(x_0)| \ge \left| \frac{t^\nu \mathbf{D}_t^\nu \psi(0)}{\Gamma(1+\nu)} - \frac{C_2 t^{\nu+\alpha\nu/2}}{\Gamma(1+\nu)} \right|$$

$$= \left| \frac{t^\nu C_0}{\Gamma(1+\nu)} - \frac{C_2 t^{\nu+\alpha\nu/2}}{\Gamma(1+\nu)} \right|.$$

This estimate together with (10.23) and (10.26) give

$$\Delta \le \frac{C_1\delta t^{-\nu}\Phi(t)}{||C_0 - C_2 t^{\alpha\nu/2}| - C_1\delta t^{-\nu}\Phi(t)|}. \tag{10.27}$$

The conditions on C_1 and δ ensure that

$$\lim_{t\to 0} \frac{C_1\delta}{||C_0 - C_2 t^{\alpha\nu/2}| - C_1\delta|} = \frac{C_1\delta}{|C_0 - C_1\delta|}.$$

We are now in the position to evaluate $|\nu - \nu_\delta|$. In the case of **STN**, keeping in mind estimate (10.27), we conclude

$$
\begin{aligned}
|\nu - \nu_\delta| &= \left| \lim_{t\to 0} \frac{\ln|1+\Delta|}{\ln t} \right| \\
&\leq \lim_{t\to 0} \frac{C_1\delta\Phi(t)t^{-\nu}}{|\ln t|||C_0 - C_2 t^{\alpha\nu/2}| - C_1\delta|} \\
&\leq \frac{C_1\delta}{|C_0 - C_1\delta|}.
\end{aligned}
$$

In the **FTN** case, we have for some positive $\gamma \in (\nu, 1)$

$$
\begin{aligned}
|\nu - \nu_\delta| &\leq \lim_{t\to 0} \frac{C_1\delta\Phi(t)t^{-\nu}}{|\ln t|||C_0 - C_2 t^{\alpha\nu/2}| - C_1\delta\Phi(t)t^{-\nu}|} \\
&\leq \lim_{t\to 0} \frac{C_1\delta C^\star t^{\gamma-\nu}}{|\ln t||C_0 - C_1\delta C^\star t^{\gamma-\nu}|} \\
&= 0.
\end{aligned}
$$

10.5 Regularized Formulation

In this section, we discuss a regularization of the reconstruction formula (10.14). Recall that this formula has been proven assuming enough smoothness of the observation data, while in practice only noisy discrete measurements are usually available. Therefore, the obtained formula should be regularized to deal with such data. Here we propose a way of combining formula (10.14) with a regularization procedure.

It is worth mentioning that the availability of only discrete observations is the more problematic issue than the presence of the noise in the continuous data $\psi_\delta(t)$, because in view of Proposition 10.3, such continuous noisy data in principle allow us to estimate the order ν rather accurately.

10.5.1 Algorithm of Reconstruction

Assume that we are able to observe the solution $u(x,t)$ of (10.1)–(10.6) at some location $x = x_0 \in \bar{\Omega}$ and at time moments t_k, $k = 1, 2, ..., N$, $0 < t_1 < t_2 < ... < t_N \leq t^\star$, but these observations are blurred by an additive noise so, that what we actually observe is

$$
\psi_{\delta,k} = u(x_0, t_k) + \delta_k, \quad k = 1, 2, ..., N.
$$

Moreover, initial condition (10.2) allows us to know the value $\psi_0 = u(x_0, 0) = u_0(x_0)$.

To be able to use such discrete noisy information in formula (10.14), a reconstruction algorithm should at first approximately recover the function $\psi(t) = u(x_0, t)$ from the values $\psi_{\delta,k}$, $k = 0, 1, ..., N$, where with a bit abuse of symbols we define

$$\psi_{\delta,0} = \psi_0 = u_0(x_0).$$

At this point, it is important to note that in view of Proposition 10.2 and Lemma 10.2 the following asymptotic holds for $t \le t^\star$:

$$\psi(t) = u_0(x_0) + O(t^\nu).$$

This tells us that our target function should be square integrable on $(0, t_N)$, $t_N \le t^\star$, with an unbounded weight $w(t) = t^{-\gamma}$, $\gamma \in (0, 1)$. Therefore, it is natural to approximate the function $\psi(t)$ by elements of the space $L^2_{t^{-\gamma}}(0, t_N)$.

Then according to the methodology of the Tikhonov regularization, the above mentioned approximate recovery of $\psi(t)$ from noisy values $\{\psi_{\delta,k}\}_{k=0}^N$ can be performed by minimizing a penalized least squares functional

$$\sum_{k=0}^{N} \left(\psi(t_k) - \psi_{\delta,k}\right)^2 + \lambda \|\psi\|^2_{L^2_{t^{-\gamma}}(0,t_N)} \to \min, \tag{10.28}$$

where λ is a regularization parameter.

Since the Jacobi polynomials $\mathcal{P}_m^{(0,-\gamma)}(t/t_N)$, $t \in (0, t_N)$, constitute an orthogonal system in $L^2_{t^{-\gamma}}(0, t_N)$, it is natural to look for the minimizer of (10.28) in the form of a linear combination of $\mathcal{P}_m^{(0,-\gamma)}(t/t_N)$. Moreover, having a series of initial guesses $\nu_1, \nu_2, ..., \nu_{\mathcal{J}}$ for the value of ν, one may incorporate the functions t^{ν_j}, $j = 1, 2, ..., \mathcal{J}$, into the basis in which minimization problem (10.28) should be solved.

Then an approximate minimizer of (10.28) can be written as

$$\psi_{\delta,\lambda}(t) = \sum_{j=1}^{\mathcal{J}} c_j t^{\nu_j} + \sum_{j=\mathcal{J}+1}^{P} c_j \mathcal{P}_{j-\mathcal{J}-1}^{(0,-\gamma)}(t/t_N), \tag{10.29}$$

where the coefficients c_j can be found from the corresponding system of linear algebraic equations written in the matrix form as follows:

$$(A^T A + \lambda \mathcal{H})\mathbf{c} = A^T \bar{\psi}_\delta,$$

where
$\mathbf{c} = (c_1, c_2, ..., c_P)^T$, $\bar{\psi}_\delta = (\psi_{\delta,0}, \psi_{\delta,1}, ..., \psi_{\delta,N})^T$, $A = \{A_{i,j}\}_{i=0,j=1}^{N\ \ P}$, $A_{ij} = e_j(t_i)$,
$\mathcal{H} = \{\mathcal{H}\}_{l,m=1}^P$, $\mathcal{H}_{l,m} = \int_0^{t_N} t^{-\gamma} e_l(t) e_m(t) dt$, and

$$e_l(t) = \begin{cases} t^{\nu_l}, & l = 1, 2, ..., \mathcal{J}, \\ \mathcal{P}^{(0,-\gamma)}_{l-\mathcal{J}-1}(t/t_N), & l = 1 + \mathcal{J}, ..., P. \end{cases}$$

Note that even after the approximate recovery of $\psi(t)$ in the form of $\psi_{\delta,\lambda}(t)$, the problem of calculating limit in reconstruction formula (10.14) is an ill-posed one and needs to be regularized.

As an approximate value of that limit one can take the quantity

$$\nu_\delta(\lambda, \tilde{t}) = \frac{\ln|\psi_{\delta,\lambda}(\tilde{t}) - \psi_0|}{\ln \tilde{t}}$$

computed at some point $t = \tilde{t}$ that is sufficiently close to zero and playing the role of a regularized parameter.

Thus, the regularized approximate value $\nu_\delta(\lambda, \tilde{t})$ of the memory order ν suggested by the proposed algorithm depends on two regularization parameters λ and $\tilde{t}$ that need to be properly chosen. Since in practice the amplitudes δ_k of the noise perturbations are usually unknown, one should rely on the so-called noise level-free regularization parameter choice rules.

The quasi-optimality criterion [32] is one of the simplest and the oldest, but still quite efficient strategy among such rules. Its version for the choice of multiple regularization parameters, such as λ and $\tilde{t}$, has been discussed in [11, 23].

To implement the quasi-optimality criterion in the present context one should consider two geometric sequences of regularization parameters values:

$$\lambda = \lambda_i = \lambda_1 q_1^{i-1}, \quad i = 1, 2, ..., M_1;$$

$$\tilde{t} = \tilde{t}_j = \tilde{t}_1 q_2^{j-1}, \quad j = 1, 2, ..., M_2;$$

$$0 < q_1, \ q_2 < 1.$$

Then the values $\nu_\delta(\lambda_i, \tilde{t}_j)$ should be computed as described above.

Next, for each $\tilde{t}_j$ one needs to find $\lambda_{i_j} \in \{\lambda_i\}_{i=1}^{M_1}$ such that

$$\left|\nu_\delta(\lambda_{i_j}, \tilde{t}_j) - \nu_\delta(\lambda_{i_j-1}, \tilde{t}_j)\right| = \min\{|\nu_\delta(\lambda_i, \tilde{t}_j) - \nu_\delta(\lambda_{i-1}, \tilde{t}_j)|, \quad i = 2, 3, ..., M_1\}.$$

Finally, $\tilde{t}_{j_0}$ is selected from $\{\tilde{t}_j\}_{j=1}^{M_2}$ such that

$$\left|\nu_\delta(\lambda_{i_{j_0}}, \tilde{t}_{j_0}) - \nu_\delta(\lambda_{i_{j_0-1}}, \tilde{t}_{j_0-1})\right|$$

$$= \min\{|\nu_\delta(\lambda_{i_j}, \tilde{t}_j) - \nu_\delta(\lambda_{i_{j-1}}, \tilde{t}_{j-1})|, \ j = 2, 3, ..., M_2\}.$$

The value $\nu_\delta(\lambda_{i_{j_0}}, \tilde{t}_{j_0})$ is the output of the proposed algorithm. In the next subsection we illustrate the performance of the algorithm by a series of numerical tests.

10.5.2 Numerical Experiments

Three different numerical examples corresponding to the final moment $T = 0.1$ and the time to measurements $t^\star = 0.012$ are treated below by the proposed algorithm.

At first, we consider problem (10.1)–(10.6) in the one-dimensional case $\Omega := (0, L)$:

$$
\begin{aligned}
\mathbf{D}_t^\nu u - a(x,t)u_{xx} + \tilde{a}(x,t)u_x - \int_0^t \mathcal{K}(t-s)b(x,s)u_{xx}(x,s)ds \\
= f(x,t,u) + g(x,t) \quad \text{in} \quad (0,L)\times(0,T), \\
u(x,0) = u_0(x), \quad x \in [0,L], \qquad (10.30) \\
u_x(0,t) = u_x(L,t) = 0, \quad t \in [0,T].
\end{aligned}
$$

In general, it is problematic to find the solution of (10.30) in the analytical form. Therefore, to generate the synthetic test data, we have to solve problem (10.30) numerically using the computational scheme described briefly below.

Introducing the space-time mesh with nodes

$$x_k = kh, \quad \tau_j = j\tau, \quad k = 0,1,\dots,\tilde{N}, \quad j = 0,1,\dots,\tilde{M}, \quad h = L/\tilde{N}, \quad \tau = T/\tilde{M},$$

and approximating the differential equation from (10.30) at each level τ_{j+1}, we derive the following finite-difference scheme:

$$
\begin{aligned}
\tau^{-\nu}\sum_{m=0}^{j+1}(u_k^{j+1-m} - u_0(x_k))\rho_m - \frac{a_k^{j+1}}{h^2}(u_{k-1}^{j+1} - 2u_k^{j+1} + u_{k+1}^{j+1}) + \frac{\tilde{a}_k^{j+1}}{2h}(u_{k+1}^{j+1} - u_{k-1}^{j+1}) \\
= \sum_{m=0}^{j}\left(b_k^m\frac{u_{k-1}^m - 2u_k^m + u_{k+1}^m}{h^2} + b_k^{m+1}\frac{u_{k-1}^{m+1} - 2u_k^{m+1} + u_{k+1}^{m+1}}{h^2}\right)\frac{\tilde{\mathcal{K}}_{m,j}}{2} \qquad (10.31) \\
+ f(x_k,\tau_j,u_k^j) + g(x_k,\tau_{j+1}), \quad k = 1,\dots,\tilde{N}-1, \quad j = 0,1,\dots,\tilde{M}-1,
\end{aligned}
$$

where we denote the finite-difference approximation of the function u at the point (x_k, τ_j) by u_k^j and put

$$a_k^{j+1} = a(x_k,\tau_{j+1}), \quad \tilde{a}_k^{j+1} = \tilde{a}(x_k,\tau_{j+1}), \quad b_k^j = b(x_k,\tau_j),$$

$$\rho_m = (-1)^m\binom{\nu}{m}, \quad \tilde{\mathcal{K}}_{m,j} = \int_{\tau_m}^{\tau_{m+1}}\mathcal{K}(\tau_{j+1}-s)ds.$$

Here we use the second-order finite-difference formulas to approximate derivatives u_x and u_{xx}; the Grünwald–Letnikov formula [7, 18, 33] to approximate the derivative $\mathbf{D}_t^\nu u$; the trapezoid-rule to approximate the integrals in the sum

$$\sum_{m=0}^{j} \int_{\tau_m}^{\tau_{m+1}} \mathcal{K}(\tau_{j+1} - s) b(x, s) u_{xx}(x, s) ds.$$

It is worth noting that an improvement in the accuracy of calculations can be achieved by Richardson extrapolation and the finite element method with mass lumping (see [7, 27]). We also use two fictitious mesh points outside the spatial domain to approximate the Neumann boundary conditions with the second order of accuracy (see, e.g. [14]).

In all our tests the noisy measurements are simulated according to (10.23), i.e.,

$$\psi_{\delta,k} = u(x_0, t_k) + C_1 \delta \Phi(t_k), \quad k = 1, 2, ..., 21,$$

where $C_1 = 0.3$, $\delta = 0.1$, and we examine the case $\Phi(t) = t|\ln t|$ corresponding to (10.24), as well as the case $\Phi(t) = t^\nu |\ln t|$ mentioned in Remark 10.5. The solution $u(x, t)$ and the space location x_0 are changing from example to example. Moreover, we consider four different distributions of the observation time moment t_k. Namely,

$$\mathbf{C1}: \quad t_k = (99 + k)\tau, \quad k = 1, 2, ..., 21,$$

$$\mathbf{C2}: \quad t_1 = 50\tau, \quad t_2 = 51\tau, \quad t_k = (99 + k)\tau, \quad k = 3, 4, ..., 21,$$

$$\mathbf{C3}: \quad t_k = (9 + k)\tau, \quad k = 1, 2, ..., 21,$$

$$\mathbf{C4}: \quad t_1 = 5\tau, \quad t_2 = 6\tau, \quad t_k = (9 + k)\tau, \quad k = 3, 4, ..., 21, \tag{10.32}$$

were $\tau = 10^{-4}$ that corresponds to $\tilde{M} = 10^3$ in scheme (10.31).

The sequences of the regularization parameters of our reconstruction algorithm are chosen as follows:

$$\lambda_i = 2^{1-i}, \quad i = 1, 2, \ldots, 60, \quad \tilde{t}_j = 2^{1-j} t_N, \quad j = 1, 2, \ldots, 10.$$

The approximate minimizer $\psi_{\delta,\lambda}(t)$ has the form (10.29) with $\mathcal{J} = 3$, $\nu_1 = 0.1$, $\nu_2 = 0.4$, $\nu_3 = 0.7$, $\gamma = 0.99$, $P = 9$ and $N = 21$, i.e., $t_N = t_{21}$.

The absolute errors $|\nu_\delta(\lambda_{i_{j_0}}, \tilde{t}_{j_0}) - \nu|$ for the analyzed examples are shown in Tables 10.1, 10.2, 10.3, 10.4, 10.5 and 10.6. The final step of the reconstruction algorithm is illustrated by Fig. 10.1 displacing the sequence $\nu_\delta(\lambda_{i_j}, \tilde{t}_j)$ and the values $\nu_\delta(\lambda_{i_{j_0}}, \tilde{t}_{j_0})$ chosen by the quasi-optimality criterion, which in the considered example (Example 10.2 below) is the best or almost the best approximation of the real ν among $\nu_\delta(\lambda_{i_j}, \tilde{t}_j)$.

Example 10.1 Consider problem (10.30) with $L = 1$ and

$$a(x,t) = \cos \pi x/2 + t, \quad \tilde{a}(x,t) = x + t, \quad b(x,t) = t^{1/3} + \sin \pi x,$$

$$\mathcal{K}(t) = t^{-1/3}, \quad u_0(x) = \cos \pi x, \quad f(x,t,u) = xt \sin(u^2),$$

$$g(x,t) = 1 + \pi^2\Big(\cos\frac{\pi x}{2} + t + \frac{3t^{2/3}\sin(\pi x)}{2} + \frac{t\pi}{3\sin \pi/3}\Big)\cos \pi x$$

$$-(x+t)\pi \sin \pi x - xt \sin\Big(\cos \pi x + \frac{t^{\nu}}{\Gamma(1+\nu)}\Big)^2.$$

In this example, we test our reconstruction algorithm for different values of the memory order $\nu = 0.05,\ 0.1,\ 0.2,\ 0.3,\ 0.4,\ 0.5,\ 0.6,\ 0.7,\ 0.8,\ 0.9$.

It is easy to verify that the function $u(x,t) = \cos \pi x + \frac{t^{\nu}}{\Gamma(1+\nu)}$ solves direct problem (10.30) with the parameters specified above. The solution is observed at the space location $x_0 = 0.70$. Then, simple calculations show that for this example the error bound provided by Proposition 10.3 in **STN** case has the value

$$\frac{C_1\delta}{||\mathcal{L}_1 u_0(x_0)|_{t=0} + f(x_0, 0, u_0(x_0)) + g(x_0, 0)| - C_1\delta|} = 0.031. \tag{10.33}$$

Although in Example 10.1 the analytic form of the solution $u(x,t)$ is known, we generate synthetic noisy data by using numerical scheme (10.31). Of course, this increases the noise level and makes the test even harder. Nevertheless, as it can be seen from Tables 10.1 and 10.2, the accuracy of our reconstruction algorithm has the order predicted by Proposition 10.3 and bound 10.33. This is remarkable, because Proposition 10.3 presupposes the availability of continuous data, while our algorithm operates only with discrete data and does not use any information about the noise level.

Example 10.2 Consider problem (10.30) with $L = 1$ and $\nu = 0.3,\ 0.9$, and

$$a(x,t) = 1, \quad \tilde{a}(x,t) = 0,$$

$$\mathcal{K} = 0, \quad b(x,t) = 1,$$

$$f(x,t,u) = 0 \quad g(x,t) = 100t(-2x^3 + 3x^2) \quad \text{and} \quad u_0(x) = -\frac{2}{3}x^3 + x^2 + 1.$$

In this example, an analytic form of the solution is unknown and we use the numerical scheme (10.31) to generate measurements $u(x_0, t)$ at the point $x_0 = 0.20$, for which

$$\frac{C_1\delta}{||\mathcal{L}_1 u_0(x_0)|_{t=0} + f(x_0, 0, u_0(x_0)) + g(x_0, 0)| - C_1\delta|} = 0.026$$

Table 10.1 The absolute error in Example 10.1 for the noise model with $\Phi(t) = t|\ln t|$

$\lvert\nu - \nu_\delta(\lambda_{i_{j_0}}, \tilde{t}_{j0})\rvert$				ν
C1	C2	C3	C4	
0.0046	0.0047	0.0031	0.0031	0.05
0.0117	0.0005	0.0044	0.0044	0.1
0.0169	0.0091	0.0096	0.0092	0.2
0.0226	0.0162	0.0119	0.0110	0.3
0.0254	0.0213	0.0134	0.0106	0.4
0.0260	0.0220	0.0144	0.0128	0.5
0.0260	0.0217	0.0148	0.0122	0.6
0.0252	0.0210	0.0140	0.0115	0.7
0.0244	0.0237	0.0151	0.0110	0.8
0.0256	0.0242	0.0187	0.0155	0.9

Table 10.2 The absolute error in Example 10.1 for the noise model with $\Phi(t) = t^\nu|\ln t|$

$\lvert\nu - \nu_\delta(\lambda_{i_{j_0}}, \tilde{t}_{j0})\rvert$				ν
C1	C2	C3	C4	
0.0325	0.0316	0.0297	0.0190	0.05
0.0341	0.0337	0.0307	0.0301	0.1
0.0417	0.0327	0.0337	0.0333	0.2
0.0467	0.0311	0.0359	0.0347	0.3
0.0500	0.0453	0.0359	0.0325	0.4
0.0519	0.0453	0.0394	0.0364	0.5
0.0477	0.0439	0.0383	0.0371	0.6
0.0453	0.0415	0.0380	0.0355	0.7
0.0411	0.0400	0.0346	0.0299	0.8
0.0367	0.0345	0.0317	0.0301	0.9

Table 10.3 The absolute error in Example 10.2 for the noise model with $\Phi(t) = t^\nu|\ln t|$

$\lvert\nu - \nu_\delta(\lambda_{i_{j_0}}, \tilde{t}_{j0})\rvert$				ν
C1	C2	C3	C4	
0.0557	0.0448	0.0429	0.0390	0.3
0.0659	0.0615	0.0524	0.0445	0.9

and, hence, condition (10.13) holds. For this example the corresponding absolute errors are listed in Tables 10.3 and 10.4. Moreover, in Fig. 10.1 we demonstrate the work of our algorithm in the final step for the order $\nu = 0.9$ and **C4** distribution. These plots show the dependence of the value $\nu_\delta(\lambda_{i_j}, \tilde{t}_j)$ (where the quantity λ_{i_j} is chosen by the quasi-optimality criterion for each $\tilde{t}_j$) on the time step j, $j = 1, 2, ...10$.

Table 10.4 The absolute error in Example 10.2 for the noise model with $\Phi(t) = t|\ln t|$

$\lvert\nu - \nu_\delta(\lambda_{i_{j_0}}, \tilde{t}_{j0})\rvert$				ν
C1	C2	C3	C4	
0.0908	0.0880	0.0221	0.0165	0.3
0.0568	0.0545	0.0424	0.0278	0.9

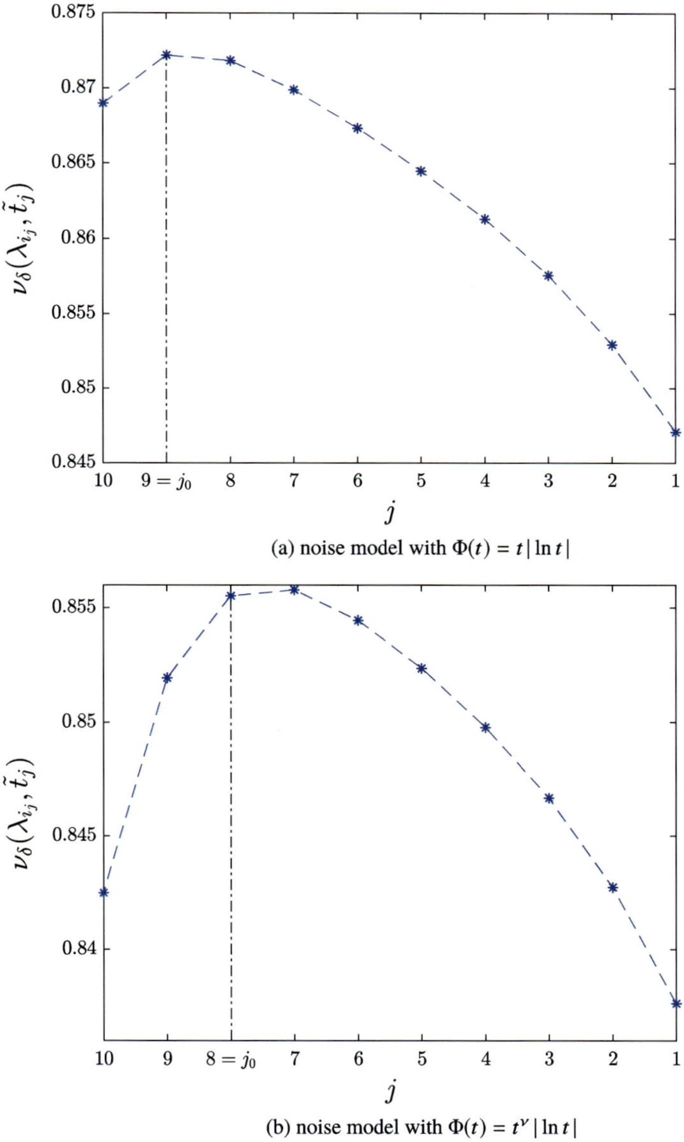

Fig. 10.1 The final step in the regularized reconstruction algorithm for Example 10.2, $\nu = 0.9$, with **a** noise model with $\Phi(t) = t|\ln t|$, **b** noise model with $\Phi(t) = t^\nu|\ln t|$

Table 10.5 The absolute error in Example 10.3 for the noise model with $\Phi(t) = t^{\nu}|\ln t|$

$\lvert\nu - \nu_\delta(\lambda_{i_{j_0}}, \tilde{t}_{j_0})\rvert$				ν
C1	C2	C3	C4	
0.0507	0.0513	0.0510	0.0496	0.2
0.0376	0.0456	0.0474	0.0485	0.5
0.0509	0.0391	0.0143	0.0009	0.9

As one can see from Fig. 10.1, in the considered example the quasi-optimality criterion provides the best (for noise model with $\Phi = t|\ln t|$) or almost the best (for noise model with $\Phi = t^{\nu}|\ln t|$) choice among the constructed approximations.

Example 10.3 In this example we consider problem (10.1)–(10.6) in the two-dimensional domain $\Omega = (0, 1) \times (0, 1)$:

$$\mathbf{D}_t^{\nu} u - u_{xx} - u_{yy} - \frac{t^{-\nu}}{\Gamma(1-\nu)} \star [u_{xx} + u_{yy}] = [\cos \pi x + \cos \pi y][\Gamma(1+\nu) + \pi^2(1+t^{\nu})$$

$$+\pi^2 t(1 + \Gamma(1+\nu)) + \frac{(1+\pi^2)t^{1-\nu}}{\Gamma(2-\nu)} + \frac{\pi^2 t^{2-\nu}}{\Gamma(3-\nu)}] \quad \text{in} \quad \Omega \times (0, T),$$

$$u(x, y, 0) = \cos \pi x + \cos \pi y, \qquad (x, y) \in [0, 1] \times [0, 1],$$

$$u_x(0, y, t) = u_x(1, y, t) = 0, \qquad t \in [0, T],\ y \in [0, 1],$$

$$u_y(x, 0, t) = u_y(x, 1, t) = 0, \qquad t \in [0, T],\ x \in [0, 1].$$

In this case, the exact solution is represented as $u(x, y, t) = (\cos \pi x + \cos \pi y)(1 + t + t^{\nu})$, and this solution will be observed in the point $(x_0, y_0) = (0.65, 0.65)$. In this point we have

$$\frac{C_1\delta}{||\mathcal{L}_1 u_0(x_0)|_{t=0} + f(x_0, 0, u_0(x_0)) + g(x_0, 0)| - C_1\delta|} = \begin{cases} 0.0373 & \text{for} \quad \nu = 0.2, \\ 0.0387 & \text{for} \quad \nu = 0.5, \\ 0.0355 & \text{for} \quad \nu = 0.9, \end{cases}$$

which means that condition (10.13) is satisfied.

Tables 10.5 and 10.6 list the outputs for this example in the case of the noises generated with noise models $\Phi(t) = t^{\nu}|\ln t|$ and $\Phi(t) = t|\ln t|$, correspondingly.

Table 10.6 The absolute error in Example 10.3 for the noise model with $\Phi(t) = t|\ln t|$

$\lvert\nu - \nu_\delta(\lambda_{i_{j_0}}, \tilde{t}_{j_0})\rvert$				ν
C1	C2	C3	C4	
0.0163	0.0162	0.0104	0.0102	0.2
0.0073	0.0089	0.0107	0.0096	0.5
0.0596	0.0490	0.0258	0.0095	0.9

10.6 Discussion and Conclusion

In this paper, we propose an approach to reconstruct the semilinear subdiffusion order. To this end, analyzing boundary value problems for the nonautonomous semilinear subdiffusion equations with memory terms in the fractional Hölder spaces, we obtain an explicit reconstruction formula for the order ν in terms of the smooth observation data for small time. Then, based on the Tikhonov regularization scheme and the quasi-optimality criterion, we construct the computational algorithm to find the order ν from noisy discrete measurements.

The computational results demonstrate that the proposed method effectively determines the unknown memory order ν. Moreover, from the methodological view point, the proposed approach can provide an efficient numerical technique for estimating the limit of the ratio of two unbounded noisy functions.

Acknowledgements This work is partially supported by the Grant H2020-MSCA-RISE-2014 project number 645672 (AMMODIT: Approximation Methods for Molecular Modelling and Diagnosis Tools). The paper has been finalized during the visit of the first, third and fourth authors to Johann Radon Institute (RICAM), Linz. The hospitality and perfect working conditions of RICAM are gratefully acknowledged.

Appendix: Proof of Proposition 10.1

First, we note that the statements **(i)–(iii)** have been obtained in the proof of Proposition 1 [22]. Moreover, statement **(v)** follows immediately from statement **(iv)**.

Hence, we are left to verify inequality **(iv)**. To this end, we first rewrite the difference $(\mathbf{D}_t^{\nu_2} U - \mathbf{D}_t^{\nu_1} U)$ as follows:

$$\mathbf{D}_t^{\nu_2} U(t) - \mathbf{D}_t^{\nu_1} U(t) = \mathbf{D}_t^{\nu_2} U(t) - \Theta_{\nu_2-\nu_1} \star \mathbf{D}_t^{\nu_2} U(t)$$

$$= \Big[1 - \frac{t^{\nu_2-\nu_1}}{\Gamma(\nu_2 - \nu_1 + 1)}\Big] \mathbf{D}_t^{\nu_2} U(t)$$

$$+\int_0^t \frac{(t-\tau)^{\nu_2-\nu_1-1}}{\Gamma(\nu_2-\nu_1)}[\mathbf{D}_t^{\nu_2}U(t)-\mathbf{D}_\tau^{\nu_2}U(\tau)]d\tau$$

$$\equiv \sum_{j=1}^{3}\mathcal{R}_j, \tag{10.34}$$

where we put

$$\mathcal{R}_1 = \frac{\Gamma(\nu_2-\nu_1+1)-1}{\Gamma(\nu_2-\nu_1+1)}\mathbf{D}_t^{\nu_2}U(t),$$

$$\mathcal{R}_2 = \frac{1-t^{\nu_2-\nu_1}}{\Gamma(\nu_2-\nu_1+1)}\mathbf{D}_t^{\nu_2}U(t),$$

$$\mathcal{R}_3 = \int_0^t \frac{(t-\tau)^{\nu_2-\nu_1-1}}{\Gamma(\nu_2-\nu_1)}[\mathbf{D}_t^{\nu_2}U(t)-\mathbf{D}_\tau^{\nu_2}U(\tau)]d\tau.$$

Now it is enough to estimate each term $\mathcal{R}_j$ separately.

• By inequalities **(i)** in Proposition 10.1,

$$\mathcal{R}_1 \leq C(\nu_2-\nu_1)\|\mathbf{D}_t^{\nu_2}U\|_{\mathrm{C}[0,T]},$$

where C is the positive constant.

• Concerning $\mathcal{R}_2$, the simple straightforward calculations give

$$|1-t^{\nu_2-\nu_1}| \leq C(\nu_2-\nu_1)t \qquad \text{if} \quad t>1,$$

$$|1-t^{\nu_2-\nu_1}| \leq C(\nu_2-\nu_1)|\ln t| \qquad \text{if} \quad t\leq 1,$$

and from this inequality, we easily draw the estimate

$$\mathcal{R}_2 \leq C(\nu_2-\nu_1)(t+|\ln t|)\|\mathbf{D}_t^{\nu_2}U\|_{\mathrm{C}[0,T]}.$$

• For $\mathcal{R}_3$, we have

$$\mathcal{R}_3 \leq C\left|\int_0^t \frac{(t-\tau)^{\nu_2-\nu_1-1+\gamma}}{\Gamma(\nu_2-\nu_1)}d\tau\right|\langle\mathbf{D}_t^{\nu_2}U\rangle_{t,[0,T]}^{(\gamma)}$$

$$\leq C(\nu_2-\nu_1)\frac{t^{\nu_2-\nu_1+\gamma}}{(\nu_2-\nu_1+\gamma)\Gamma(1+\nu_2-\nu_1)}\langle\mathbf{D}_t^{\nu_2}U\rangle_{t,[0,T]}^{(\gamma)}.$$

Summarizing the above estimates, we obtain statement **(iv)**.

References

1. M. Abramowitz, I.A. Stegun, *Handbook of Mathematical Functions with Formulas, Graphs and Mathematical Tables*, vol. 55, Applied Mathematics Series (National Bureau of Standards, Washington, 1964)
2. R.A. Adams, *Sobolev Spaces* (Academic Press, New York, 1975)
3. R.L. Bagley, P. Torvik, A theoretical basis for the application of fractional calculus to viscoelastisity. J. Rheol. **27**, 201–210 (1983)
4. M. Caputo, Models of flux in porous media with memory. Water Resour. Res. **36**, 693–705 (2000)
5. M. Caputo, J.M. Carcione, M.A.B. Botelho, Modeling extreme-event precursors with the fractional diffusion equation. Frac. Calc. Appl. Anal. **18**, 208–222 (2015)
6. M. Caputo, W. Plastino, Diffusion in porous layers with memory. Geophys. J. Int. **158**, 385–396 (2004)
7. K. Diethelm, N.J. Ford, A.D. Freed, Yu. Luchko, Algorithms for the fractional calculus: a selection of numerical methods. Comput. Methods Appl. Mech. Eng. **194**, 743–773 (2005)
8. W.G. Glöckle, T.F. Nonnenmacher, A fractional calculus approach to self-similar protein dynamics. Biophys. J. **68**, 46–53 (1995)
9. N. Engheia, On the role of fractional calculus in electromagnetic theory. IEEE Antennas Propag. Mag. **39**, 35–46 (1997)
10. S. Fedotov, N. Korable, Subdiffusion in an external potential: anomalous effects hiding behind normal behavior. Phys. Rev E. Stat. Nonlin. Soft Matter. Phys. **91**(4), 042112-1–042112-7 (2015)
11. M. Fornasier, V. Naumova, S.V. Pereverzyev, Parameter choice strategies for multipenalty regularization. SIAM J. Numer. Anal. **52**(4), 1770–1794 (2014)
12. G. Iaffaldano, M. Caputo, S. Martino, Experimental and theoretical memory diffusion of water in sand. Hydrol. Earth. Syst. Sci. Discuss. **2**, 1329–1357 (2005)
13. Y. Hatano, N. Hatano, Dispersive transport of ions in column experiments: an explanation of long tailed profiles. Water Resour. Res. **34**, 1027–1033 (1998)
14. M. Huntul, D. Lesnic, T. Johansson, Determination of an additive time- and space-dependent coefficient in the heat equation. Int. J. Numer. Methods Heat Fluid Flow **28**(6), 1352–1373 (2018)
15. Y. Hatano, J. Nakagawa, Sh Wang, M. Yamamoto, Determination of order in fractional diffusion equation. J. Math-for-Ind. **5A**, 51–57 (2013)
16. J. Janno, Determination of the order of fractional derivative and a kernel in an inverse problem for a generalized time fractional diffusion equation. Electron. J. Differ. Equ. **2016** (2016)
17. J. Janno, N. Kinash, Reconstruction of an order of derivative and a source term in a fractional diffusion equation from final measurements. Inverse Probl. **34**(2), 025007 (2018)
18. A.A. Kilbas, H.M. Srivastava, J.J. Trujillo, *Theory and Applications of Fractional Differential Equations*, vol. 204, North-Holland Mathematics Studies (Elsevier Science B.V., Amsterdam, 2006)
19. M. Krasnoschok, V. Pata, N. Vasylyeva, Semilinear subdiffusion with memory in the one-dimensional case. Nonlinear Anal. **165**, 1–17 (2017)
20. M. Krasnoschok, V. Pata, N. Vasylyeva, Solvability of linear boundary value problems for subdiffusion equations with memory. J. Integr. Equ. Appl. **30**(3), 417–445 (2018)
21. M. Krasnoschok, V. Pata, N. Vasylyeva, Semilinear subdiffusion with memory in multidimensional domains. Math. Nachr. **292**(7), 1490–1513 (2019)
22. G. Li, D. Zhang, X. Jia, M. Yamamoto, Simultaneous inversion for the space-dependent diffusion coefficient and the fractional order in the time-fractional diffusion equation. Inverse Probl. **29**(6), 065014 (2013)
23. S. Lu, S.V. Pereverzyev, *Regularization Theory for Ill-posed Problems. Selected topics*, vol. 58, Inverse and Ill-posed Problems Series (De Gruyter, Berlin, 2013)
24. M.M. Meerschart, A. Sikorskii, *Stochastic Models for Fractional Calculus*, vol. 43 (Walter De Grayter, Berlin, 2011)

25. M. Raberto, E. Scalas, F. Mainardi, Waiting-time and returns in high-frequency financial data: and empirical study. J. Phys. A. **314**(1), 749–755 (2002)
26. Z. Ruan, W. Zhang, Z. Wang, Simultaneous inversion of the fractional order and the space-dependent source term for the time-fractional diffusion equation. Appl. Math. Comput. **328**, 365–379 (2018)
27. S.V. Siryk, A note on the application of the Guermond-Pasquetti mass lumping correction technique for convection-diffusion problems. J. Comput. Phys. **376**, 1273–1291 (2019)
28. F. Shen, W. Tan, Y. Zhao, T. Masuoka, The Rayleigh-Stokes problem for a heated generalized second grade fluid with fractional derivative model. Nonlinear Anal. Real World Appl. **7**, 1072–1080 (2006)
29. I.M. Sokolov, J. Klafter, From diffusion to anomalous diffusion: a century after Einstein's Brownian motion. Chaos **15**, 1–7 (2005)
30. G. Szegö, *Orthogonal Polynomials*, 4th edn. (AMS, Providence, 1975)
31. C. Sun, G. Li, X. Jia, Numerical inversion for the multiple fractional orders in the multiterm TFDE. Adv. Math. Phys. 3204959 (2017)
32. A.N. Tikhonov, V.B. Glasko, Use of the regularization methods in nonlinear problems. USSR Comput. Math. Math. Phys. **5**, 93–107 (1965)
33. B. Yu, X. Jiang, H. Qi, An inverse problem to estimate an unknown order of a Riemann-Liouville fractional derivative for a fractional Stokes' first problem for a heated generalized second grade fluid. Acta Mech. Sin. **31**(2), 153–161 (2015)
34. Y. Zhang, C.T. Green, E.M. LaBolle, R.M. Neupauer, H.G. Sun, Bounded fractional diffusion in geological media: definition and Lagrangian approximation. Water Resour. Res. **52**, 8561–8577 (2016)
35. Y.X. Zhang, J. Jia, L. Yan, Bayesian approach to a nonlinear inverse problem for a time-space fractional diffusion equation. Inverse Prob. **34**, 125002 (2018)

Chapter 11
On the Singular Value Decomposition of n-Fold Integration Operators

Ronny Ramlau, Christoph Koutschan and Bernd Hofmann

Abstract In theory and practice of inverse problems, linear operator equations $Tx = y$ with compact linear forward operators T having a non-closed range $\mathcal{R}(T)$ and mapping between infinite dimensional Hilbert spaces plays some prominent role. As a consequence of the ill-posedness of such problems, regularization approaches are required, and due to its unlimited qualification spectral cut-off is an appropriate method for the stable approximate solution of corresponding inverse problems. For this method, however, the singular system $\{\sigma_i(T), u_i(T), v_i(T)\}_{i=1}^{\infty}$ of the compact operator T is needed, at least for $i = 1, 2, ..., N$, up to some stopping index N. In this note we consider n-fold integration operators $T = J^n$ $(n = 1, 2, ...)$ in $L^2([0, 1])$ occurring in numerous applications, where the solution of the associated operator equation is characterized by the nth generalized derivative $x = y^{(n)}$ of the Sobolev space function $y \in H^n([0, 1])$. Almost all textbooks on linear inverse problems present the whole singular system $\{\sigma_i(J^1), u_i(J^1), v_i(J^1)\}_{i=1}^{\infty}$ in an explicit manner. However, they do not discuss the singular systems for J^n, $n \geq 2$. We will emphasize that this seems to be a consequence of the fact that for higher n the eigenvalues $\sigma_i^2(J^n)$ of the associated ODE boundary value problems obey transcendental equations, the complexity of which is growing with n. We present the transcendental equations for $n = 2, 3, ...$ and discuss and illustrate the associated eigenfunctions and some of their properties.

R. Ramlau (✉)
Institute for Industrial Mathematics, Johannes Kepler University Linz,
and Johann Radon Institute for Computational and Applied Mathematics (RICAM), Altenberger Straße 69, 4040 Linz, Austria
e-mail: ronny.ramlau@jku.at

C. Koutschan
Johann Radon Institute for Computational and Applied Mathematics (RICAM),
Austrian Academy of Sciences, Altenberger Straße 69, 4040 Linz, Austria
e-mail: christoph.koutschan@ricam.oeaw.ac.at

B. Hofmann
Faculty of Mathematics, Chemnitz University of Technology,
09107 Chemnitz, Germany
e-mail: bernd.hofmann@mathematik.tu-chemnitz.de

J. Cheng et al. (eds.), *Inverse Problems and Related Topics*,
Springer Proceedings in Mathematics & Statistics 310,
https://doi.org/10.1007/978-981-15-1592-7_11

Keywords Integration operators · Singular systems · N-fold integration · Boundary value problems · Symbolic determinant

11.1 Introduction

For the stable approximate solution of the ill-posed linear operator equation

$$Tx = y, \qquad y \in \mathcal{R}(T), \tag{11.1}$$

with a compact linear operator T mapping between the infinite dimensional Hilbert spaces X and Y with norms $\|\cdot\|$ and inner products $\langle\cdot,\cdot\rangle$ and range $\mathcal{R}(T) \neq \overline{\mathcal{R}(T)} = Y$ the *spectral cut-off* method is appropriate due to its unlimited qualification which avoids saturation of the method (cf., e.g., [6, Example 4]). However, the use of spectral cut-off requires the knowledge of the singular system $\{\sigma_i(T), u_i(T), v_i(T)\}_{i=1}^{\infty}$ of the compact operator T, at least for $i = 1, 2, ..., N$, up to some stopping index N, which plays the role of a regularization parameter and occurs in case of noisy data $y^\delta \in Y$ obeying the noise model $\|y - y^\delta\| \le \delta$ with noise level $\delta > 0$ inside the regularization procedure

$$x_N^\delta := \sum_{i=1}^{N} \frac{1}{\sigma_i(T)} \langle y^\delta, v_i(T)\rangle u_i(T)$$

(cf., e.g., [4, p.36]).

In this note, we restrict our considerations with respect to Eq. (11.3) to the family of *Riemann–Liouville fractional integral operators* $T := J^\alpha$ defined for all exponents $0 < \alpha < \infty$ as *compact* operators

$$[J^\alpha x](s) := \int_0^s \frac{(s-t)^{\alpha-1}}{\Gamma(\alpha)} x(t)\,dt, \qquad 0 \le s \le 1, \tag{11.2}$$

mapping in the separable infinite dimensional Hilbert space $X = Y := L^2([0, 1])$ of quadratic integrable Lebesgue-measurable real functions over the unit interval $[0, 1]$, which are of particular interest in the mathematical literature. Namely the linear operator equation

$$J^\alpha x = y, \qquad y \in \mathcal{R}(J^\alpha), \tag{11.3}$$

is solved in a unique manner by applying the α-fold Riemann–Liouville fractional derivative D^α to the right-hand side y, which is assumed to belong to the range $\mathcal{R}(J^\alpha)$ of the operator J^α. For the left inverse D^α of J^α there exists the explicit formula

$$[D^\alpha y](t) = \frac{d}{dt} \int_0^t \frac{(t-s)^{-\alpha}}{\Gamma(1-\alpha)} y(s)\,ds, \qquad 0 \le t \le 1,$$

in the case $0 < \alpha < 1$. If $\alpha = n + \widetilde{\alpha}$ with positive integer n and $0 < \widetilde{\alpha} < 1$ we have $D^\alpha y = D^{\widetilde{\alpha}} y^{(n)}$.

The following facts are well-known from the literature (see, for example, [1, 2, 5]):

Fact 11.1 For all real numbers $0 < \alpha < \infty$ the linear convolution operators J^α mapping in $L^2([0, 1])$ are *injective* and *compact*, and so are the adjoint operators

$$[(J^\alpha)^* z](t) := \int_t^1 \frac{(s-t)^{\alpha-1}}{\Gamma(\alpha)} z(s)\,ds, \qquad 0 \le t \le 1, \tag{11.4}$$

too. Hence, the range $\mathcal{R}(J^\alpha)$ is a *dense* and *non-closed* subset of $L^2([0, 1])$. Consequently, the operator Eq. (11.3) is *ill-posed of type II* in the sense of Nashed [7]. The linear Volterra integral operators J^α are linear Fredholm integral operators with quadratically integrable kernel and hence Hilbert–Schmidt operators whenever $\frac{1}{2} < \alpha < \infty$.

Fact 11.2 For all real numbers $0 < \alpha < \infty$ the operator J^α possesses a singular system $\{\sigma_i(J^\alpha), u_i(J^\alpha), v_i(J^\alpha)\}_{i=1}^\infty$ with the uniquely determined ordered singular values $\sigma_1(J^\alpha) > \sigma_2(J^\alpha) > \ldots > 0$, where $\lim_{i\to\infty} \sigma_i(J^\alpha) = 0$, and two orthonormal systems $\{u_i(J^\alpha)\}_{i=1}^\infty$ and $\{v_i(J^\alpha)\}_{i=1}^\infty$, which are both complete in $L^2([0, 1])$, such that for $i = 1, 2, \ldots$

$$J^\alpha u_i(J^\alpha) = \sigma_i(J^\alpha)\, v_i(J^\alpha) \quad \text{and} \quad (J^\alpha)^* v_i(J^\alpha) = \sigma_i(J^\alpha)\, u_i(J^\alpha)\,. \tag{11.5}$$

The focus of the present note is on the case $\alpha = n$ with natural numbers $n = 1, 2, \ldots$, where $D^n y = y^{(n)}$ coincides with the n-fold generalized derivative of the Sobolev space function $y \in \mathcal{R}(J^n) \subset H^n([0, 1])$. We try to answer the frequently asked question why the complete *singular system* $\{\sigma_i(J^n), u_i(J^n), v_i(J^n)\}_{i=1}^\infty$ of J^n is made explicit only for $n = 1$ in many textbooks and papers, but for $n \ge 2$ such a detailed discussion is mostly avoided. One of the reasons may be that there are no nice explicit formulas for the singular systems. We describe them by implicit transcendental equations which become increasingly unhandy as n grows. Therefore we employ symbolic computation (in form of the computer algebra system Mathematica) to derive some of the formulas presented here. Also the arbitrary-precision arithmetic that is available through such a system is crucial for finding some of the numerical approximations.

11.2 Singular Value Asymptotics of Riemann–Liouville Fractional Integral Operators

Many authors have discussed upper and lower bounds for the singular values $\sigma_i(J^\alpha)$ of J^α aimed at deriving a singular value asymptotics with respect to the Riemann–Liouville fractional integral operators mapping in $L^2([0,1])$ in the case of specific exponents α and exponent intervals. However, in the paper [9] we find the complete asymptotics:

Proposition 11.1 *For all* $0 < \alpha < \infty$ *there exist constants*

$$0 < \underline{c}(\alpha) \le \overline{c}(\alpha) < \infty$$

such that

$$\underline{c}(\alpha)\, i^{-\alpha} \le \sigma_i(J^\alpha) \le \overline{c}(\alpha)\, i^{-\alpha}.$$

As a consequence of Proposition 11.1 the *degree of ill-posedness* (cf. [3]) of the operator Eq. (11.3) is α and grows with the level of integration. Abel integral equations ($0 < \alpha < 1$) are *weakly ill-posed* and the problem of n-fold differentiation with $\alpha = n \in \mathbb{N}$ and

$$\underline{c}(n)\, i^{-n} \le \sigma_i(J^n) \le \overline{c}(n)\, i^{-n}, \qquad i = 1, 2, ... \tag{11.6}$$

is *mildly ill-posed*. No *severely (exponentially) ill-posed* problem occurs in the context of Eq. (11.3).

11.3 The Boundary Value Problem for the Singular Value Decomposition

By deriving $u_i(J^n)$ $(i \in \mathbb{N})$ from the well-known equations

$$\begin{aligned} [J^n u_i(J^n)]^{(n)} &= u_i(J^n) &&= \sigma_i(J^n)[v_i(J^n)]^{(n)}, \\ [(J^n)^* v_i(J^n)]^{(n)} &= (-1)^n v_i(J^n) &&= \sigma_i(J^n)[u_i(J^n)]^{(n)}, \end{aligned}$$

one can verify the singular system of J^n from the following boundary value problem of an ordinary differential equation of order $2n$:

$$\begin{cases} \lambda\, u^{(2n)}(t) + (-1)^{(n+1)} u(t) = 0, \qquad 0 < t < 1\,, \\ u(1) = u'(1) = ... = u^{(n-1)}(1) = 0\,, \\ u^{(n)}(0) = u^{(n+1)}(0) = ... = u^{(2n-1)}(0) = 0\,. \end{cases} \tag{11.7}$$

More precisely, we are searching for all eigenvalues $\lambda > 0$ such that the system (11.7) possesses nontrivial solutions $0 \neq u \in L^2([0, 1])$. According to Proposition 11.1, there will be with $\lambda = \lambda_i$ an infinite sequence $\lambda_1 > \lambda_2 > \ldots > 0$ of ordered eigenvalues $\lambda_i := (\sigma_i(J^n))^2$ which are the i-largest eigenvalues of both operators $(J^n)^* J^n$ and $J^n(J^n)^*$. Moreover, with $u := u_i(J^n)$ there will be an associated orthonormal eigensystem $\{u_i(J^n)\}_{i=1}^{\infty}$ which leads with $v_i(J^n) := \frac{1}{\sigma_i(J^n)} J^n u_i(J^n)$ to the orthonormal eigensystem $\{v_i(J^n)\}_{i=1}^{\infty}$. Thus the singular system $\{\sigma_i(J^n), u_i(J^n), v_i(J^n)\}_{i=1}^{\infty}$ of J^n is complete.

The computation of the eigensystem follows a schema listed in the algorithm below that has been frequently used in the literature for the case $n = 1$ (for results see Sect. 11.4) and can be applied to any larger integer $n \in \mathbb{N}$. This approach is based on the zeros ν of the characteristic polynomial

$$p_n(\nu) = \lambda \nu^{2n} + (-1)^{n+1} \tag{11.8}$$

of the homogeneous differential equation $\lambda\, u^{(2n)}(t) + (-1)^{(n+1)} u(t) = 0$ of order $2n$ occurring in the boundary value problem (11.7). It is clear that these zeros ν obey the equation

$$\lambda \nu^{2n} = (-1)^n. \tag{11.9}$$

For fixed $\lambda > 0$, the solutions u of this ODE are characterized by a corresponding fundamental system

$$\left\{\varphi_k(t) \;\middle|\; k = 0, \ldots, 2n-1\right\}$$

such that

$$u(t) = \sum_{k=0}^{2n-1} \gamma_k \varphi_k(t). \tag{11.10}$$

Each non-zero coefficient vector $\gamma = (\gamma_0, \gamma_1, \ldots, \gamma_{2n-1})^T \in \mathbb{R}^{2n}$ represents a nontrivial solution of the ODE. Taking into account the required initial and terminal conditions of the boundary value problem (11.7), the vector γ must satisfy the linear system $A_n(\lambda)\gamma = 0$ with a singular quadratic matrix

$$A_n(\lambda) = \begin{pmatrix} \varphi_0(1) & \varphi_1(1) & \cdots & \varphi_{2n-1}(1) \\ \varphi_0'(1) & \varphi_1'(1) & \cdots & \varphi_{2n-1}'(1) \\ \vdots & \vdots & & \vdots \\ \varphi_0^{(n-1)}(1) & \varphi_1^{(n-1)}(1) & \cdots & \varphi_{2n-1}^{(n-1)}(1) \\ \varphi_0^{(n)}(0) & \varphi_1^{(n)}(0) & \cdots & \varphi_{2n-1}^{(n)}(0) \\ \vdots & \vdots & & \vdots \\ \varphi_0^{(2n-1)}(0) & \varphi_1^{(2n-1)}(0) & \cdots & \varphi_{2n-1}^{(2n-1)}(0) \end{pmatrix} \in \mathbb{R}^{2n \times 2n}. \tag{11.11}$$

This means that only those $\lambda > 0$ for which

$$\det(A_n(\lambda)) = 0 \tag{11.12}$$

yield non-zero vectors γ such that u in (11.10) is non-trivial. It can be seen that only a countable set $\{\lambda_i\}_{i=1}^{\infty}$ of values $\lambda > 0$ satisfies (11.12). This set consists of the eigenvalues of $J_n^* J_n$ with associated eigenfunctions

$$u_i(t) = \sum_{k=0}^{2n-1} \gamma_k^{(i)} \varphi_{i,k}(t),$$

where $\{\varphi_{i,0}, \ldots, \varphi_{i,2n-1}\}$ is the fundamental system associated to the eigenvalue λ_i, and the vector $\gamma^{(i)} = \left(\gamma_0^{(i)}, \ldots, \gamma_{2n-1}^{(i)}\right)^T \in \mathbb{R}^{2n}$ satisfies the linear system $A_n(\lambda_i)$ $\gamma^{(i)} = 0$ and is normalized such that $\|u_i\|_{L^2([0,1])} = 1$.

Now we are ready to formulate the algorithm for obtaining the desired eigenvalues and eigensystems.

Algorithm 11.1

(i) Compute, by solving Eq. (11.9), the $2n$ zeros of the characteristic polynomial $p_n(\nu)$ of the ODE occurring in problem (11.7).
(ii) Construct the fundamental system $\left\{\varphi_k(t) \;\middle|\; k = 0, \ldots, 2n-1\right\}$ of the ODE for arbitrary $\lambda > 0$.
(iii) Form the $(2n \times 2n)$-matrix $A_n(\lambda)$ that expresses the initial and terminal conditions occurring in (11.7).
(iv) Determine the eigenvalues λ_i, $i = 1, 2, \ldots$, of $J_n^* J_n$ by solving the equation $\det(A_n(\lambda_i)) = 0$.
(v) Calculate the eigenfunctions $u_i = \sum_{k=0}^{2n-1} \gamma_k^{(i)} \varphi_{i,k}(t)$ for all $i = 1, 2, \ldots$, such that $\|u_i\|_{L^2(0,1)} = 1$.

Although this algorithm seems to be straightforward, we will see that only steps (i)–(iii) can be done explicitly.

Proposition 11.2 *The zeros of the characteristic polynomial* (11.8) *are given by*

$$\nu_k = \left(\frac{1}{\lambda}\right)^{\frac{1}{2n}} \exp\left(i\,\frac{[n]_2\pi + 2\pi k}{2n}\right), \qquad k = 0, \ldots, 2n-1, \tag{11.13}$$

where $[n]_2 := n \bmod 2$ *for* $n \in \mathbb{N}$.

Proof For n even, we have to solve the equation $\nu^{2n} = 1/\lambda$. Its zeros are given as

$$\nu_k = \left(\frac{1}{\lambda}\right)^{\frac{1}{2n}} e^{i\,\frac{2\pi k}{2n}}, \qquad k = 0, \ldots, 2n-1. \tag{11.14}$$

For n odd, the equation reads

$$\nu^{2n} = \frac{-1}{\lambda} = \frac{e^{i\pi}}{\lambda} \tag{11.15}$$

having the roots

$$\nu_k = \left(\frac{1}{\lambda}\right)^{\frac{1}{2n}} e^{i\frac{\pi+2\pi k}{2n}}, \qquad k = 0, \ldots, 2n-1. \tag{11.16}$$

As roots of the characteristic polynomial, the ν_k are either real, or if one root is complex, then its conjugate is also a root. In detail, we have the assertions of the following proposition.

Proposition 11.3 *For n even, there exist two real zeros*

$$\nu_0 = \lambda^{-\frac{1}{2n}} \tag{11.17}$$

$$\nu_n = -\lambda^{-\frac{1}{2n}}. \tag{11.18}$$

Additionally, we have

$$\overline{\nu_k} = \nu_{2n-k}, \qquad k = 1, \ldots, n-1. \tag{11.19}$$

If n is odd, all zeros are complex and satisfy

$$\overline{\nu_k} = \nu_{2n-(k+1)}, \qquad k = 0, \ldots, n-1. \tag{11.20}$$

Proof If n is even, the values of ν_k for $k = 0$ and $k = n$ are evident. Moreover, we have for $k = 1, \ldots, n-1$

$$\overline{\nu_k} = \overline{\lambda^{-\frac{1}{2n}} e^{i\frac{\pi k}{n}}} = \lambda^{-\frac{1}{2n}} e^{-i\frac{\pi k}{n}} = \lambda^{-\frac{1}{2n}} e^{i\pi\frac{2n-k}{n}} = \nu_{2n-k}. \tag{11.21}$$

In the case that n is odd, the zeros are given by

$$\nu_k = \lambda^{-\frac{1}{2n}} e^{\frac{i\pi}{2n}(1+2k)}, \tag{11.22}$$

and as there is no $k \in \mathbb{N}$ such that $1 + 2k$ equals zero or a multiple of $2n$, there are no real roots. Additionally, we have

$$\begin{aligned} \lambda^{\frac{1}{2n}} \overline{\nu_k} &= e^{\overline{\frac{i\pi}{2n}(1+2k)}} = e^{-\frac{i\pi}{2n}(1+2k)+2\pi i} = e^{\frac{i\pi}{n}(2n-k-\frac{1}{2})} \\ &= e^{\frac{i\pi}{n}(\frac{1}{2}+(2n-(k+1))} = \lambda^{\frac{1}{2n}} \nu_{2n-(k+1)}. \end{aligned}$$

It is well known that a complex root and its complex conjugate create a pair of real fundamental solutions of an ODE. Specifically, a complex root $\nu_k = \alpha_k \pm i\beta_k$ with multiplicity one creates the two real fundamental solutions

$$e^{\alpha_k \cdot t}\cos(\beta_k \cdot t), \qquad e^{\alpha_k \cdot t}\sin(\beta_k \cdot t). \tag{11.23}$$

For what follows, let us denote the roots of the characteristic polynomial $p_n(\lambda)$ by $\nu_k^{(e)}$ if n is even and by $\nu_k^{(o)}$ if n is odd. Then we obtain the following result.

Proposition 11.4 *Let* $\nu_k^{(e,o)} = \alpha_k^{(e,o)} + i\beta_k^{(e,o)}$ *denote the roots of the characteristic polynomial* (11.8) *and*

$$\{\varphi_0,\ \varphi_1, \ldots, \varphi_{2n-1}\} \tag{11.24}$$

the fundamental system of the ODE occurring in (11.7). *Then we have:*

(a) *If n is even then, for $k = 1, \ldots, n-1$, the system (11.24) is characterized as*

$$\begin{aligned}
\varphi_0(t) &= e^{(\lambda^{-1/2n})\cdot t}\\
\varphi_1(t) &= e^{(-\lambda^{-1/2n})\cdot t}\\
\varphi_{2k}(t) &= e^{\alpha_k^{(e)}\cdot t}\cos(\beta_k^{(e)}\cdot t)\\
\varphi_{2k+1}(t) &= e^{\alpha_k^{(e)}\cdot t}\sin(\beta_k^{(e)}\cdot t).
\end{aligned}$$

(b) *If n is odd then, for $k = 1, \ldots, n-1$, the system (11.24) is characterized as*

$$\begin{aligned}
\varphi_{2k}(t) &= e^{\alpha_k^{(o)}\cdot t}\cos(\beta_k^{(o)}\cdot t)\\
\varphi_{2k+1}(t) &= e^{\alpha_k^{(o)}\cdot t}\sin(\beta_k^{(o)}\cdot t).
\end{aligned}$$

Proof Taking into account (11.23), the proof follows by the characterization of the roots and their complex conjugates in Proposition 11.3.

11.4 The Onefold Integration Operator

Along the lines outlined above in the algorithm and frequently presented in the literature one finds the singular system for $n = 1$, i.e. for the simple integration operator J^1, from the ODE system

$$\begin{cases}
\lambda\, u''(t) + u(t) = 0, & 0 < t < 1\,,\\
u(1) = 0\,,\\
u'(0) = 0\,.
\end{cases}$$

The explicit structure of this singular system is outlined in the following proposition.

Proposition 11.5 *For $n = 1$ we have the explicitly given singular system*

$$\left\{\sigma_i = \tfrac{2}{(2i-1)\pi},\, u_i(t) = \sqrt{2}\cos\!\left(\left(i-\tfrac{1}{2}\right)\pi t\right),\, v_i(t) = \sqrt{2}\sin\!\left(\left(i-\tfrac{1}{2}\right)\pi t\right)\right\}_{i=1}^{\infty}$$

of the operator J^1 mapping in the Hilbert space $L^2([0,1])$. Hence, formula (11.6) applies in the form

$$\frac{1}{\pi}\, i^{-1} \le \sigma_i(J^1) \le \frac{2}{\pi}\, i^{-1}, \qquad i = 1, 2, \ldots .$$

11.5 The Twofold Integration Operator

11.5.1 General Assertions

In the case $n = 2$, i.e., for the twofold integration operator J^2, the ODE-system (11.7) attains the form:

$$\begin{cases} \lambda\, u^{(4)}(t) - u(t) = 0, \qquad 0 < t < 1\,, \\ u(1) = u'(1) = 0\,, \\ u''(0) = u'''(0) = 0\,. \end{cases} \tag{11.25}$$

For the eigenfunctions $u = u_i(J^2)$ it is a necessary condition that they satisfy the homogeneous fourth-order differential equation in (11.25), which implies the ansatz structure via the corresponding fundamental system as

$$u(t) = \gamma_1 \exp\left(\frac{t}{\lambda^{1/4}}\right) + \gamma_2 \exp\left(-\frac{t}{\lambda^{1/4}}\right) + \gamma_3 \sin\left(\frac{t}{\lambda^{1/4}}\right) + \gamma_4 \cos\left(\frac{t}{\lambda^{1/4}}\right).$$

To obtain such $u \neq 0$, the linear (4×4)-system of equations

$$A_2(\mu) \cdot (\gamma_1, \gamma_2, \gamma_3, \gamma_4)^T = (0, 0, 0, 0)^T \quad \text{with} \quad \mu := \lambda^{-1/4}$$

must have a singular matrix $A_2(\mu)$, which means that

$$\det \begin{pmatrix} e^{\mu} & e^{-\mu} & \sin(\mu) & \cos(\mu) \\ e^{\mu} & -e^{-\mu} & \cos(\mu) & -\sin(\mu) \\ 1 & 1 & 0 & -1 \\ 1 & -1 & -1 & 0 \end{pmatrix} = 4\big(\cos(\mu)\cosh(\mu) + 1\big) = 0\,.$$

This leads to the following proposition:

Proposition 11.6 *The eigenvalues λ of the operator $(J^2)^* J^2$ are the solutions of the nonlinear transcendental equation*

$$\cos\left(\frac{1}{\lambda^{1/4}}\right) \cdot \cosh\left(\frac{1}{\lambda^{1/4}}\right) + 1 = 0. \tag{11.26}$$

In the next subsection we motivate the fact that the sequence $\{\lambda_i\}_{i=1}^{\infty}$ of solutions to (11.26) is of the form $\lambda_i = \frac{1}{((i-\frac{1}{2})\pi + \varepsilon_i)^4}$ $(i = 1, 2, \ldots)$, where the sequence $\{\varepsilon_i\}_{i=1}^{\infty}$ tends to zero exponentially fast. This gives evidence that the singular values $\sigma_i(J^2)$ are very close to

$$\frac{1}{\big((i+\frac{1}{2})\pi\big)^2} \quad (i = 1, 2, ...)$$

for sufficiently large i, which is in accordance with the assertion of Proposition 11.1 in the case $\alpha := 2$.

11.5.2 On the Zeros of the Function $f(z) = \cos(z)\cosh(z) + 1 = 0$

To our knowledge there does not exist a closed form for the zeros of the transcendental equation $f(z) = \cos(z)\cosh(z) + 1 = 0$. From the form of the equation it becomes apparent that there are infinitely many zeros, whose distribution is approximately π-periodic. Applying Newton's method to $f(z)$, we find the following numeric values for the first few positive roots:

$$\begin{aligned}
z_1 &= 1.87510406871196116644530824107821 4...\\
z_2 &= 4.69409113297417457643639177801981 2...\\
z_3 &= 7.85475743823761256486100858276457 0...\\
z_4 &= 10.9955407348754669906673491078547 0...\\
z_5 &= 14.1371683910464705809170468125517 7...
\end{aligned}$$

The almost-periodic behavior of $\{z_i\}_{i=1}^{\infty}$ suggests to write

$$z_i = (i - \tfrac{1}{2})\pi + \varepsilon_i, \tag{11.27}$$

where the sequence $\{\varepsilon_i\}_{i=1}^{\infty}$ tends to zero exponentially fast. Our first goal is to derive a bound on the absolute value of ε_i, thereby proving the claimed asymptotic behavior of $\{\varepsilon_i\}_{i=1}^{\infty}$.

For this purpose, consider the function

$$g(z) := f(z) - 1 = \cos(z)\cosh(z),$$

whose zeros are at the positions

$$\zeta_i := (i - \tfrac{1}{2})\pi \quad (i = 1, 2, ...).$$

The locations where the graph of $g(z)$ intersects the line $y = -1$ are exactly the zeros of $f(z)$. From $g''(z) = -2\sin(z)\sinh(z)$ we see that $g(z)$ is convex if $\sin(z) < 0$ and that $g(z)$ is concave when $\sin(z) > 0$.

Lemma 11.1 *If $i \geq 1$ is an odd integer, then* $0 < \varepsilon_i < 2\,\mathrm{e}^{-\zeta_i}$.

Proof If i is odd, then $g'(\zeta_i) = -\cosh(\zeta_i)$. Hence g has a negative slope at ζ_i and therefore $\varepsilon_i > 0$. Since g is concave in the interval $\big(\zeta_i - \frac{\pi}{2}, \zeta_i + \frac{\pi}{2}\big)$, it follows that in this interval the tangent to g at ζ_i is above g. This tangent intersects the line $y = -1$ at $\zeta_i + 1/\cosh(\zeta_i)$, which yields the desired upper bound on ε_i:

$$\varepsilon_i < \frac{1}{\cosh(\zeta_i)} = \frac{2}{e^{\zeta_i} + e^{-\zeta_i}} < \frac{2}{e^{\zeta_i}}.$$

Lemma 11.2 *If $i \geq 2$ is an even integer, then* $0 < -\varepsilon_i < 4\,e^{-\zeta_i}$.

Proof In this case, $g'(\zeta_i) = \cosh(\zeta_i)$. Hence g is increasing, which means that its graph intersects $y = -1$ left to ζ_i, thus $\varepsilon_i < 0$. Moreover, g is convex in the interval $\left(\zeta_i - \frac{\pi}{2}, \zeta_i + \frac{\pi}{2}\right)$, and therefore the tangent to g at ζ_i is below g. Unfortunately, the intersection between this tangent and $y = -1$ does not deliver an upper bound on $|\varepsilon_i|$.

Instead, we define $\xi_i := \zeta_i - 4\,e^{-\zeta_i}$ and show that $g(\xi_i) < -1$, yielding the claimed bound on ε_i. Equivalently, we show $-g(\xi_i) > 1$:

$$\begin{aligned}
-g(\xi_i) &= \sin\left(4\,e^{-\zeta_i}\right)\cosh(\xi_i) \\
&= \tfrac{1}{2}\sin\left(4\,e^{-\zeta_i}\right)\left(e^{\xi_i} + e^{-\xi_i}\right) \\
&> \tfrac{1}{2}\sin\left(4\,e^{-\zeta_i}\right)e^{\xi_i} \\
&= \tfrac{1}{2}\sin\left(4\,e^{-\zeta_i}\right)e^{\zeta_i}\exp\left(-4\,e^{-\zeta_i}\right) \\
&= \frac{\sin(4x)}{2x}\,e^{-4x} \quad \text{with } x = e^{-\zeta_i}.
\end{aligned}$$

The function $h(x) = \left(\sin(4x)/(2x)\right)e^{-4x}$ is monotonically decreasing in the interval $\left(0, e^{-\zeta_1}\right)$ with $h(0) = 2$ and $h\left(e^{-\zeta_2}\right) \approx 1.92899$. In particular, we have $h(x) > 1$ in this interval and therefore $h(e^{-\zeta_i}) > 1$ for all $i \geq 2$, which implies our claim on $g(\xi_i)$. The asserted bound on ε_i follows.

Remark 11.1 The factor 4 in the previous lemma is because of our crude estimate; actually we have for even and odd i

$$|\varepsilon_i| \sim 2\,e^{-\zeta_i} \quad \text{for } i \to \infty.$$

Remark 11.2 Analogous statements can be made about the negative roots of the function $f(z)$; they follow immediately by symmetry since f is an even function.

Instead of a bound on ε_i, we can also derive an exact expression for it in the form of an infinite series. Plugging the representation (11.27) into the equation $f(z_i) = 0$, one obtains

$$(-1)^i \sin(\varepsilon_i)\cosh\left(\left(i - \tfrac{1}{2}\right)\pi + \varepsilon_i\right) = -1,$$

or equivalently

$$\sin(\varepsilon_i) + (-1)^i \frac{2}{w_i^{-1} + w_i} = 0 \qquad \text{with } w_i = \exp\left(-\left(i - \tfrac{1}{2}\right)\pi - \varepsilon_i\right).$$

We expand the left-hand side as a geometric series in w_i^2, which gives

$$\sin(\varepsilon_i) + 2\,(-1)^i \sum_{k=0}^{\infty} (-1)^k w_i^{2k+1}.$$

Next, we write $(-1)^i w_i = x_i \cdot \exp(-\varepsilon_i)$ with $x_i = (-1)^i \exp\bigl(-\bigl(i+\frac{1}{2}\bigr)\pi\bigr)$, and perform Taylor expansion with respect to ε_i:

$$\sum_{j=0}^{\infty} \frac{(-1)^j}{(2j+1)!} \varepsilon_i^{2j+1} + 2 \sum_{k=0}^{\infty} (-1)^k x_i^{2k+1} \sum_{j=0}^{\infty} \frac{(-2k-1)^j}{j!} \varepsilon_i^j.$$

Formally speaking, this is a bivariate power series in the variables x_i and ε_i. Making an ansatz for ε_i, i.e., substituting for ε_i a power series in x_i with undetermined coefficients,

$$\varepsilon_i = \sum_{i=1}^{\infty} a_i x_i^i,$$

we obtain a univariate series:

$$(a_1+2)x_i + (a_2 - 2a_1)x_i^2 + \left(a_3 - 2a_2 - \tfrac{1}{6}a_1^3 + a_1^2 - 2\right) x_i^3 \\ + \left(a_4 - 2a_3 - \tfrac{1}{2}a_1^2 a_2 + 2a_1 a_2 - \tfrac{1}{3}a_1^3 + 6a_1\right) x_i^4 + \ldots$$

Coefficient comparison with respect to x_i then allows us to compute the unknown coefficients a_i; note that in the coefficient of x_i^k the indeterminate a_k appears linearly and can therefore be easily computed from the previous ones:

$$a_1 = -2,\ a_2 = -4,\ a_3 = -\tfrac{34}{3},\ a_4 = -\tfrac{112}{3},\ a_5 = -\tfrac{2006}{15},\ a_6 = -\tfrac{1516}{3},\ \ldots$$

Remark 11.3 We have computed the first 100 coefficients a_i symbolically, but we were not able to identify a nice closed form for them. They do not satisfy a (nice) linear recurrence equation with polynomial coefficients, either. Also in the OEIS [8], we could not find any information about these numbers.

11.5.3 Eigenfunctions

With the acquired knowledge on the eigenvalues λ of the operator $(J^2)^* J^2$, we are able to derive the corresponding eigenfunctions u_i, at least numerically. Recall the fundamental system

$$\gamma_1 \exp\left(\frac{t}{\lambda^{1/4}}\right) + \gamma_2 \exp\left(-\frac{t}{\lambda^{1/4}}\right) + \gamma_3 \sin\left(\frac{t}{\lambda^{1/4}}\right) + \gamma_4 \cos\left(\frac{t}{\lambda^{1/4}}\right).$$

Table 11.1 Eigenvalues and multipliers of the fundamental system

i	λ_i	γ_1	γ_2	γ_3	γ_4
1	0.08089068	0.13295224	0.86704776	−0.73409551	1.00000000
2	0.00205965	−0.00923366	1.00923370	−1.01846730	1.00000000
3	0.00026271	0.00038775	0.99961225	−0.99922450	1.00000000
4	0.00006841	−0.00001678	1.00001680	−1.00003360	1.00000000
5	0.00002504	0.00000072	0.99999928	−0.99999855	1.00000000

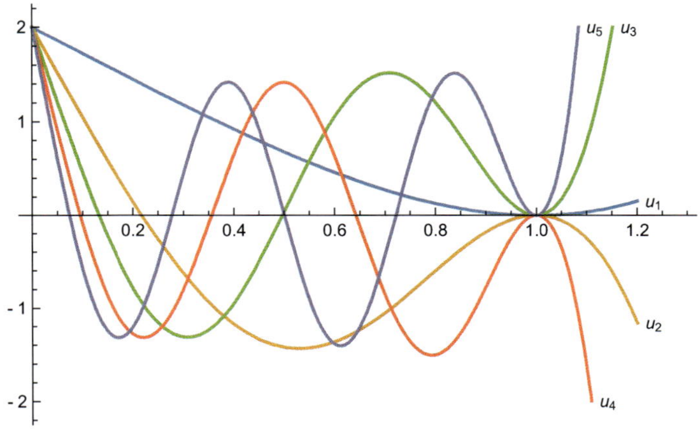

Fig. 11.1 Eigenfunctions $u_1, \ldots, u_5$ for $n = 2$

By plugging the computed values for λ_i, $1 \le i \le 5$, into the matrix $A_2(\lambda)$, we can determine the constants $\gamma_1, \gamma_2, \gamma_3, \gamma_4$. The results are shown in Table 11.1 (after the normalization $\|u_i\|_{L^2([0,1])} = 1$) and the eigenfunctions themselves are plotted in Fig. 11.1.

11.6 The *n*-fold Integration Operator

With the notation $\omega_k = \exp\big(\frac{i\pi}{2n}(2k + [n]_2)\big)$, the roots ν_k of the characteristic polynomial (11.8), $p_n(\nu){=}\lambda\nu^{2n} + (-1)^{n+1}$, can be written as $\nu_k = \lambda^{-\frac{1}{2n}}\omega_k$, $k = 0, \ldots, 2n-1$. Let $z = \lambda^{-\frac{1}{2n}}$ and write the fundamental system of the ODE in (11.7) in terms of the complex exponential functions $\varphi_k(t) = e^{\omega_k z t}$, then the matrix A_n given in (11.11) attains the following form

$$A_n(z) = \left(a_{j,k}^{(n)}(z)\right)_{0\le j,k\le 2n-1} \quad \text{with} \quad a_{j,k}^{(n)}(z) = \begin{cases} \omega_k^j z^j e^{\omega_k z}, & j < n, \\ \omega_k^j z^j, & j \ge n. \end{cases}$$

We want to determine the values of $z \neq 0$ for which $\det(A_n(z)) = 0$. Hence the common factor z^j from the jth row can be removed. In order to obtain an explicit expression for the determinant of $A_n(z)$, we first study the more general matrix

$$M_n = \left(m_{j,k}^{(n)}\right)_{0\le j,k\le 2n-1} \quad \text{with} \quad m_{j,k}^{(n)} = \begin{cases} \omega_k^j z_k, & j < n, \\ \omega_k^j, & j \ge n, \end{cases}$$

where $z_0, \dots, z_{2n-1}$ are indeterminates. The matrix looks as follows:

$$M_n = \begin{pmatrix} z_0 & z_1 & z_2 & \cdots & z_{2n-1} \\ \omega_0 z_0 & \omega_1 z_1 & \omega_2 z_2 & \cdots & \omega_{2n-1} z_{2n-1} \\ \omega_0^2 z_0 & \omega_1^2 z_1 & \omega_2^2 z_2 & \cdots & \omega_{2n-1}^2 z_{2n-1} \\ \vdots & \vdots & \vdots & & \vdots \\ \omega_0^{n-1} z_0 & \omega_1^{n-1} z_1 & \omega_2^{n-1} z_2 & \cdots & \omega_{2n-1}^{n-1} z_{2n-1} \\ \omega_0^n & \omega_1^n & \omega_2^n & \cdots & \omega_{2n-1}^n \\ \vdots & \vdots & \vdots & & \vdots \\ \omega_0^{2n-1} & \omega_1^{2n-1} & \omega_2^{2n-1} & \cdots & \omega_{2n-1}^{2n-1} \end{pmatrix}.$$

For $z_0 = \dots = z_{2n-1} = 1$, the matrix M_n equals exactly the Vandermonde matrix $V(\omega_0, \dots, \omega_{2n-1})$, which for even n is the Fourier matrix, since in this case the ω_k are precisely the $2n$th complex roots of unity.

For symbolic indeterminates $z_0, \dots, z_{2n-1}$, the determinant of M_n is a polynomial in $z_0, \dots, z_{2n-1}$ that is homogeneous of degree n and linear in each variable z_k. Let $I \subset \{0, 1, \dots, 2n-1\}$ be an index set with $|I| = n$. We aim at computing the coefficient of the monomial $\prod_{k\in I} z_k$ in $\det(M_n)$. This corresponds to setting $z_k = 0$ for all $k \in C := \{0, 1, \dots, 2n-1\} \setminus I$. By permuting its columns, the matrix M_n can be transformed into a block matrix of the form

$$\left(\begin{array}{c|c} V\big((\omega_k)_{k\in I}\big) \cdot \operatorname{diag}\big((z_k)_{k\in I}\big) & 0 \\ \hline * & V\big((\omega_k)_{k\in C}\big) \cdot \operatorname{diag}\big((\omega_k^n)_{k\in C}\big) \end{array}\right).$$

Moving all columns whose index is an element in the set I to the first n positions requires $\sum_{k\in I} k - \frac{1}{2}n(n-1)$ swaps of neighboring columns. Hence for the coefficient of the monomial $\prod_{k\in I} z_k$ in $\det(M_n)$ one obtains:

$$(-1)^{\sum_{k\in I} k - \frac{1}{2}n(n-1)} \cdot \det V\big((\omega_k)_{k\in I}\big) \cdot \det V\big((\omega_k)_{k\in C}\big) \cdot \prod_{k\in C} \omega_k^n. \tag{11.28}$$

From the definition of ω_k it follows immediately that $\omega_k^n = (-1)^k$ if n is even, and $\omega_k^n = i \cdot (-1)^k$ if n is odd. Hence the term $\prod_{k\in C} \omega_k^n$ turns into $i^{[n]_2 \cdot n} \cdot (-1)^{\sum_{k\in C} k}$. The sign in (11.28) can now be determined by the parity of

$$\sum_{k\in I} k - \frac{n(n-1)}{2} + \sum_{k\in C} k = \sum_{k=0}^{2n-1} k - \frac{n(n-1)}{2}$$
$$= \frac{2n(2n-1)}{2} - \frac{n(n-1)}{2}$$
$$= \frac{n(3n-1)}{2} = n(n-1) + \frac{n(n+1)}{2}.$$

In addition, by employing the well-known formula for the Vandermonde determinant, Eq. (11.28) simplifies to

$$(-1)^{n(n+1)/2}\, i^{[n]_2\cdot n} \Bigg(\prod_{\substack{k,\ell\in I\\ k<\ell}} (\omega_\ell - \omega_k)\Bigg)\Bigg(\prod_{\substack{k,\ell\in C\\ k<\ell}} (\omega_\ell - \omega_k)\Bigg) =: c_I^{(n)}. \tag{11.29}$$

Hence the determinant of M_n can be written as follows:

$$\det(M_n) = \sum_{\substack{I\subset\{0,\ldots,2n-1\}\\ |I|=n}} c_I^{(n)} \cdot \prod_{k\in I} z_k. \tag{11.30}$$

Note that for computational purposes the Vandermonde products in the definition of $c_I^{(n)}$ can always be expressed in terms of $2n$th roots of unity, even if n is odd. In this case one has to extract the factor i^{n-1} which simplifies in combination with the term $i^{[n]_2\cdot n}$, namely $i^{n-1}\cdot i^{[n]_2\cdot n} = i$.

Theorem 11.1 *Let* $\omega_k = \exp\big(\frac{i\pi}{2n}(2k+[n]_2)\big)$ *and* $c_I^{(n)}$ *be as before, and define* $\alpha_I := \mathrm{Re}\big(\sum_{k\in I}\omega_k\big)$ *and* $\beta_I := \mathrm{Im}\big(\sum_{k\in I}\omega_k\big)$. *Then*

$$\det(A_n(z)) = z^{n(2n-1)} \cdot \sum_{\substack{I\subset\{0,\ldots,2n-1\}\\ |I|=n}} c_I^{(n)} \cosh(\alpha_I z)\cos(\beta_I z). \tag{11.31}$$

Proof We show that $\det(A_n(z))$ is basically a special case of (11.30) (up to the factor $z^{n(2n-1)}$ which comes from the common row factors z^j that were omitted in the definition of M_n). Recall that the variables z_k in the matrix M_n have replaced the exponential functions $e^{\omega_k z}$ that appear in the matrix $A_n(z)$. Thus, with $z_k = e^{\omega_k z}$ one obtains

$$\prod_{k\in I} z_k = \exp\Bigg(z\cdot\sum_{k\in I}\omega_k\Bigg) = \exp\big(z\cdot(\alpha_I + i\beta_I)\big).$$

As before, let $C := \{0,\ldots,2n-1\}\setminus I$ be the set complement of I. From $\sum_{k\in I}\omega_k + \sum_{k\in C}\omega_k = \sum_{k=0}^{2n-1}\omega_k = 0$, it follows immediately that

$$\alpha_C = -\alpha_I \quad\text{and}\quad \beta_C = -\beta_I.$$

Moreover, define

$$\bar{I} := \{2n - [n]_2 - k \bmod 2n \mid k \in I\},$$

which corresponds to reflecting the set $\{\omega_k\}_{k \in I}$ across the real axis. Then it follows that

$$\alpha_{\bar{I}} = \alpha_I \quad \text{and} \quad \beta_{\bar{I}} = -\beta_I.$$

Apparently, for those sets I for which $I = \bar{I}$ holds, one has $\beta_I = 0$. In such cases, we can combine the exponentials that correspond to I and to C, using the identity

$$e^{\alpha z} + e^{-\alpha z} = 2\cosh(\alpha z).$$

Otherwise, if $I \neq \bar{I}$, then $I, \bar{I}, C, \bar{C}$ are four pairwise distinct sets, and their corresponding exponentials can be combined as follows:

$$e^{(\alpha+i\beta)z} + e^{(\alpha-i\beta)z} + e^{(-\alpha+i\beta)z} + e^{(-\alpha-i\beta)z} = 4\cosh(\alpha z)\cos(\beta z).$$

The asserted formula follows by observing that $c_I^{(n)} = c_{\bar{I}}^{(n)} = c_C^{(n)} = c_{\bar{C}}^{(n)}$ (which is obvious from symmetry arguments).

Example 11.1 For $n = 2$, we deduce again the transcendental equation satisfied by $z = \lambda^{-\frac{1}{4}}$, but now using Theorem 11.1:

- For $I = \{0, 2\}$ one has $I = \bar{I}$ and $C = \bar{C} = \{1, 3\}$; it follows that

$$\begin{aligned} c_I^{(n)} &= (-1)^3 \cdot (\omega_2 - \omega_0) \cdot (\omega_3 - \omega_1) = -4i, \\ \alpha_I &= \beta_I = 0. \end{aligned}$$

- For $I = \{0, 1\}$ one has $\bar{I} = \{0, 3\}$, $C = \{2, 3\}$ and $\bar{C} = \{1, 2\}$; it follows that

$$\begin{aligned} c_I^{(n)} &= -(\omega_1 - \omega_0)(\omega_3 - \omega_2) = -(i-1)(-i+1) = -2i, \\ \alpha_I &= \mathrm{Re}(\omega_0 + \omega_1) = 1, \\ \beta_I &= \mathrm{Im}(\omega_0 + \omega_1) = 1. \end{aligned}$$

Summing over all possible subsets $I \subset \{0, 1, 2, 3\}$, we get

$$2 \cdot (-4i)\cosh(0)\cos(0) + 4 \cdot (-2i)\cosh(z)\cos(z) = -8i \cdot \big(1 + \cosh(z)\cos(z)\big),$$

which is in accordance with our previous result (Proposition 11.6).

Corollary 11.1 *In the case $n = 3$, i.e., for the threefold integration operator J^3, the eigenvalues λ of the operator $(J^3)^* J^3$ are the solutions of the nonlinear transcendental equation*

Table 11.2 Numerical approximations of the first five zeros z_i of the transcendental equations in Corollary 11.1 (with $z = \lambda^{-1/6}$) and Corollary 11.2 (with $z = \lambda^{-1/8}$)

	$n = 3$ (Corollary 11.1)	$n = 4$ (Corollary 11.2)
z_1	2.2247729764011889	2.5902718684989891
z_2	4.8026572459190195	5.0106222998859963
z_3	7.8476475910871745	7.8970686069935174
z_4	10.9951601546635699	10.9949247590502524
z_5	14.1371941952108977	14.1366518856561214

$$8\cos\big(\lambda^{-\frac{1}{6}}\big) + \cos\big(2\lambda^{-\frac{1}{6}}\big) + 2\cos\big(\lambda^{-\frac{1}{6}}\big)\cosh\big(\sqrt{3}\lambda^{-\frac{1}{6}}\big) + 16\cos\big(\tfrac{1}{2}\lambda^{-\frac{1}{6}}\big)\cosh\big(\tfrac{\sqrt{3}}{2}\lambda^{-\frac{1}{6}}\big) + 9 = 0.$$

Corollary 11.2 *In the case $n = 4$, i.e., for the fourfold integration operator J^4, the eigenvalues λ of the operator $(J^4)^*J^4$ are the solutions of the nonlinear transcendental equation (which is obtained from Theorem 11.1 after further simplifications):*

$$\begin{aligned}
&\cos\big(\sqrt{2}\lambda^{-\frac{1}{8}}\big) + \cosh\big(\sqrt{2}\lambda^{-\frac{1}{8}}\big) + 2\cos\big(\lambda^{-\frac{1}{8}}\big)\cosh\big(\lambda^{-\frac{1}{8}}\big)\\
&+ 3\big(\cos\big(\sqrt{2}\lambda^{-\frac{1}{8}}\big) + \cosh\big(\sqrt{2}\lambda^{-\frac{1}{8}}\big)\big)\cos\big(\lambda^{-\frac{1}{8}}\big)\cosh\big(\lambda^{-\frac{1}{8}}\big)\\
&+ 8\big(\cos\big(\lambda^{-\frac{1}{8}}\big) + \cosh\big(\lambda^{-\frac{1}{8}}\big)\big)\cos\big(\tfrac{1}{\sqrt{2}}\lambda^{-\frac{1}{8}}\big)\cosh\big(\tfrac{1}{\sqrt{2}}\lambda^{-\frac{1}{8}}\big)\\
&+ 4\sqrt{2}\sin\big(\lambda^{-\frac{1}{8}}\big)\sin\big(\tfrac{1}{\sqrt{2}}\lambda^{-\frac{1}{8}}\big)\cosh\big(\tfrac{1}{\sqrt{2}}\lambda^{-\frac{1}{8}}\big)\\
&- 4\sqrt{2}\cos\big(\tfrac{1}{\sqrt{2}}\lambda^{-\frac{1}{8}}\big)\sinh\big(\lambda^{-\frac{1}{8}}\big)\sinh\big(\tfrac{1}{\sqrt{2}}\lambda^{-\frac{1}{8}}\big)\\
&+ 2\sqrt{2}\sin\big(\lambda^{-\frac{1}{8}}\big)\sin\big(\sqrt{2}\lambda^{-\frac{1}{8}}\big)\cosh\big(\lambda^{-\frac{1}{8}}\big)\\
&- 2\sqrt{2}\cos\big(\lambda^{-\frac{1}{8}}\big)\sinh\big(\lambda^{-\frac{1}{8}}\big)\sinh\big(\sqrt{2}\lambda^{-\frac{1}{8}}\big) + 6 = 0.
\end{aligned}$$

The transcendental equations given in Corollaries 11.1 and 11.2 can be used to compute accurate approximations to the eigenvalues λ_i, using Newton's method, for example. In Table 11.2 the values for $\lambda_i^{-1/6}$ (resp. $\lambda_i^{-1/8}$) for $i = 1, \ldots, 5$ are displayed. As in the case $n = 2$, one observes that the ith value is close to $(2i-1)\frac{\pi}{2}$ (we give an explanation of this phenomenon in the next section). The corresponding eigenfunctions u_i are shown in Figs. 11.2 and 11.3.

11.7 Distribution of the Eigenvalues λ_i

In Theorem 11.1 we have stated that the determinant of $A_n(z)$ can be written as a sum of expressions of the form $c \cdot \cosh(\alpha z) \cdot \cos(\beta z)$ with $c \neq 0$ and $\alpha, \beta \in \mathbb{R}$. The location of the zeros of this determinant is mostly governed by the summand

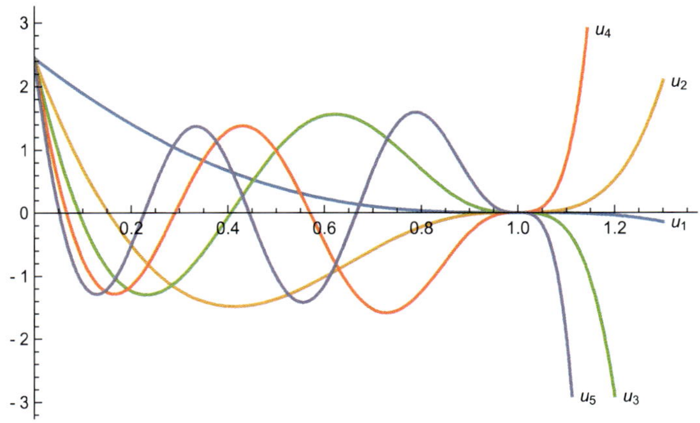

Fig. 11.2 Eigenfunctions $u_1, \ldots, u_5$ for $n = 3$

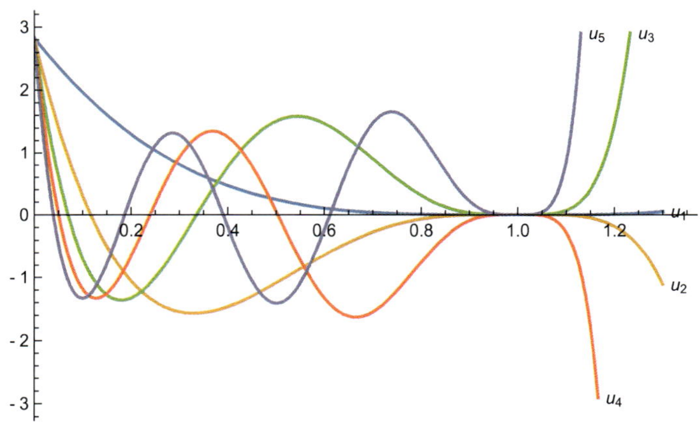

Fig. 11.3 Eigenfunctions $u_1, \ldots, u_5$ for $n = 4$

whose $\cosh(\alpha z)$ term has the fastest asymptotic growth, i.e., whose scaling factor α is largest. Recall that the α's are obtained as the real parts of sums (with n summands) of different $\omega_k = \exp\big(\frac{i\pi}{2n}(2k + [n]_2)\big)$. Figure 11.4 shows all possible pairs (α, β) when $n = 8$; note that there are $\binom{16}{8} = 12870$ sums, some of which add up to the same values.

If n is even, then obviously $\{i, -i\} \subseteq \{\omega_k \mid k = 0, \ldots, 2n-1\}$. Due to the "correction term" $[n]_2$, the same is true for odd n. Hence, there are exactly $n - 1$ numbers among the ω_k with strictly positive real part. By choosing them, plus one out of $\{i, -i\}$ (w.l.o.g. we choose i), their sum will clearly exhibit the largest real part among all n-sums of the ω_k.

We now study this sum in more detail. For even n we obtain (by combining pairs of complex conjugates):

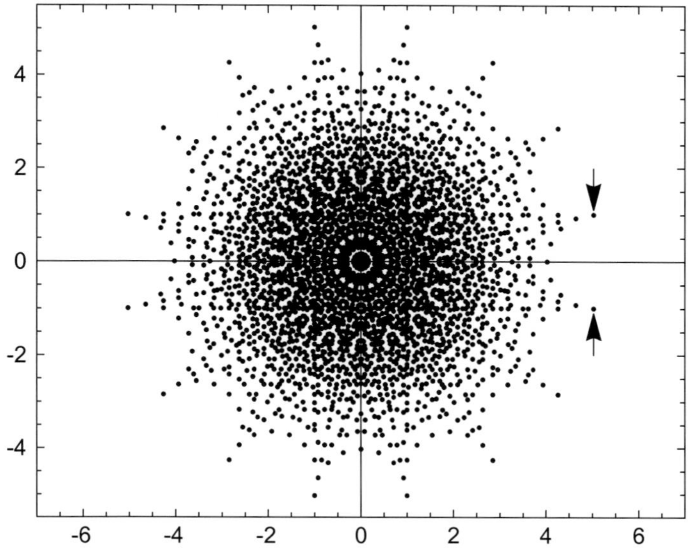

Fig. 11.4 The set $\left\{\sum_{k\in I}\omega_k \,\middle|\, I\subseteq\{0,\ldots,2n-1\}, |I|=n\right\}$ plotted in the complex plane (for $n=8$); the two arrows point to those two numbers which dominate the asymptotics, i.e., with largest real part

$$\sum_{k=-n/2+1}^{n/2}\omega_k = \sum_{k=-n/2+1}^{n/2}\exp\left(i\frac{2\pi k}{2n}\right) = i+1+\sum_{k=1}^{n/2-1}2\,\mathrm{Re}\left(\exp\left(i\frac{2\pi k}{2n}\right)\right)$$
$$= i+1+\sum_{k=1}^{n/2-1}2\cos\left(\frac{2\pi k}{2n}\right) = i+\cot\left(\frac{\pi}{2n}\right).$$

For odd n we obtain

$$\sum_{k=-(n-1)/2}^{(n-1)/2}\omega_k = \sum_{k=-(n-1)/2}^{(n-1)/2}\exp\left(i\frac{2\pi k+\pi}{2n}\right) = i+\sum_{k=0}^{(n-1)/2-1}2\,\mathrm{Re}\left(\exp\left(i\frac{2\pi k+\pi}{2n}\right)\right)$$
$$= i+\sum_{k=0}^{(n-1)/2-1}2\cos\left(\frac{2\pi k+\pi}{2n}\right) = i+\cot\left(\frac{\pi}{2n}\right).$$

Note that the result is the same for even and for odd n. Hence the behavior of $\det(A_n(z))$ is dominated by the term $c\cdot\cosh(\alpha z)\cdot\cos(\beta z)$ with $\alpha=\cot\left(\frac{\pi}{2n}\right)$ and $\beta=1$, as z goes to infinity. This implies that for large z the zeros of $\det(A_n(z))$ tend to the zeros of $\cos(z)$, which is exactly what was observed in Sect. 11.5.2 for the special case $n=2$, and in Sect. 11.6 for the cases $n=3$ and $n=4$.

11.8 Conclusions

In the paper, we have further investigated the Singular Value Decomposition of the n-fold integration operators J^n. We have presented an algorithm that can be used to compute the eigenvalues and eigenfunctions of $(J^n)^* J^n$. Out of the 5 steps of the algorithm, the first three can be done explicitely, but for the last two it seem impossible to solve them explicitely. The reason lies in the fact that the computation of the eigenvalues requires to solve a transcendental equation. However, for the cases $n = 2, 3, 4$ we computed some of the eigenvalues as well as their eigenfunctions approximately. It needs to be mentioned that the numerical computation of the eigenvalues requires a computer program that allows computations in arbitrary precision—systems like MATLAB failed to deliver reasonable approximations to the eigenvalues.

Acknowledgements Ronny Ramlau was supported by the Austrian Science Fund (FWF): SFB F68-N36 and DK W1214. Christoph Koutschan was supported by the Austrian Science Fund (FWF): P29467-N32 and F5011-N15. Bernd Hofmann was supported by German Research Foundation (DFG): HO 1454/12-1.

References

1. H.W. Engl, *Integralgleichungen* (Springer, Wien/New York, 1997)
2. R. Gorenflo, S. Vessella, *Abel Integral Equations* (Springer, Berlin, 1991)
3. B. Hofmann, L. von Wolfersdorf, Some results and a conjecture on the degree of ill-posedness for integration operators with weights. Inverse Probl. **21**(2), 427–433 (2005)
4. A. Kirsch, *An Introduction to the Mathematical Theory of Inverse Problems*, 2nd edn. (Springer, New York, 2011)
5. R. Kress, *Linear Integral Equations* (Springer, Berlin, 1989)
6. P. Mathé, Saturation of regularization methods for linear ill-posed problems in Hilbert spaces. SIAM J. Numer. Anal. **42**(3), 968–973 (2004)
7. M.Z. Nashed, A new approach to classification and regularization of ill-posed operator equations. in *Inverse and Ill-Posed Problems*, Sankt Wolfgang, 1986, eds. by H.W. Engl, C.W. Groetsch (Academic Press, Boston, MA, 1987), pp. 53–75
8. N.J.A. Sloane, The on-line encyclopedia of integer sequences. OEIS Foundation Inc., http://oeis.org
9. V.K. Tuan, R. Gorenflo, Asymptotics of singular values of fractional integral operators. Inverse Probl. **10**(4), 949–955 (1994)

Chapter 12
The Kurdyka–Łojasiewicz Inequality as Regularity Condition

Daniel Gerth and Stefan Kindermann

Abstract We show that a Kurdyka–Łojasiewicz (KL) inequality can be used as regularity condition for Tikhonov regularization with linear operators in Banach spaces. In fact, we prove the equivalence of a KL inequality and various known regularity conditions (variational inequality, rate conditions, and others) that are utilized for postulating smoothness conditions to obtain convergence rates. Case examples of rate estimates for Tikhonov regularization with source conditions or with conditional stability estimates illustrate the theoretical result.

Keywords KL-inequality · Convex Tikhonov-regularization · Ill-posed problem · Convergence rates · Regularity condition

12.1 Introduction

In the theory of the regularization of ill-posed inverse problems, it is well-known that the behavior of regularization methods essentially depends on the interplay of the forward operator with the true solution. Over time, several conditions have been developed that, usually formulated as assumptions, allow for a more or less precise description of the regularization process. In this paper, we will connect the set of smoothness conditions discussed in the recent papers [23, 27] to a Kurdyka–Łojasiewicz (KL) inequality. The KL inequality, which we introduce in detail in Sect. 12.3, has been utilized in various branches of mathematics since its discovery in the 1960s. Hence, it may open new perspectives to inverse problems.

Before going into detail, we introduce the setting of our paper. We consider operator equations

D. Gerth (✉)
Faculty of Mathematics, Chemnitz University of Technology, 09107 Chemnitz, Germany
e-mail: daniel.gerth@mathematik.tu-chemnitz.de

S. Kindermann
Industrial Mathematics Institute, Johannes Kepler University Linz, 4040 Linz, Austria
e-mail: kindermann@indmath.uni-linz.ac.at

J. Cheng et al. (eds.), *Inverse Problems and Related Topics*,
Springer Proceedings in Mathematics & Statistics 310,
https://doi.org/10.1007/978-981-15-1592-7_12

$$Ax = y \tag{12.1}$$

where A is a bounded linear operator mapping from an infinite-dimensional Banach space X to an infinite-dimensional Hilbert space Y. We assume that the range $\mathcal{R}(A)$ of A is not closed in Y, $\mathcal{R}(A) \neq \overline{\mathcal{R}(A)}$, such that A is not continuously invertible and hence (12.1) is ill-posed. We assume that only noisy data y^δ is available with $||y - y^\delta|| \leq \delta$ for $\delta > 0$. Due to the ill-posedness of (12.1) and the noisy data, we employ the Tikhonov-type regularization

$$T_\alpha^\delta(x) = \frac{1}{2}||Ax - y^\delta||^2 + \alpha J(x) \tag{12.2}$$

to determine a stable approximation to the true solution $x^\dagger$ for which $Ax^\dagger = y$ holds. In (12.2), $\alpha > 0$ is the regularization parameter and $J : \mathcal{D}(J) \subset X \to \mathbb{R}$ the penalty functional. The minimizer of (12.2) is the regularized solution, i.e.,

$$x_\alpha^\delta = \arg\min_{x \in \mathcal{D}(A)} T_\alpha^\delta(x). \tag{12.3}$$

By omitting the superscript δ, we denote noise-free data and variables, i.e.,

$$x_\alpha = \arg\min_{x \in \mathcal{D}(A)} T_\alpha \quad \text{with} \quad T_\alpha(x) = \frac{1}{2}||Ax - y||^2 + \alpha J(x). \tag{12.4}$$

In order to guarantee existence and stability of the approximations x_α^δ and x_α, respectively, we impose the following standard assumptions (see, e.g., [23, 32]) on the penalty functional J throughout the paper:

Assumption 12.1 The functional $J : X \to [0, \infty]$ is a proper, convex functional defined on a Banach space X, which is lower semicontinuous with respect to weak (or weak*) sequential convergence. Additionally, we assume that J is a stabilizing (weakly coercive) functional, i.e., the sublevel sets $[J \leq c] := \{x \in X : J(x) \leq c\}$ of J are, for all $c \geq 0$, weakly (or weakly*) sequentially compact. Moreover, we assume that at least one solution $x^\dagger$ of (12.1) with finite penalty value $J(x^\dagger) < \infty$ exists and that the subgradient $\partial J(x^\dagger)$ exists.

With the basic regularization properties covered as consequence of Assumption 12.1, we move directly to the discussion of convergence rates. In Banach space regularization, the Bregman distance

$$B_\xi(z, x) := J(x) - J(z) - \langle \xi, x - z \rangle \geq 0, \quad x \in X, \quad \xi \in \partial J(z) \subset X^*,$$

where the subgradient ξ is an element of the subdifferential $\partial J(z)$ of J in the point $z \in X$, has become a popular choice to measure the speed of convergence of the approximate solution to the true solution $x^\dagger$. In this paper, we follow the approach of [23] and consider the Bregman distance

$$B_{\xi_\alpha^\delta}(x_\alpha^\delta, x^\dagger)$$

with subgradient taken at the approximate solutions. Note that the Bregman distance is not symmetric in its arguments. Our task is to find an index function φ, i.e., a monotonically increasing function $\varphi : [0, \infty) \to [0, \infty)$ with $\varphi(0) = 0$ that is continuous (possibly only in a neighborhood of 0), such that

$$B_{\xi_\alpha^\delta}(x_\alpha^\delta, x^\dagger) \le \varphi(\delta). \tag{12.5}$$

It is well-known that no uniform function φ exists *for all* $x^\dagger \in X$, and that φ has to take into account the interplay between the operator A, the solution $x^\dagger$, and the penalty functional J, in combination with an appropriate choice of the regularization parameter $\alpha > 0$ in (12.2) and (12.4), respectively. Many conditions have been developed that control this interplay and yield convergence rates (12.5). It is the aim of this paper to show the equivalence of most of the known conditions, and more important, we add another equivalent condition in form of the KL-inequality.

12.2 Convergence Rate Theory for Convex Tikhonov Regularization

For the complete statement of our equivalence results, we also need Flemming's distance function [14–16]:

$$D(r) := \sup_{x \in X} \left(J(x^\dagger) - J(x) - r\|Ax - Ax^\dagger\| \right).$$

Theorem 12.1 *The following statements are equivalent:*

(a) (J-rate) There is an index function Ψ_1 such that

$$J(x^\dagger) - J(x_\alpha) \le \Psi_1(\alpha) \quad \text{for all} \quad \alpha > 0. \tag{12.6}$$

(b) (T-rate) There is an index function Ψ_2 such that

$$\frac{1}{\alpha}\left(T_\alpha(x^\dagger) - T_\alpha(x_\alpha)\right) \le \Psi_2(\alpha) \quad \text{for all} \quad \alpha > 0. \tag{12.7}$$

(c) (Variational inequality) There is an index function Ψ_3 such that

$$J(x^\dagger) - J(x) \le \Psi_3(\|Ax^\dagger - Ax\|) \qquad \text{for all} \quad x \in X. \tag{12.8}$$

(d) (Distance function) There is an index function Ψ_4 such that

$$D(\tfrac{1}{r}) \le \Psi_4(r) \qquad \forall r > 0. \tag{12.9}$$

(e) *(Dual T-rate) There exists an index function Ψ_5 such that for all $\alpha > 0$ a $z \in Y$ exist with $x^* = \partial J(x^\dagger)$ and*

$$J^*(A^*z) - J(x^*) - (x^\dagger, A^*z - x^*)_{X,X^*} + \alpha \frac{1}{2}\|z\|^2 \leq \Psi_5(\alpha), \qquad (12.10)$$

(f) *(KL-inequality) There exists a concave index function φ such that $(\partial\varphi)^{-1}(z)z$ is nonincreasing with $\lim_{z\to\infty}(\partial\varphi)^{-1}(z)z = 0$, with*

$$\left\|\partial\left(\varphi \circ \left(T_\alpha(x^\dagger) - T_\alpha(x_\alpha)\right)\right)\right\| \geq \frac{1}{k}. \qquad (12.11)$$

Proof In the proof we provide the formula for converting the various index functions: In [23, Prop. 2.4] the equivalence of (a) and (b) was shown:

$$\text{(a)} \Rightarrow \text{(b)} : \Psi_2 \leq \Psi_1 \qquad \text{(b)} \Rightarrow \text{(a)} : \Psi_1 \leq 2\Psi_2.$$

Also in [23, Prop 3.3] it was shown that

$$\text{(c)} \Rightarrow \text{(b)} : \Psi_2(\alpha) \leq \sup_{t>0}\left(\Phi_3(t) - \frac{t^2}{2\alpha}\right).$$

It follows that Ψ_2 is increasing and by continuity of Ψ_3, it can be shown that $\Psi_2(0) = 0$. We now show (b) $\Rightarrow$ (c): From (12.7), it follows, for all x and all α,

$$J(x^\dagger) - J(x_\alpha) - \frac{1}{2\alpha}\|Ax_\alpha - Ax^\dagger\|^2 \leq \frac{1}{\alpha}T_\alpha(x^\dagger, Ax^\dagger) - \frac{1}{\alpha}T_\alpha(x_\alpha, Ax^\dagger) \leq \Psi_2(\alpha).$$

Thus, from the optimality of x_α, we find

$$J(x^\dagger) - J(x) \leq \Psi_2(\alpha) + \frac{1}{2\alpha}\|Ax - Ax^\dagger\|^2.$$

Taking the infimum over α yields the variational inequality (12.8) with the function Φ_3

$$\text{(b)} \Rightarrow \text{(c)} : \Psi_3(\alpha) = \inf_{t>0}\left(\Psi_2(t) + \frac{\alpha^2}{2t}\right).$$

If follows easily that Ψ is an index function.

Moreover, (d) $\Leftrightarrow$ (c) by results of Flemming [16, Lemma 3.4] [14, Thm. 12.32], with

$$\text{(d)} \Rightarrow \text{(c)} : \Psi_3(\alpha) = \inf_{r>0}\left(\Psi_4(r) + r\alpha\right)$$

and

$$\text{(c)} \Rightarrow \text{(d)} : \Psi_4(\alpha) = \inf_{r>0}\left(\Psi_3(t) - \alpha t\right).$$

Concerning (f), we remark that by duality we may rewrite the Tikhonov functional as

$$\frac{1}{\alpha}\left(T_\alpha(x^\dagger) - T_\alpha(x_\alpha)\right) = J(x^\dagger) - \tfrac{1}{\alpha} T_\alpha(x_\alpha)$$
$$= J(x^\dagger) - \tfrac{1}{\alpha}\sup_p \left[-\tfrac{1}{2}\|p\|^2 + (p, Ax^\dagger) - \alpha J^*\left(\frac{1}{\alpha}A^* p\right)\right]$$
$$= \inf_p \left[J(x^\dagger) + J^*\left(\frac{1}{\alpha}A^* p\right) - \tfrac{1}{\alpha}(p, Ax^\dagger) + \tfrac{1}{2\alpha}\|p\|^2\right].$$

Young's inequality yields $J(x^\dagger) = (x^\dagger, x^*) - J^*(x^*)$, and by setting $z = \frac{1}{\alpha}p$ it is clear that (f) is just a reformulation of (b): (Note that the infimum over p is attained).

$$\text{(b)} \Leftrightarrow \text{(f)} : \Psi_5 = \Psi_2.$$

Similar formulas were actually already used by Flemming [14].

The essential equivalence of the KL inequality (g) is one of the main issues in this paper and will be shown in later sections in Theorem 12.3. □

Hence, any of the conditions in Theorem 12.1 implies the other ones. These conditions imply a certain decay rate for the approximation error in the Bregman distance. This subsequently yields convergence rate for the total error measured in the Bregman distance. Not only this, but we immediately obtain errors in the strict metric and a Tikhonov rate (These results were obtained or follow easily from [23, Thm. 2.8, Prop. 3.7]):

Theorem 12.2 *Let any of the equivalent assumptions in Theorem 12.1 hold.*

1. *(Bregman rate) There is a constant C such that, for all* $\alpha > 0$,

$$B_{\xi_\alpha^\delta}(x_\alpha^\delta, x^\dagger) \le C \inf_\alpha \left(\frac{\delta^2}{\alpha} + \Psi_2(\alpha)\right). \tag{12.12}$$

2. *(strict metric rate) There is a constant C such that for all* $\alpha > 0$

$$\begin{aligned} \left|J(x^\dagger) - J(x_\alpha^\delta)\right| &\le C\left(\Psi_2(\alpha) + \frac{\delta^2}{\alpha}\right), \\ \|Ax_\alpha^\delta - Ax^\dagger\|^2 &\le C\left(\alpha\Psi_2(\alpha) + \delta^2\right). \end{aligned} \tag{12.13}$$

3. *(Tikhonov rate) There is a constant C such that, for all* $\alpha > 0$,

$$|J(x^\dagger) - \frac{1}{\alpha}T_\alpha^\delta(x_\alpha^\delta)| \le C\left(\Psi_2(\alpha) + \frac{\delta^2}{\alpha}\right). \tag{12.14}$$

Moreover, defining the companion $\Theta(\alpha)$ as

$$\Theta(\alpha) := \sqrt{\alpha \Psi_2(\alpha)}, \tag{12.15}$$

the a-priori choice

$$\alpha_* = \alpha_*(\delta) := \left(\Theta^2\right)^{-1}\left(\frac{\delta^2}{2}\right) = \Theta^{-1}\left(\frac{\delta}{\sqrt{2}}\right) \tag{12.16}$$

obtained by equilibrating the error decomposition (12.12) yields the following convergence rate:

Corollary 12.1 *Let any of the equivalent assumptions in Theorem 12.1 hold. Then with the choice* (12.16) *we obtain the convergence rates*

$$B_{\xi_\alpha^\delta}(x_\alpha^\delta, x^\dagger) \leq 2\Psi_2\left(\Theta^{-1}\left(\frac{\delta}{\sqrt{2}}\right)\right). \tag{12.17}$$

Note that the same rates holds for the analog error measures in (12.13) and (12.14).

12.3 The Łojasiewicz-Inequality and Its Consequences

In this section we give a brief overview over the Kurdyka–Łojasiewicz (KL) inequality and some of its implications. A main reason for our interest in this inequality is its broad spectrum of applications in several mathematical disciplines. This may open new interconnections for inverse problems. We start with a short and certainly incomplete overview of the KL inequality.

Łojasiewicz showed that for any real analytic function $f : D(f) \subset \mathbb{R}^n \to \mathbb{R}$ there is $\theta \in [0, 1)$ such that

$$\frac{|f(x) - f(\bar{x})|^\theta}{\|\nabla f(x)\|}$$

remains bounded around any critical point $\bar{x}$, i.e., $\nabla f(\bar{x}) = 0$ [29, 30]. Kurdyka [28] later generalized the result to C^1 functions whose graphs belong to an o-minimal structure. A further generalization to nonsmooth subanalytic functions was given in [5]. It can also be formulated in (general) Hilbert spaces, see, e.g., [12, 22], and has applications, for example, in PDE analysis (see, for example, [21, 24, 33]), neural networks [17] and complexity theory [31]. First approaches towards inverse problems were made in [19, 20]. In the optimization literature, the KL inequality has emerged as a powerful tool to characterize the convergence properties of iterative algorithms; see, e.g., [1, 2, 5, 7, 8, 18, 19].

It is known that the KL inequality immediately yields a measure for the distance between the level-sets of a function, which, under some additional assumptions, directly yields convergence rates for the noise-free Tikhonov functional (12.4). To show the generality of the KL inequality, we temporarily consider the problem

$$f(x) \to \min_{x \in X}$$

where X is a complete metric space with metric $d(x, y)$ and $f : X \to \mathbb{R} \cup \{\infty\}$ is lower semicontinuous. To formulate the result in this abstract setting, we use the following notation.

Definition 12.1 We denote by

$$[t_1 \leq f \leq t_2] := \{x \in X : t_1 \leq f(x) \leq t_2\} \tag{12.18}$$

the level-set of f for the levels $t_1 \leq t_2$. With slight abuse of notation we write, for fixed $x \in X$, $[f(x)] := [f = f(x)]$. Furthermore, for any $x \in X$, the distance of x to a set $S \subset X$ is denoted by

$$dist(x, S) := \inf_{y \in S} d(x, y). \tag{12.19}$$

With this we recall the Hausdorff distance between sets,

$$D(S_1, S_2) := \max\{\sup_{x \in S_1} dist(x, S_2), \sup_{x \in S_2} dist(x, S_1)\}. \tag{12.20}$$

The KL inequality is directly linked to certain index functions, which we specify below.

Definition 12.2 A concave function $\varphi : [0, \bar{r}) \to \mathbb{R}$ is called desingularizuation function or smooth index function if $\varphi \in C(0, \bar{r}) \cap C^1(0, \bar{r})$, $\varphi(0) = 0$, and $\varphi'(x) > 0$ for all $x \in (0, \bar{r})$. We denote the set of all such φ with $\mathcal{K}(0, \bar{r})$.

Now we are ready to cite the main inspiration for our work. It is taken from [6]. In comparison to the original result we have omitted a third equivalence to the concept of metric regularity, see [26]. Note that we replaced f with $f - \inf f$.

Proposition 12.1 ([6, Corollary 4]) *Let $f : X \to \mathbb{R} \cup \{\infty\}$ be a lower semicontinuous function defined on a complete metric space and $\varphi \in \mathcal{K}(0, r_0)$. Assume that $[\inf f < f < r_0 - \inf f] \neq \emptyset$. Then the following assumptions are equivalent.*

(a) For all $r_1, r_2 \in (\inf f, r_0)$

$$D([f \leq r_1 - \inf f], [f \leq r_2 - \inf f]) \leq k|\varphi(r_1 - \inf f) - \varphi(r_2 - \inf f)|. \tag{12.21}$$

(b) For all $x \in [0 < f < r_0]$

$$|\nabla(\varphi \circ (f - \inf f))|(x) \geq \frac{1}{k}, \tag{12.22}$$

where $|\nabla f|(x) := \limsup_{\tilde{x} \to x} \frac{\max(f(x) - f(\tilde{x}), 0)}{d(x, \tilde{x})}$ is the strong slope.

Now we return to X being a Banach space and consider the Tikhonov functional $f = T_\alpha(x)$. Due to the convexity of the penalty J, we can write Proposition 12.1 in the following way, where

$$\|\partial f(x)\|_- := \inf_{p \in \partial f(x)} \|p\|_{X^*} = \operatorname{dist}(0, \partial f(x)) = |\nabla f|(x) \tag{12.23}$$

is the *remoteness* of the subdifferential of f in x; see also [3, 4].

Corollary 12.2 *Let either A be injective or J be strictly convex. Then, for the Tikhonov functional $T_\alpha(x)$ from* (12.4)*, the following are equivalent for a smooth index function $\varphi \in \mathcal{K}(0, \tilde{r})$, $x \in [T_\alpha(x_\alpha) \le T_\alpha(x) \le \tilde{r}]$, and $0 < k < \infty$.*

(a)

$$\|x - x_\alpha\| \le k\varphi(T_\alpha(x) - T_\alpha(x_\alpha)),$$

(b)

$$\varphi'(T_\alpha(x) - T_\alpha(x_\alpha))\, \|\partial T_\alpha(x)\|_- \ge \frac{1}{k}. \tag{12.24}$$

Proof Due to Assumption 12.1 minimizers of $T_\alpha(x)$ exist, and due to the injectivity of A or strict convexity of J the minimizers are unique. Hence it is plain to see from the definition of the Hausdorff-metric (12.20) that

$$\|x - x_\alpha\| \le D([T_\alpha(x)], [T_\alpha(x_\alpha)]),$$

and we obtain (a). For (semi)-convex functions, the strong slope coincides with $\|\partial T_\alpha(x)\|_-$ ([6, Remark 12]), from which the remainder follows.

We close this section by mentioning two obstacles in the application of Corollary 12.2. Firstly, it should be noted that a functional $f = g + h$ does not necessarily fulfill a KL inequality although both g and h do so. It is therefore not clear how to properly treat such a sum functional. While a partial answer is given in [19, Thm. 3.11], we can not apply the results since they require an invertible operator A. We will sketch in Sect. 12.6 that the Tikhonov functional (12.4) behaves differently than it would be expected from the sum of its parts. The second issue in applying Corollary 12.2 lies in the fact that it only holds in the noise-free case. To the best of the authors knowledge, there are no results on how the KL inequality behaves under noisy data. It is, however, out of the scope of this paper to close this gap.

12.4 The KL-Regularity Condition

Due to the equivalences of Theorem 12.1, it is sufficient to connect one of the conditions (a)–(e) with the KL inequality, and (b) appears to be most simple.

Theorem 12.3 *The following are equivalent:*

(a) There is a $\varphi \in \mathcal{K}(0,\infty)$ *such that* $(\partial\varphi)^{-1}(z)z$ *is nonincreasing with* $\lim_{z\to\infty}(\partial\varphi)^{-1}(z)z = 0$ *and a constant k such that*

$$\|\partial\left(\varphi\circ\left(T_\alpha(x^\dagger) - T_\alpha(x_\alpha)\right)\right)\| \geq \frac{1}{k}. \tag{12.25}$$

(b) There is an index function Ψ *such that*

$$\frac{1}{\alpha}\left(T_\alpha(x^\dagger) - T_\alpha(x_\alpha)\right) \leq \Psi(\alpha) \ \ \textit{for all} \ \ \alpha > 0. \tag{12.26}$$

The functions φ *and* Ψ *are connected via* $\Psi(t) = \frac{1}{t}(\partial\varphi)^{-1}\left(\frac{1}{tk\|[\partial J](x^\dagger)\|_-}\right)$.

Proof First, we observe that in our context, where x_α is the minimizer of the Tikhonov functional and $x^\dagger$ is the point of interest, the KL inequality (12.25) can be written as

$$\partial\varphi\left(T_\alpha(x^\dagger, y) - T_\alpha(x_\alpha, y)\right)\operatorname{dist}(0, [\partial T_\alpha(x^\dagger, y)](x^\dagger)) \geq \frac{1}{k}, \tag{12.27}$$

where

$$[\partial T_\alpha(x^\dagger, y)](x^\dagger) = A^*(Ax^\dagger - y) + \alpha[\partial J](x^\dagger) = \alpha[\partial J](x^\dagger).$$

By concavity, $\partial\varphi$ is monotonically decreasing and thus (12.27) leads to

$$T_\alpha(x^\dagger) - T_\alpha(x_\alpha) \leq \partial\varphi^{-1}\left(\frac{1}{k\alpha\|\partial J(x^\dagger)\|_-}\right).$$

Dividing both sides by $\alpha > 0$ yields (b) with

$$\Psi(\alpha) = \tfrac{1}{\alpha}\partial\varphi^{-1}\left(\frac{1}{k\alpha\|\partial J(x^\dagger)\|_-}\right).$$

This function is an index function by assumptions.

On the other hand, we write (b) as

$$T_\alpha(x^\dagger) - T_\alpha(x_\alpha) \leq \alpha\Psi(\alpha),$$

and by defining

$$\bar{\Theta}(\alpha) := \frac{\Psi(\frac{1}{\alpha})}{\alpha}$$

we have

$$\Delta T \leq \bar{\Theta}\left(\frac{1}{\alpha}\right).$$

As $\Psi(\frac{1}{\alpha})$ is nonincreasing so is $\bar{\Theta}$, hence

$$\bar{\Theta}^{-1}(\Delta T) \geq \frac{1}{\alpha}.$$

Finally, identifying $\partial\varphi = \bar{\Theta}^{-1}$ and noting that $\|\partial T_\alpha(x^\dagger)\| \sim \alpha$, we get the KL inequality (12.27) up to constants. As $\bar{\Theta}^{-1}$ is nonincreasing, φ is concave. Note that $\partial\varphi^{-1}(z)z = \Psi(\frac{1}{z})$ such that the stated condition on φ follow as Ψ is an index function.

It is interesting that in the proof we stumbled upon the companion function Θ from (12.15). Namely, we have $\Theta^2(\alpha) = \bar{\Theta}(\frac{1}{\alpha})$. The proof also reveals the identification

$$\Theta^2(\alpha) = (\partial\varphi)^{-1}\left(\frac{c}{\alpha}\right).$$

Equation (12.16) for the a priori choice ($\Theta^2(\alpha_*) \sim \delta^2$) of the regularization parameter then reads

$$\alpha_* = \frac{1}{\partial\varphi(\delta^2)}, \tag{12.28}$$

and we obtain the formal convergence rate

$$B_{\xi_\alpha^\delta}(x_\alpha^\delta, x^\dagger) \leq \partial\varphi(\delta^2)(\partial\varphi)^{-1}\left(c\partial\varphi(\delta^2)\right) \sim \partial\varphi(\delta^2)\delta^2.$$

Since $\varphi \in \mathcal{K}(0, r_0)$ is by definition concave, it holds that

$$\partial\varphi(\delta^2)\delta^2 \leq \varphi(\delta^2),$$

which follows from the property of the "subgradient" of concave functions, where the inequality is reversed compared to convex ones:

$$\partial\varphi(x)(0 - x) + \varphi(x) \geq \varphi(0) = 0.$$

12.5 Relation to Conditional Stability Estimates

We illustrate how the KL-theory quite directly yields convergence rates in case that a conditional stability estimate holds. Note that such estimates are a very useful tool in, e.g., parameter identification problems in partial differential equations; for examples, see, e.g., [9, 11, 25, 34]. The use of conditional stability estimates for rate estimates was in particular investigated by Cheng and Yamamoto in the seminal article [10].

Consider the Tikhonov functional

$$T_\alpha^\delta(x) = \frac{1}{2}\|Ax - y^\delta\|_Y^2 + \frac{1}{2}\alpha\|x\|_Z^2, \tag{12.29}$$

where $A : Z \to Y$ and $y = Ax^\dagger$ with $\|x^\dagger\|_Z \le M_0$. We furthermore assume that the Hilbert space $Z \hookrightarrow X$ is continuously embedded into a Banach space X, and there we assume a conditional stability estimate to hold: for some $0 \le \gamma < 1$ we assume that

$$\|f_1 - f_2\|_X \le C(M)\|Af_1 - Af_2\|_Y^\gamma \qquad \forall \|f_1\|_Z, \|f_2\|_Z \le M. \tag{12.30}$$

Cheng and Yamamoto have considered a more general setup and verified convergence rates. For simplicity and to show the relation between conditional stability estimates and the KL inequality, we deem it sufficient to consider only the Hölder-type stability estimate (12.30). Note that the case $0 \le \gamma < 1$ is not artificial. Let for example X a Hilbert space and $Z := \{x \in X : x = (A^*A)^\mu w\}$ for some $0 < \mu < \infty$ with norm $\|x\|_Z = \|(A^*A)^{-\mu}x\|_X = \|w\|_X$. Then it is well known (e.g., [13]), that (12.30) holds with $\gamma = \frac{2\mu}{2\mu+1}$ and $C(M) = M^{\frac{1}{2\mu+1}}$.

Here we illustrate the approach of Cheng and Yamamoto via the KL-inequality. To this end, we extend the Tikhonov functional (12.29) as follows to X:

$$\bar{T}_\alpha^\delta(x) := \begin{cases} T_\alpha^\delta(x) & \text{if } x \in Z, \\ \infty & \text{if } x \in X, x \notin Z. \end{cases}$$

In order to verify a KL-inequality as in Proposition 12.1 and Corollary 12.2, we verify first the KL-inequality (12.24) for $\bar{T}_\alpha^\delta$ for $x \in X$ with

$$T_\alpha^\delta(x_\alpha^\delta) \le T_\alpha^\delta(x) \le T_\alpha^\delta(x_\alpha^\delta) + C_0\delta^2 =: \tilde{r}, \tag{12.31}$$

where $1 < C_0 < \infty$ and $x_\alpha^\delta := \arg\min_{x \in X} T_\alpha^\delta(x)$, which we assume to exist. Note that this means in particular $\bar{T}_\alpha^\delta(x) < \infty$, i.e., $x \in Z$. In this case (12.24) reads

$$\varphi'(T_\alpha^\delta(x) - T_\alpha^\delta(x_\alpha))\|\partial\bar{T}_\alpha^\delta(x)\|_- \ge \frac{1}{k}.$$

In the following we write A^* for the adjoint of A in the space Z.

By [3, Prop 3.1] the strong slope or the remoteness can be characterized by the directional derivative $(\bar{T}_\alpha^\delta)'$,

$$\begin{aligned} \|\partial\bar{T}_\alpha^\delta(x)\|_- &= \sup_{\bar{T}_\alpha^\delta(z) < \bar{T}_\alpha^\delta(x)} \frac{-(\bar{T}_\alpha^\delta)'(x, z-x)}{\|x - z\|_X}, \qquad \text{which simplifies to} \\ &= \sup_{T_\alpha^\delta(z) < T_\alpha^\delta(x)} \frac{\left(\nabla T_\alpha^\delta(x), x - z\right)_Z}{\|x - z\|_X}, \end{aligned} \tag{12.32}$$

where $\nabla T_\alpha^\delta(x)$ is the usual gradient in the space Z:

$$\nabla T_\alpha^\delta(x) = A^*(Ax - y^\delta) + \alpha x.$$

The optimality condition for x_α^δ reads

$$A^*Ax_\alpha^\delta + \alpha x_\alpha^\delta = A^*y^\delta.$$

After some algebraic manipulation exploiting this identity, we obtain

$$\begin{aligned} T_\alpha^\delta(x) - T_\alpha^\delta(x_\alpha^\delta) &= \frac{1}{2}\|Ax - y^\delta\|^2 + \frac{1}{2}\alpha\|x\|_Z^2 - \left(\frac{1}{2}\|Ax_\alpha^\delta - y^\delta\|^2 + \alpha\frac{1}{2}\|x_\alpha^\delta\|_Z^2\right) \\ &= \frac{1}{2}\|Ax - Ax_\alpha^\delta\|^2 + \alpha\frac{1}{2}\|x_\alpha^\delta - x\|_Z^2. \end{aligned} \tag{12.33}$$

The difference $T_\alpha^\delta(x) - T_\alpha^\delta(x_\alpha^\delta)$ clearly depends on the choice of α. Therefore, we now consider the a priori choice

$$\alpha = c_\alpha \delta^2 \tag{12.34}$$

with positive constant c_α, which is the choice in [10].

This parameter choice, together with (12.33) and (12.31) implies

$$\frac{1}{2}\|x_\alpha^\delta - x\|_Z^2 \le \frac{C_0}{c_\alpha}. \tag{12.35}$$

Also, together with the optimality of the Tikhonov functional, $T_\alpha^\delta(x_\alpha^\delta) \le T_\alpha^\delta(x^\dagger)$, the parameter choice yields

$$\|x_\alpha^\delta\|_Z^2 \le \frac{1}{c_\alpha} + M_0^2,$$

and hence with (12.35) we obtain

$$\|x\|_Z^2 \le \|x_\alpha^\delta\|_Z^2 + \|x_\alpha^\delta - x\|_Z^2 \le \frac{1 + 2C_0}{c_\alpha} + M_0^2.$$

We set $M_1 := \frac{1+2C_0}{c_\alpha} + M_0^2$ and note that $\|x_\alpha^\delta\|_Z^2 \le M_1$. Now we can apply the conditional stability estimate (12.30). Namely, using additionally the first-order optimality condition and (12.32), we have

$$\begin{aligned} &\frac{1}{2}\|Ax - Ax_\alpha^\delta\|^2 + \alpha\frac{1}{2}\|x_\alpha^\delta - x\|_Z^2 = \frac{1}{2}(A^*A(x - x_\alpha^\delta) + \alpha(x - x_\alpha^\delta), x - x_\alpha^\delta)_Z \\ &= \frac{1}{2}(A^*(Ax - y^\delta) + \alpha x, x - x_\alpha^\delta)_Z = \frac{1}{2}(\nabla T_\alpha^\delta(x), x - x_\alpha^\delta)_Z \\ &\le \frac{1}{2}\|\partial \bar{T}_\alpha^\delta(x)\|_- \|x - x_\alpha^\delta\|_X \le \frac{1}{2}\|\partial \bar{T}_\alpha^\delta(x)\|_- C(M_1)\|A(x - x_\alpha^\delta)\|^\gamma \\ &\le 2^{\frac{\gamma}{2}} C(M_1)\|\partial \bar{T}_\alpha^\delta(x)\|_- \left(\frac{1}{2}\|Ax - Ax_\alpha^\delta\|^2 + \alpha\frac{1}{2}\|x_\alpha^\delta - x\|_Z^2\right)^{\frac{\gamma}{2}}. \end{aligned} \tag{12.36}$$

With some abuse of notation we write $2^{\frac{\gamma}{2}} C(M_1) = C(M_1)$. Thus,

$$\left(\frac{1}{2}\|Ax - Ax_\alpha^\delta\|^2 + \alpha\frac{1}{2}\|x_\alpha^\delta - x\|_Z^2\right)^{1-\frac{\gamma}{2}} \leq C(M_1)\|\partial \bar{T}_\alpha^\delta(x)\|_-,$$

and consequently

$$\left(T_\alpha^\delta(x) - T_\alpha^\delta(x_\alpha^\delta)\right)^{1-\frac{\gamma}{2}} \leq C(M_1)\|\partial T_\alpha^\delta(x)\|_{X'}.$$

We have thus found a KL-inequality (12.24) with

$$\varphi'(t) := \frac{1}{t^{1-\frac{\gamma}{2}}}, \quad \text{i.e.,} \quad \varphi(t) = t^{\frac{\gamma}{2}}.$$

We now apply Proposition 12.1 and Corollary 12.2 to $\bar{T}_\alpha^\delta$ (which agrees with T_α^δ for the relevant arguments) and obtain

$$\begin{aligned}\|x^\dagger - x_\alpha^\delta\|_X &\leq C(M_1)|\varphi\left(T_\alpha^\delta(x^\dagger) - T_\alpha^\delta(x_\alpha^\delta)\right) - \varphi\left(T_\alpha^\delta(x_\alpha^\delta) - T_\alpha^\delta(x_\alpha^\delta)\right)| \\ &\leq C(M_1)\varphi\left(T_\alpha^\delta(x^\dagger) - T_\alpha^\delta(x_\alpha^\delta)\right).\end{aligned}$$

We have

$$\begin{aligned}T_\alpha^\delta(x^\dagger) - T_\alpha^\delta(x_\alpha^\delta) &= \frac{1}{2}\|Ax^\dagger - Ax_\alpha^\delta\|^2 + \alpha\frac{1}{2}\|x_\alpha^\delta - x^\dagger\|_Z^2. \\ &\leq \|Ax^\dagger - y^\delta\|^2 + \|Ax_\alpha^\delta - y^\delta\|^2 + \alpha(\|x_\alpha^\delta\|_Z^2 + \|x^\dagger\|_Z^2) \\ &\leq 2\left(\|Ax^\dagger - y^\delta\|^2 + \alpha\|x^\dagger\|_Z^2\right) \leq 2\delta^2 + 2M_0^2\alpha\end{aligned}$$

due to the optimality of $T_\alpha^\delta(x_\alpha^\delta)$. Hence, with the choice (12.34) we obtain that

$$T_\alpha^\delta(x^\dagger) - T_\alpha^\delta(x_\alpha^\delta) \leq 2(1 + M_0^2 c_\alpha))\delta^2.$$

Now, as long as $1 + M_0^2 c_\alpha < C_0$, $x^\dagger$ is in the admissible set of (12.31), and we have the convergence rate

$$\|x^\dagger - x_\alpha^\delta\|_X \leq C(M_1)\varphi(2(1 + M_0^2 c_\alpha)\delta^2) \sim \delta^\gamma.$$

This is the same rate as obtained by Cheng and Yamamoto in this setting.

12.6 Example: Tikhonov Regularization

Due to the (partial) equivalence of the KL-inequality with the conditions of [23], their examples apply in our case as long as Ψ is a power function. Therefore, we will not go through all of those examples again, but focus on the most prominent one, which is classical Tikhonov regularization

$$T_\alpha(x) := \|Ax - y\|^2 + \alpha\|x\|^2, \tag{12.37}$$

where $A : X \to Y$ is a linear operator between Hilbert spaces X and Y and $\|\cdot\|$ denotes the norm in the respective spaces.

As is well known, the convergence behavior of Tikhonov-regularization (12.37) depends on the specific solution $x^\dagger$, and we employ here source conditions of the type

$$x^\dagger = (A^*A)^\mu w, \quad \|w\| \le 1, \quad \mu > 0. \tag{12.38}$$

While the treatment of more general source conditions $x^\dagger = \phi(A^*A)w$ is possible within our framework (see [23]), it shall be sufficient here to treat only the classical setting (12.38).

We recall from [20] that the residual $\|Ax - Ax^\dagger\|^2$ fulfills a KL inequality with

$$\varphi(t) \sim t^{\frac{\mu}{2\mu+1}} \tag{12.39}$$

if

$$x - x^\dagger = (A^*A)^\mu w, \quad \|w\| \le 1, \quad \mu > 0, \tag{12.40}$$

i.e., both x and $x^\dagger$ lie in the source set (12.38). This will become important again later. For now we simply apply the theory from [23] in the case $0 < \mu < \frac{1}{2}$ and demonstrate that the KL inequality and Corollary 12.1 yield convergence in the Bregman distance. Before starting, we summarize some results from [23, Sect. 4.1]. Namely, we have for (12.37) and under (12.38) that

$$\|x^\dagger\|^2 - \|x_\alpha\|^2 \le \alpha^{2\mu} \tag{12.41}$$

and

$$\|Ax_\alpha - Ax^\dagger\|^2 \le \alpha^{2\mu+1}. \tag{12.42}$$

Then we have from (12.41) and (12.42) that

$$\Delta T = \alpha(\|x^\dagger\|^2 - \|x_\alpha\|^2) - \|Ax_\alpha - Ax^\dagger\|^2 \sim \alpha^{2\mu+1}.$$

Because $\nabla T_\alpha(x^\dagger) = \alpha\|x^\dagger\|$, the KL inequality requires

$$\partial\varphi(\alpha^{2\mu+1})\alpha \ge c,$$

and it is easy to see that we even have equality for

$$\varphi(t) = t^{\frac{2\mu}{2\mu+1}} \tag{12.43}$$

with derivative

$$\partial\varphi(t) = ct^{-\frac{1}{2\mu+1}}, \quad (\partial\varphi)^{-1}(t) = t^{-(2\mu+1)}.$$

This function satisfies the condition in Theorem 12.3. From this, we obtain

$$\Psi(\alpha) = \frac{(\partial\varphi)^{-1}\left(\frac{1}{\alpha}\right)}{\alpha} = \frac{\alpha^{2\mu+1}}{\alpha} = \alpha^{2\mu}.$$

This yields, according to (12.28)

$$\alpha^* \sim \frac{1}{\partial\varphi(\delta^2)} \sim \delta^{\frac{2}{2\mu+1}},$$

and the convergence rate is given by

$$B_{\xi_\alpha^\delta}(x_\alpha^\delta, x^\dagger) \sim \delta^{\frac{4\mu}{2\mu+1}}.$$

Identifying $\|x_\alpha^\delta - x^\dagger\|^2 = B_{\xi_\alpha^\delta}(x_\alpha^\delta, x^\dagger)$, we obtain the well-known rate

$$\|x_\alpha^\delta - x^\dagger\| \sim \delta^{\frac{2\mu}{2\mu+1}}.$$

Note that Corollary 12.2 does not apply directly since it would yield a convergence rate $\|x^\dagger - x_\alpha^\dagger\| \le c\delta^{\frac{4\mu}{2\mu+1}}$, which is clearly off the correct rate by a square in the exponent. We will now sketch a likely explanation for this.

Comparing the functionals (12.37) and (12.29), it appears that similar techniques should lead to a KL inequality. This is indeed the case, and we obtain for the classical Tikhonov functional (12.37)

$$T_\alpha(x) - T_\alpha(x_\alpha) = \frac{1}{2}\|Ax - Ax_\alpha\|^2 + \alpha\frac{1}{2}\|x_\alpha - x\|^2.$$

We follow the next steps to arrive at the equivalent of (12.36), which reads

$$\begin{aligned} \frac{1}{2}\|Ax - Ax_\alpha\|^2 + \alpha\frac{1}{2}\|x_\alpha - x\|^2 &\le \frac{1}{2}(\nabla T_\alpha(x), x - x_\alpha) \\ &\le \frac{1}{2}\|\partial\bar{T}_\alpha(x)\|_{-}\|x - x_\alpha\|_X. \end{aligned} \tag{12.44}$$

The conditional stability estimate (12.30) no longer holds, but the source condition (12.40) yields an alternative. Namely, using the interpolation inequality

$$\|(A^*A)^r x\| \le \|(A^*A)^q x\|^{\frac{r}{q}} \|x\|^{1-\frac{r}{q}}, \tag{12.45}$$

for all $q > r \ge 0$, we see that

$$\|x_\alpha - x\| = \|(A^*A)^\mu w\| \le \|A(x - x_\alpha)\|^{\frac{2\mu}{2\mu+1}} \|w\|^{\frac{1}{2\mu+1}}.$$

Inserting this into (12.44), and following the argument after (12.36), we obtain

$$(T_\alpha(x) - T_\alpha(x_\alpha))^{1-\frac{\mu}{2\mu+1}} \le C\|\partial \bar{T}_\alpha(x)\|_-,$$

which yields a KL inequality with $\varphi'(t) \sim t^{\frac{\mu}{2\mu+1}-1}$ or

$$\varphi(t) \sim t^{\frac{\mu}{2\mu+1}}. \tag{12.46}$$

Comparing this with the previous results, we see that we have the same function φ as for the residual functional (12.39), but this φ is only the square root of the function from (12.43) that we derived earlier in this section. Note that $\|\cdot\|^2$ fulfill a KL inequality with $\varphi(t) = \sqrt{t}$. The discrepancy is due to the local character of the KL inequality for ill-posed problems. From the optimality condition of the classical Tikhonov functional (12.37) it follows that x_α (and x_α^δ, respectively) are always in the range of $A^* = (A^*A)^{\frac{1}{2}}$. Therefore, while $x^\dagger$ may fulfill the source condition (12.38) for arbitrary $0 < \mu < \infty$, the source condition (12.40) with $x \in \{x_\alpha, x_\alpha^\delta\}$ only holds for $\mu = \frac{1}{2}$, and we can only apply Corollary 12.2 in this case. Indeed, using the well-known a priori choice $\alpha \sim \delta^{\frac{2}{2\mu+1}} = \delta$, we have $T_\alpha(x^\dagger) - T_\alpha(x_\alpha^\delta) \sim \delta^2$, which yields via Corollary 12.2 with $\varphi(t)$ from (12.39) with $\mu = \frac{1}{2}$ the convergence rate $\|x_\alpha^\delta - x^\dagger\| \sim \sqrt{\delta}$. Therefore, the different index functions φ (12.43) and (12.39) are no contradiction.

Acknowledgements Part of this research was started during a visit of the second author at the Chemnitz University of Technology. S.K. would like to thank the Faculty of Mathematics in Chemnitz and especially Bernd Hofmann for their great hospitality. D.G. would like to thank Prof. Masahiro Yamamoto for his hospitality during his stay in Tokio, where the author first learned of the KL inequality. D.G. was supported by Deutsche Forschungsgemeinschaft (DFG), project GE3171/1-1.

References

1. P.-A. Absil, R. Mahony, B. Andrews, Convergence of the iterates of descent methods for analytic cost functions. SIAM J. Optim. **16**, 531–547 (2005)
2. H. Attouch, J. Bolte, On the convergence of the proximal algorithm for nonsmooth functions involving analytic features. Math. Program. **116**, 5–16 (2009)
3. D. Azé, J.N. Corvellec, Characterizations of error bounds for lower semicontinuous functions on metric spaces. ESAIM Control Optim. Calc. Var. **10**, 409–425 (2004)
4. D. Azé, J.N. Corvellec, Nonlinear error bounds via a change of function. J. Optim. Theory Appl. **172**, 9–32 (2017)
5. J. Bolte, A. Daniilidis, A. Lewis, The Łojasiewicz inequality for nonsmooth subanalytic functions with applications to subgradient dynamical systems. SIAM J. Opt. **17**, 1205–1223 (2007)
6. J. Bolte, A. Daniilidis, O. Ley, L. Mazet, Characterizations of Lojasiewicz inequalities: subgradient flows, talweg, convexity. Trans. Am. Math. Soc. **382**, 3319–3363 (2010)

7. J. Bolte, S. Sabach, M. Teboulle, Proximal alternating linearized minimization for nonconvex and nonsmooth problems. Math. Program. **146**, 459–494 (2014)
8. R.I. Boţ, E.R. Csetnek, Proximal-gradient algorithms for fractional programming. Optimization **66**, 1383–1396 (2017)
9. A.L. Bukhgeim, J. Cheng, M. Yamamoto, Stability for an inverse boundary problem of determining a part of a boundary. Inverse Probl. **15**, 1021–1032 (1999)
10. J. Cheng, M. Yamamoto, One new strategy for a priori choice of regularizing parameters in Tikhonov's regularization. Inverse Probl. **16**, L31–L38 (2000)
11. J. Cheng, M. Yamamoto, Conditional stabilizing estimation for an integral equation of first kind with analytic kernel. J. Integr. Equ. Appl. **12**, 39–61 (2000)
12. R. Chill, The Łojasiewicz-Simon gradient inequality in Hilbert spaces, in *Proceedings of the 5th European-Maghrebian Workshop on Semigroup Theory, Evolution Equations, and Applications* 2006, ed. by M.A. Jendoubi, pp. 25–36
13. H.W. Engl, M. Hanke, A. Neubauer, *Regularization of Inverse Problems* (Kluwer Academic Publishers, Dordrecht, 1996)
14. J. Flemming, Generalized Tikhonov regularization and modern convergence rate theory in Banach spaces. Ph.D. thesis, Shaker Verlag, Aachen, 2012
15. J. Flemming, A converse result for Banach space convergence rates in Tikhonov-type convex regularization of ill-posed linear equations. J. Inverse Ill Posed Probl. **26**, 639–646 (2018)
16. J. Flemming, Existence of variational source conditions for nonlinear inverse problems in Banach spaces. J. Inverse Ill Posed Probl. **26**, 277–286 (2018)
17. M. Forti, P. Nistri, M. Quincampoix, Convergence of neural networks for programming problems via a nonsmooth Lojasiewicz inequality. IEEE Trans. Neural Netw. **17**, 1471–1486 (2006)
18. P. Frankel, G. Garrigos, J. Peypouquet, Splitting methods with variable metric for Kurdyka–Łojasiewicz functions and general convergence rates. J. Optim. Theory Appl. **165**, 874–900 (2015)
19. G. Garrigos, L. Rosasco, S. Villa, Convergence of the forward-backward algorithm: beyond the worst case with the help of geometry (2017). arXiv:1703.09477
20. D. Gerth, Using Landweber method to quantify source conditions—a numerical study. J. Inverse Ill Posed Probl. **27**, 367–383 (2019)
21. A. Haraux, A hyperbolic variant of Simon's convergence theorem, in *Evolution Equations and Their Applications in Physical and Life Sciences (Bad Herrenalb, Germany, 1998)*. (Dekker, New York, 2001), pp. 255–264
22. A. Haraux, M.A. Jendoubi, The Łojasiewicz gradient inequality in the infinite-dimensional Hilbert space framework. J. Funct. Anal **260**, 0022–1236 (2011)
23. B. Hofmann, S. Kindermann, P. Mathé, Penalty-based smoothness conditions in convex variational regularization. J. Inverse Ill Posed Probl. **27**, 283–300 (2019)
24. S.-Z. Huang, *Gradient Inequalities. With Applications to Asymptotic Behavior and Stability of Gradient-Like Systems*. Mathematical Surveys and Monographs, vol. 126 (American Mathematical Society, Providence, RI, 2006)
25. O.Y. Imanuvilov, M. Yamamoto, Global uniqueness and stability in determining coefficients of wave equations. Comm. Part. Differ. Equ. **26**, 1409–1425 (2001)
26. A.D. Ioffe, Metric regularity—a survey Part 1: theory. J. Aust. Math. Soc. **2**, 188–243 (2016)
27. S. Kindermann, Convex Tikhonov regularization in Banach spaces: new results on convergence rates. J. Inverse Ill Posed Probl. **24**, 341–350 (2016)
28. K. Kurdyka, On gradients of functions definable in o-minimal structures. Ann. Inst. Fourier **48**, 769–783 (1998)
29. S. Łojasiewicz, Ensembles semi-analytiques, lecture notes, Institut des Hautes Etudes Scientifiques Bures-sur-Yvette (Seine-et-Oise), France, 1965. http://perso.univ-rennes1.fr/michel.coste/Lojasiewicz.pdf
30. S. Łojasiewicz, Une propriété topologique des sous-ensembles analytiques reéls, in *Les Équations aux Dériveés Partielles* (Éditions du centre National de la Recherche Scientifique, Paris, 1963)

31. Y. Nesterov, B.T. Polyak, Cubic regularization of Newton method and its global performance. Math. Program. Ser. A **108**(1), 177–205 (2006)
32. T. Schuster, B. Kaltenbacher, B. Hofmann, K. Kazimierski, *Regularization Methods in Banach Spaces* (De Gruyter, Berlin, 2012)
33. L. Simon, Asymptotics for a class of non-linear evolution equations, with applications to geometric problems. Ann. Math. **118**, 525–571 (1983)
34. M. Yamamoto, On ill-posedness and a Tikhonov regularization for a multidimensional inverse hyperbolic problem. J. Math. Kyoto Univ. **36**, 825–856 (1996)

Chapter 13
Value Function Calculus and Applications

Kazufumi Ito

Abstract In this paper the sensitivity analysis is discussed for the parameter-dependent optimization and constraint optimization. The sensitivity of the optimality value function with respect to the change in parameters plays a significant role in the inverse problems and the optimization theory, including economics, finance, the Hamilton–Jacobi theory, the inf-sup duality and the topological design and the bi-level optimization. We develop the calculus for the value function and present its applications in the variational calculus, the bi-level optimization and the optimal control and optimal design, shape calculus and inverse problems.

Keywords Parameter dependent optimization · Constrained optimization · Optimal value function · Sensitivity analysis · Value function calculus and applications

13.1 Introduction

Optimization and sensitivity analysis are key aspects of successful process design. Optimizing a process involves maximization of the project value and plant performance the minimization of the project cost, and facilitates the selection of the best design, for example. Specifically, we consider the optimization problem in which cost-functional and constraint are parameter dependent and thus the optimal value function is a functional over the parameters. In this paper we discus the sensitivity of the optimality value function in the parameter dependent optimization with respect to the change in parameters. It plays an essential role in developing robust algorithms and effective optimization methods. References [1, 7, 8, 13] and references therein.

We call such problems the value function calculus, i.e., find the derivative of the value function with respect to the parameter. Especially we develop the value function calculus for the constrained optimization in which the constraint is parameter depen-

K. Ito (✉)
Department of Mathematics, North Carolina State University, Raleigh, USA
e-mail: kito@ncsu.edu

J. Cheng et al. (eds.), *Inverse Problems and Related Topics*,
Springer Proceedings in Mathematics & Statistics 310,
https://doi.org/10.1007/978-981-15-1592-7_13

dent. The value function calculus is a very important aspect of the implicit function calculus and plays a significant role in inverse problems, parametric design and the optimization theory, including economics, finance, the Hamilton–Jacobi theory, the min-max problems. The existence of derivatives and the formula for the derivative are discussed. Applications of the calculus is discussed, including the variational calculus, the bi-level optimization and the optimal control and optimal design, shape calculus and inverse problems.

First, consider the parameter-dependent optimization problem;

$$\min \quad F(x, p) \quad x \in C. \tag{13.1}$$

Define the value function

$$V(p) = \inf_{x \in C} F(x, p) \tag{13.2}$$

where C is a closed (convex) set in Banach space X For given $p \in \mathcal{P}$, a closed convex subset of a Banach space Q, we assume minimizers $x = x(p) \in C$ of (13.1) exist and thus $V(p) = F(x(p), p)$. Throughout the paper, $S(p) = \{x(p)\}$ denote a solution set of (13.1) at p and $x(p) \in S(p)$ denotes an arbitrary minimizer. Recall that if $x \to F(x, p)$ is weakly lower semicontinuous and $F(x, p)$ is coercive on $x \in C$, given p, then there exists a solution to (13.1), given p. If the derivative $V'(\bar{p})$ at $p = \bar{p}$ exists, then it will be shown that

$$V'(\bar{p}) = F_p(x(\bar{p}), \bar{p}). \tag{13.3}$$

The existence of the derivative and the characterization by the Danskin theorem [2] are established. An application of the value function calculus in the bi-level optimization of the form

$$\max \quad V(p) + \Psi(p) \text{ over } p \in \mathcal{P}. \tag{13.4}$$

For the general bi-level optimization

$$\max \quad J(x(p)) + \Psi(p) \text{ over } p \in \mathcal{P}. \tag{13.5}$$

where $x(p) \in C$ minimizes $F(x, p)$ over $x \in C$, given $p \in \mathcal{P}$ is analyzed. One considers the regularized bi-level optimization of the form

$$\min \quad J(x) + \frac{F(x, p) - V(p)}{\epsilon} \text{ over } (x, p) \in C \times \mathcal{P}. \tag{13.6}$$

It is a single level optimization problem over (x, p). We show under appropriate conditions as $\epsilon \to 0^+$ solution pair (x_ϵ, p^ϵ) for (13.6) converges to a solution pair (x, p) to (13.5). Also, one can derive and analyze the necessary optimality condition for he original problem (13.5). Then, concrete applications are presented for nonsmooth variational problems $x \to F(x, p)$, $x \in C$. Moreover, we discuss the constraint bi-level optimization of the form

$$\max \quad J(x(p)) + \Psi(p) \text{ over } p \in \mathcal{P}. \tag{13.7}$$

where $x(p) \in C$ minimizes $F(x, p)$ subject to the equity constraint $E(x, p) = 0$ in Y^*, given $p \in \mathcal{P}$.

Next we consider the implicit function calculus for

$$V(p) = F(x, p) \text{ subject to } E(x, p) = 0 \text{ in } Y^*$$

based on the Lagrangian functional

$$L(x, p, \lambda) = F(x, p) + \langle \lambda, E(x, p) \rangle_{Y \times Y^*}$$

where Y^* is the dual space a Banach space Y. That is, under appropriate conditions we show

$$V'(p) = L_p(x(p), p, \lambda), \tag{13.8}$$

assuming $x(p) \in X$ satisfies $E(x(p), p) = 0$ and $\lambda \in Y$ satisfies the adjoint equation

$$E_x(x(p), p)^* \lambda + F_x(x(p), p) = 0.$$

Then, we consider the parameter-dependent constraint optimization, i.e., the equality constraint:

$$V(p) = \inf_{x \in C} F(x, p) \text{ subject to } E(x, p) = 0 \text{ in } Y^*$$

and the inequality constraint:

$$V(p) = \inf_{x \in C} F(x, p) \text{ subject to } G(x, p) \leq 0 \text{ in } Z.$$

where Z is the Hilbert lattice space. Also, for the general mathematical programing

$$V(p) = \inf_{x \in C} F(x, p) \text{ subject to } E(x, p) = 0, \quad G(x, p) \leq 0.$$

The value function calculus based on the Lagrange multiplier theory is developed, i.e.,

$$V'(p) = L_p(x(p), \lambda, \mu), \tag{13.9}$$

where L is the Lagrangian defined by

$$L(x, \lambda, \mu) = F(x, p) + \langle \lambda, E(x, p) \rangle_{Y \times Y^*} + (\mu, G(x, p))_Z, \tag{13.10}$$

where $x(p) \in X$ is a minimizer and (x, λ, μ) satisfies the optimality condition

$$\begin{cases} E_x(x(p), p)^*\lambda + G_x(x(p), p)^*\mu + F_x(x(p), p) = 0 \\ E(x(p), p) = 0 \\ \mu = \max(0, \mu + G(x(p), p)). \end{cases}$$

Concrete applications, including the optimal control problem, shape optimization and inverse problems are described and analyzed.

In conclusion the value calculus (13.3), (13.8) and (13.9) are very useful tool to analyze the sensitivity of the value function for the parameter dependent (optimization) problems. We apply the value function calculus to various examples and demonstrate the applicability throughout the paper. There are many potential applications including the ones we present in the paper for a wide class area and problems, design optimization, topological optimization, shape optimization, economics and finance, inverse medium, optimal control, Hamilton–Jacobi theory, duality theory, max-min optimization. In inverse problems a parameter calibration problem based on ground truths can be formulated as parameter optimization. For multi-objective optimization the parameter represents the weight distribution for multi-objectives. The parameter represents the regularization weights in the Tikhonov regularization method in the inverse, control and design problems. Based on the sensitivity analysis the balance principle for regularization parameter selection [10, 11] and the homotopy algorithms [15]. Moreover, we refer to [1–3, 5, 7, 8, 13, 16] and references therein for numerical optimization, sensitivity analysis and its applications.

An outline of our presentation is as follows. In Sect. 13.2 we discuss the unconstrained case and establish the differentiability of the value function $V(p)$ and if $x(\bar{p})$ is unique at $\bar{p}$. We apply it to the max-min optimization. For non-uniqueness case we derive the Danskin theorem [2] and a local supper differentiability of the value function $V(p)$. In Sect. 13.3 we apply the results in Sect. 13.2 to the bi-level optimization (13.5)–(13.7). In Sect. 13.4 we discuss the implicit function calculus and derive under appropriate conditions we show (13.8). In Sect. 13.5 we consider the parameter-dependent mathematical programming and derive (13.9) based on the Lagrange multiplier theory. We then apply it to the optimal control problem and the shape optimization problems.

13.2 Unconstrained Case

In this section we discuss the unconstrained case (13.2).

$$\min F(x, p) \text{ over } x \in C$$

where C is a closed convex set in X. First, we have

Lemma 2.1 (Weak Continuity) *Assume $S(p) = \{x(p)\} \in U$, a bounded set in X in a neighborhood of $\bar{p}$, and that $x \to F(x, \bar{p})$ is weakly sequentially lower semicontinuous, and $p \to F(x, p)$ is continuous uniformly on U.*

(a) If $p \to F(x, p)$ is weakly upper semicontinuous for $x \in U$, then $p \to V(p)$ is weakly upper semicontinuous.
(b) Every weak accumulation point $\hat{x} \in C$ of $x(p)$ is a minimizer of $F(x, \bar{p})$ over $x \in C$ and $V(p) \to V(\bar{p})$ as $p \to \bar{p}$.
(c) If $\bar{x} = x(\bar{p})$ is unique, then $x(p)$ is weakly convergent to $x(\bar{p})$ as $p \to \bar{p}$.
(d) Moreover, If $x \to F(x, \bar{p})$ is uniformly convex, then $|x(p) - \bar{x}| \to 0$ as $p \to \bar{p}$.

Proof (a) For all $x \in U$

$$V(p) \le F(x, p) = F(x, p) - F(x, \bar{p}) + F(x, \bar{p}).$$

Thus,

$$\limsup_{p \to \bar{p}} V(p) \le F(x, \bar{p}).$$

Since $x \in U$ is arbitrary, $\limsup_{p \to \bar{p}} V(p) \le V(\bar{p})$.
(b) Since $x(p)$ is bounded in X one can assume there exists a sequence $\{x(p_n)\}$ that converges weakly to $\hat{x}$ for some $\hat{x} \in C$. It follows from assumptions that

$$\lim_{n \to \infty} F(x_n, p_n) \le \lim_{n \to \infty} F(\bar{x}, p_n) \to F(\bar{x}, \bar{p}) = V(\bar{p})$$

and

$$\lim_{n \to \infty} F(x_n, p_n) = \lim_{n \to \infty} (F(x_n, p_n) - F(x_n, \bar{p})) + \lim_{n \to \infty} F(x_n, \bar{p}) \ge F(\hat{x}, \bar{p})$$

Thus, $F(\hat{x}, \bar{p}) \le F(\bar{x}, \bar{p})$ and hence $\hat{x}$ is a minimizer of (13.1) at $\bar{p}$ and $V(p_n) \to V(\bar{p})$ as $p_n \to \bar{p}$. If $\bar{x}$ is unique, then $p \to x(p)$ is weakly continuous at $\bar{p}$.
(d) There exists a nonnegative monotone function ϕ on R^+ satisfying $\phi(0)=0$ for $\xi \in \partial F(\bar{x}, \bar{p})$ such that

$$F(x(p), \bar{p}) - F(\bar{x}, \bar{p}) - (\xi, x - \bar{x}) \ge \phi(|x(p) - \bar{x}|)$$

where

$$F(x(p), p) \to F(\bar{x}, \bar{p}) \text{ since } V(p) \to V(\bar{p}).$$ □

Next, we show the existence of the derivative of $V(p)$ under a minimum assumption on F.

Definition 2.1 (1) $V(p)$ is directionally differentiable in a direction $\dot{p}$ at p if

$$\lim_{t \to 0^+} \frac{V(p + t\,\dot{p}) - V(p)}{t} \text{ exits.}$$

(2) $V(p)$ is G-differentiable at p if all direction $\dot{p}$

$$\lim_{|t|\to 0} \frac{V(p+t\,\dot{p})-V(p)}{t} = V'(p)\dot{p} \text{ exits.}$$

We assume that given $x \in U$, $p \to F(x,p)$ is G-differentiable. First note that for $\bar{p} \in \mathcal{P}$ and $p = \bar{p} + t\dot{p} \in \mathcal{P}$ with $t \in R$ and an increment $\dot{p}$ we have the comparisons:

$$F(x(p),p) \le F(x(\bar{p}),p), \quad \text{and} \quad F(x(\bar{p}),\bar{p}) \le F(x(p),\bar{p}).$$

Thus, we have

$$\begin{aligned} F(x(p),p)-F(x(\bar{p}),\bar{p}) &= F(x(p),p)-F(x(\bar{p}),p)+F(x(\bar{p}),p) \\ &\quad -F(x(\bar{p}),\bar{p}) \le F(x(\bar{p}),p)-F(x(\bar{p}),\bar{p}) \\ F(x(p),p)-F(x(\bar{p}),\bar{p}) &= F(x(p),\bar{p})-F(x(\bar{p}),\bar{p})+F(x(p),p) \\ &\quad -F(x(p),\bar{p}) \ge F(x(p),p)-F(x(p),\bar{p}) \end{aligned} \tag{13.11}$$

and hence

$$\begin{aligned} F(x(p),p)-F(x(p),\bar{p}) &\le F(x(p),p)-F(x(\bar{p}),\bar{p}) \\ &\le F(x(\bar{p}),p)-F(x(\bar{p}),\bar{p}). \end{aligned} \tag{13.12}$$

Therefore for $t > 0$ and all direction $\dot{p}$, we have

$$\begin{aligned} &\frac{F(x(\bar{p}+t\dot{p}),\bar{p}+t\dot{p})-F(x(\bar{p}+t\dot{p}),\bar{p})}{t} \le \frac{F(x(\bar{p}+t\dot{p}),\bar{p}+t\dot{p})-F(x(\bar{p}),\bar{p})}{t} \\ &\le \frac{F(x(\bar{p}),\bar{p}+t\dot{p})-F(x(\bar{p}),\bar{p})}{t} \\ &\frac{F(x(\bar{p}),\bar{p})-F(x(\bar{p}),\bar{p}-t\dot{p})}{t} \le \frac{F(x(\bar{p}),\bar{p})-F(x(\bar{p}-t\dot{p}),\bar{p}-t\dot{p})}{t} \\ &\le \frac{F(x(\bar{p}-t\dot{p}),\bar{p})-F(x(\bar{p}-t\,\dot{p}),\bar{p}-t\dot{p})}{t}. \end{aligned} \tag{13.13}$$

Based on (13.12) we have

Theorem 2.2 *Let $d \in X^*$ be a local supper differential of V at $\bar{p}$, i.e., $d \in D^+V(\bar{p})$ is defined by*

$$D^+V(\bar{p})(\dot{p}) = \{d : \limsup_{t\to 0} \frac{V(p+t\,\dot{p})-V(\bar{p})-t\,(d,\dot{p})}{|t|} \le 0\}.$$

Then, $F_p(\bar{x},\bar{p}) \in D^+V(\bar{p})$ for all minimizer $\bar{x}$ at $\bar{p}$.

The following Corollary follows from Lemma 4.28 [13].

Corollary 2.3 *Suppose* $(x, p) \to F(x, p)$ *is convex, then* $p \to V(p)$ *is convex and thus* $F_p(\bar{x}, \bar{p}) \in D^- V(\bar{p})$ *and G-derivative of* V *at* $\bar{p}$ *exits with* $V'(\bar{p}) = F_p(\bar{x}, \bar{p})$.

Also, based on (13.13) we have

Theorem 2.4 (Sensitivity) *Assume*
(H1) there exists the continuous graph $p \to x(p) \in C$ *in a neighborhood of* $(x(\bar{p}), \bar{p}) \in C \times P$ *and thus*

$$V(p) = F(x(p), p).$$

(H2) in a neighborhood of $(x(\bar{p}), \bar{p})$

$$(x, p) \in C \times \mathcal{P} \to F_p(x, p) \text{ is continuous.}$$

Then, the G-derivative of V *at* $\bar{p}$ *exists and is given by*

$$V'(\bar{p})\dot{p} = F_p(x(\bar{p}), \bar{p})\dot{p}. \tag{13.14}$$

Conversely, if $V(p)$ *is differentiable at* $\bar{p}$ *then* (13.14) *holds for all minimizers* $x(\bar{p})$.

Corollary 2.5 *In addition to the assumptions of Lemma 2.1 we assume that as* $x(p + t\dot{p}) \to x(\bar{p})$ *weakly and*

$$\liminf_{t \to 0^+} \frac{F(x(\bar{p} + t\,\dot{p}), \bar{p} + t\,\dot{p}) - F(x(\bar{p} + t\,\dot{p}), \bar{p})}{t} \geq F_p(x(\bar{p}), \bar{p})\dot{p}, \tag{13.15}$$

or sufficiently assume

$$\begin{aligned} &\liminf_{t \to 0^+} F_p(x(p + t\,\dot{p}, \bar{p})\dot{p} \geq F_p(x(\bar{p}), \bar{p})\dot{p} \\ &\lim_{t \to 0^+} \frac{F(x(p + t\,\dot{p}), p) - F(x(p + t\,\dot{p}), \bar{p})}{t} - F_p(x(p + t\,\dot{p}), \bar{p})\dot{p} = 0 \end{aligned} \tag{13.16}$$

for all $\dot{p}$. *Then,* $p \to V(p)$ *is directionally differentiable at* $\bar{p}$. *If* $x(\bar{p})$ *is unique then* $V(p)$ *is differentiable at* $\bar{p}$ *and* (13.14) *holds.*

Proof It follows from (13.13) and that for $p = \bar{p} + t\,\dot{p}$

$$\begin{aligned} \frac{F(x(p), p) - F(x(\bar{p}), \bar{p})}{t} &= F_p(x(p), \bar{p})\dot{p} \\ &+ \frac{F(x(p), p) - F(x(p), \bar{p})}{t} - F_p(x(p), \bar{p})\dot{p}. \end{aligned}$$

□

Corollary 2.6 *If* $x(\bar{p}) = \bar{x}$ *is unique and* X *is finite dimensional, then* $p \in \mathcal{P} \to V(p)$ *is differentiable at* $\bar{p}$ *and* (13.14) *holds.*

Lemma 2.1 and Corollary 2.5 provide very minimum assumptions for the value function calculus

$$V'(p)\dot{p} = F_p(x(\bar{p}), \bar{p})\dot{p}.$$

If minimizer $x(\bar{p})$ at $\bar{p}$ is not unique, then we have

Theorem 2.7 (Danskin theorem) *Under the assumptions of Lemma 2.1 and Corollary 2.5 there exists minimizers $\bar{x}^+$ and $\bar{x}^-$ at $\bar{p}$ such that*

$$\lim_{t\to 0^+} \frac{V(\bar{p}+t\,\dot{p}) - V(\bar{p})}{t} = F_p(\bar{x}^+, \bar{p})\dot{p} = \inf_{\bar{x}\in x(\bar{p})} F_p(\bar{x}, \bar{p})\dot{p}$$

$$\lim_{t\to 0^+} \frac{V(\bar{p}) - V(\bar{p}-t\,\dot{p})}{t} = F_p(\bar{x}^-, \bar{p})\dot{p} = \sup_{\bar{x}\in x(\bar{p})} F_p(\bar{x}, \bar{p})\dot{p}$$

where inf and sup is taken over all minimizes $x(\bar{p})$.

Proof It follows from the left hand side of inequality of (13.13) and (13.15) that for all weak cluster points x of $x(p_n) = x(\bar{p}+t_n\,\dot{p})$ as $t \to 0^+$ and all minimizers $\bar{x}$ at $\bar{p}$

$$\lim_{t_n\to 0^+} \frac{F(x(\bar{p}+t_n\,\dot{p}), \bar{p}+t_n\,\dot{p}) - F(x(\bar{p}+t_n\dot{p}), \bar{p})}{t_n} \geq F_p(\bar{x}, \bar{p})\dot{p}$$

Since $\bar{x}$ is a minimizer at $\bar{p}$ and from the right hand side of inequality (13.13)

$$\limsup_{t\to 0^+} \frac{V(p+t\,\dot{p}) - V(\bar{p})}{t} \leq \inf_{\bar{x}\in x(\bar(p))} F_p(\bar{x}, \bar{p})\dot{p},$$

the first claim holds. Similarly, $x(p_n) = x(\bar{p}-t_n\,\dot{p}) \to x$ weakly as $t_n \to 0^-$

$$\lim_{t_n\to 0^+} \frac{F(x(\bar{p}-t_n\,\dot{p}), \bar{p}) - F(x(\bar{p}-t_n\,\dot{p}), \bar{p})}{t_n} \leq F_p(\bar{x}, \bar{p})\dot{p}$$

and

$$\liminf_{t\to 0^+} \frac{V(\bar{p}) - V(p-t\,\dot{p})}{t} \leq \sup_{\bar{x}\in x(\bar(p))} F_p(\bar{x}, \bar{p})\dot{p},$$

and thus the second claim holds. □

Remark We also refer the Danskin theorem, e.g., see [2, 4, 6]. Equation (13.16) for (13.15) is an explicit sufficient condition which one can prove directly for a given example.

Example 1 (Stefan problem) Consider the enthalpy formulation of the stationary two phase stefan problem

$$\min \quad F(\theta, g) = \int_\Omega \frac{\kappa}{2}\,|\nabla\theta|^2 + \gamma(\theta)\,dx - \int_{\partial\Omega} g\theta ds_x$$

where the enthalpy $E = \gamma(\theta)$ is a monotone function of the temperature θ that represents the total heat energy and κ is the thermal conductivity. $g \in L^2(\partial\Omega)$ is the thermal flux at the boundary $\partial\Omega$. There exists the unique solution $\theta = \theta(g) \in X = H^1(\Omega)$, given $g \in L^2(\partial\Omega)$. Since $\theta(g) \to \theta(\bar{g})$ weakly in X and

$$\lim_{t\to 0} \frac{F(\theta(g), g) - F(\theta(g), \bar{g})}{t} = (\theta(\bar{g}), \dot{g})_{\partial\Omega}. \tag{13.17}$$

It follows from Corollary 2.6 that

$$V'(\bar{g})\dot{g} = -\int_{\partial\Omega} \theta(\bar{p})\dot{g}\, ds_x.$$

Example 2 (Non-Newtonian Mechanics) For $1 < q < \infty$ let $X = W_0^{1,q}(\Omega)$ and $Q = \{p \in L^\infty(\Omega) : 0 < \underline{p} \le p \le \bar{p}\}$. For $u \in X$ and $p \in Q$ consider

$$F(u, p) = \int_\Omega (\frac{1}{q} p(x)|\nabla u|^q - f u)\, dx. \tag{13.18}$$

Lemma 2.1 holds and the minimizer $u(p) \in X$ is unique and uniformly bounded. The necessary and sufficient condition $u(p) \in X$ is given by

$$-\nabla \cdot (p(x)|\nabla u(p)|^{q-2}\nabla u(p)) = f. \tag{13.19}$$

In this case

$$\frac{F(u(p), p) - F(u(p), \bar{p})}{t} = F_p(u(p), \bar{p})\dot{p} = (\dot{p}, \frac{1}{q}|\nabla u(p)|^q).$$

Since $V(p) \to V(\bar{p})$ as $p \to \bar{p}$,

$$\int_\Omega \frac{1}{q} p(x)|\nabla u(p)|^q\, dx \to \int_\Omega \frac{1}{q} \bar{p}(x)|\nabla u(\bar{p})|^q\, dx$$

For $\alpha\,\bar{p} + \dot{p} > 0$ for some α and since the norm is weakly lower-semicontinuous

$$\lim_{t\to 0}(\alpha\bar{p} + \dot{p}, \frac{1}{q}|\nabla u(p)|^q) \ge (\alpha\bar{p} + \dot{p}, \frac{1}{q}|\nabla u(\bar{p})|^q)$$

which implies (13.15) and $V'(p)\dot{p} = \int_\Omega \frac{1}{q}\dot{p}(x)|\nabla u(p)|^q\, dx.$

In addition for the case $x(\bar{p})$ is not necessary unique, then the following lemma holds.

Lemma 2.9 (Sensitivity) *Suppose* $V(p) = \inf_{x\in C} F(x, p)$ *where* C *is a closed convex set. If* $p \to F(x, p)$ *is concave, then* $p \to V(p)$ *is concave and* $-F_p(\bar{x}, \bar{p}) \in \partial(-V)(\bar{p})$.

Proof Since for $0 \le t \le 1$

$$F(x, t\,p_1 + (1-t)\,p_2) \ge t\,F(x, p_1) + (1-t)\,F(t, p_2)$$

we have

$$V(t\,p_1 + (1-t)\,p_2) \ge t\,V(p_1) + (1-t)\,V(p_2).$$

and thus V is concave. Since

$$V(p) - V(\bar{p}) = F(x, p) - F(\bar{x}, p) + F(\bar{x}, p) - F(\bar{x}, \bar{p}) \le F(\bar{x}, p) - F(\bar{x}, \bar{p})$$

we have

$$-F_p(\bar{x}, \bar{p}) \in \partial(-V)(\bar{p}).$$ □

From Corollary 2.2 and Theorem 2.3, we have

Corollary 2.10 (Value function optimization) *Consider the problem of minimizing the cost functional* $J(p) = V(p) + G(p)$ *over* $p \in \mathcal{K}$. *Assume* $G(p) = G_0(p) + G_1(p)$ *with concave* G_0 *and continuously differentiable and* G_1 *is concave and* $\mathcal{K}$ *a a closed convex set. The necessary optimality condition for a maximizer* p^* *is that*

$$0 \in F_p(x(p^*), p^*) + G_0'(p^*) - \partial(-G_1)(p^*)$$

if p^* *is an interior point of* $\mathcal{K}$. *In general* $p \to F(x(p), p)$ *is differentiable at* p^* *we have*

$$F_p(x(p^*), p^*)(p - p^*) + (G_0'(p^*), p - p^*) + G_1(p) - G_1(p^*) \ge 0 \text{ for all } p \in \mathcal{K}.$$

Example 2 (Continued) Consider the topological optimization: Minimize the compliance

$$\min \int_\Omega f u(p)\,dx \text{ over } Q$$

where $u(p) \in X = W_0^{1,q}(\Omega)$ is a minimizer of (13.18). Since from (13.19)

$$\frac{1}{q}\int_\Omega (p|\nabla u(p)|^q - f u(p))\,dx = 0,$$

it is equivalently formulated as

$$\min \quad -V(p) + \frac{\alpha}{2}|p|^2 \text{ over } \mathcal{K}$$

Since $p \to F(u,p)$ is linear, $p \to V(p)$ is concave and thus a maximizer exists. The necessary optimality is given by

$$\int_\Omega (\alpha\,(p-p^*) - \frac{1}{q}(p(x)-p^*(x))|\nabla u(p^*)|^q)\,dx \le 0 \text{ for all } p \in \mathcal{K}.$$

Thus,

$$\alpha\, p^* = \frac{1}{q}\max(\min(|\nabla u(p^*)|^q, \bar{p}, \underline{p}).$$

13.3 Bi-level Optimization

In this section we apply the results in Sect. 13.2 to bi-level optimizations. Consider the bi-level optimization of the form

$$\min \quad J(x(p),p) \text{ over } p \in \mathcal{K} \tag{13.20}$$

where $x(p)$ minimizes $F(x,p)$ over $x \in C$.

Consider the penalty formulation of the bi-level optimization: for $\epsilon > 0$

$$\min \quad J(x,p) + \frac{F(x,p) - V(p)}{\epsilon} \text{ over } (x,p) \in C \times \mathcal{K}. \tag{13.21}$$

For the existence we assume $p \in \mathcal{K} \to V(p) = F(x(p),p)$ is weakly upper continuous and $(x,p) \in C \times \mathcal{K} \to F(x,p)$ and $J(x,p)$ is weakly lower semi continuous.

Theorem (Convergence) *Every weak cluster point $(\hat{x},\hat{p})$ of (x^*,p^*) of* (13.20) *as $\epsilon \to 0^+$ is a minimizer of the bi-level optimization* (13.20).

Proof Note that

$$J(x^*,p^*) \le J(x(p),p) \text{ for all } p \in \mathcal{K}$$

and

$$F(x^*,p^*) - F(x(p^*),p^*) \le \epsilon\,(J(x(p^*),p^*) - J(x^*,p^*)).$$

Thus, if (x^*,p^*) converges weakly to $(\hat{x},\hat{p})$ as $\epsilon \to 0^+$, we have $F(\hat{x},\hat{p}) \le V(\hat{p})$ and hence $V(\hat{p}) = F(\hat{x},\hat{p})$. Since $J(\hat{x},\hat{p}) \le J(x(p),p)$ for all $p \in \mathcal{K}$. Thus the claim holds. □

Corollary 3.1 (Bi-level optimization) *Assume F and J are C^1 and $p \to V(p)$ is differentiable at p^*. Then, the necessary optimality given by*

$$\begin{cases} (F_x(x(p^*), p^*), x - x(p^*)) \geq 0 \; over \; x \in \mathcal{C} \\ (J_x(x^*, p^*) + \dfrac{F_x(x^*, p^*)}{\epsilon}, x - x^*) \geq 0 \; over \; x \in \mathcal{C} \\ (J_p(x^*, p^*) + \dfrac{F_p(x^*, p^*) - F_p(x(p^*), p^*)}{\epsilon}, p - p^*) \geq 0 \; over \; p \in \mathcal{K} \end{cases} \tag{13.22}$$

Proof It follows from Theorem 2,5 that the value function calculus $\frac{dF(x(p),p)}{dp} = F_p(x(p), p)$ holds. Thus, $(x^*, p^*, x(p^*))$ satisfies the system of variational inequalities (13.22). □

For non constraint case $\mathcal{C} = X$ (13.22) is equivalent to the system for $(x^*, p^*, x(p^*))$

$$\begin{cases} \dfrac{F_x(x^*, p^*) - F_x(x(p^*), p^*)}{\epsilon} + J_x(x^*, p^*) = 0 \\ (\dfrac{F_p(x^*, p^*) - F_p(x(p^*), p^*)}{\epsilon} + J_p(x^*, p^*), p - p^*) \geq 0 \\ F_x(x(p^*), p^*) = 0 \end{cases} \tag{13.23}$$

As will be shown in the following examples we have much relaxed conditions for the existence of minimizers for the penalty formulation and for the necessary optimality it does not need to use the second derivatives of F. As $\epsilon \to 0^+$ we obtain a formal necessary optimality condition to the original bi-level optimization (13.20). Under a regularity assumption on a minimizer $(\bar{x}, \bar{p})$ of the original bi-level optimization it can be shown that it is the necessary optimality condition for (13.20).

Example 3.2 Consider the linear quadratic case.

$$F(x, p) = \frac{1}{2}(Ax, x) - (Bp + f, x) \text{ and } J(x, p) = J(x) + \frac{\alpha}{2}|p|^2,$$

If we define $\mu = \dfrac{x^* - x(p^*)}{\epsilon}$ then (13.23) is written as

$$\begin{cases} A\mu + J'(x^*) = 0, \quad Ax(p^*) = Bp^* + f, \quad \mu = \dfrac{x^* - x(p^*)}{\epsilon} \\ (\alpha\, p^* + B^*\mu, p - p^*) = 0 \text{ for all } p \in \mathcal{K}. \end{cases}$$

Or, equivalently it is for $(\mu, \bar{x} = x(p^*))$

$$\begin{cases} A\mu + J'(\bar{u} + \epsilon\, \mu) = 0, \quad A\bar{x} = Bp^* + f \\ (\alpha\, p^* + B^*\mu, p - p^*) = 0 \text{ for all } p \in \mathcal{K}. \end{cases} \tag{13.24}$$

Also, note that the necessary optimality condition for (13.20) is given by

$$\begin{cases} A\mu + J'(\bar{x}) = 0, \quad A\bar{x} = Bp^* + f \\ \\ (\alpha\, p^* + B^*\mu,\, p - \bar{p}) = 0 \text{ for all } p \in \mathcal{K}, \end{cases}$$

which coincides with (13.24) with $\epsilon = 0$.

Example 3.3 (Control of Obstacle problem and American option) We consider

$$\min \quad F(u, p) = \int_\Omega \frac{1}{2}|\nabla u|^2\, dx - (Bp + f, u) \text{ over } u \le \psi$$

given p, where $B \in \mathcal{P} \to H^{-1}(\Omega)$ is a linear operator. Consider a bi-level optimization of the form

$$\min J(u, p) = \frac{1}{2}(|u(p) - u_d|^2 + \alpha\, |p|^2) \text{ over } p.$$

Let (u^*, p^*) is an optimal pair of the regularized problem (13.21) for $\epsilon > 0$ and $A = -\Delta$. For $\bar{u} = u(p^*)$ there exist Lagrange multipliers $\bar{\lambda} \ge 0$ and $\lambda^* \ge 0$ [12, 13] such that

$$A\bar{u} - Bp^* - f + \bar{\lambda} = 0, \quad J'(u^*) + \tfrac{Au^* - Bp^* - f + \lambda^*}{\epsilon} = 0.$$

$$\bar{\lambda} = \max(0, \bar{\lambda} + \bar{u} - \psi), \quad \lambda^* = \max(0, \lambda^* + u^* - \psi).$$

Thus, we have

$$A\frac{u^* - \bar{u}}{\epsilon} + J'(u^*) + \frac{\lambda^* - \bar{\lambda}}{\epsilon} = 0.$$

Let $\mu = \dfrac{u^* - \bar{u}}{\epsilon}$, $q = \dfrac{\lambda^* - \bar{\lambda}}{\epsilon}$ and $u^* = \bar{u} + \epsilon\mu$. Then, we obtain the system for $(\mu, p^*, \bar{x})$

$$\begin{cases} A\mu + J'(\bar{u} + \epsilon\mu) + q = 0, \\ \\ \alpha\, p^* + B^*\mu = 0 \\ \\ A\bar{u} + \bar{\lambda} = Bp^* + f, \;\; \bar{\lambda} = \max(0, \bar{\lambda} + \bar{u} - \psi). \end{cases} \tag{13.25}$$

Note that

$$\begin{cases} q = 0 & \text{on } \{u^* - \psi < 0\} \cap \{\bar{u} - \psi < 0\} \\ \mu = 0 & \text{on } \{u^* - \psi = 0\} \cap \{\bar{u} - \psi = 0\} \\ \mu > 0 \text{ and } q \ge 0 & \text{on } \{u^* - \psi = 0\} \cap \{\bar{u} - \psi < 0\} = I_1, \\ \mu < 0 \text{ and } q \le 0 & \text{on } \{u^* - \psi < 0\} \cap \{\bar{u} - \psi = 0\} = I_2 \end{cases}$$

Since $(q, \mu) \geq 0$ it follows from the first equation of (13.25) that $\mu \in X$ is bounded uniformly in $\epsilon > 0$. Thus, $u^* - \bar{u} \to 0$ in X. Let $(\bar{u}, \lambda, p)$ is the optimal triple for the original problem. We assume the strict complementarity:

$$\int_\Omega \bar{\lambda}\, dx > 0 \quad \text{and} \quad \int_\Omega (\bar{u} - \psi)^-\, dx < 0. \tag{13.26}$$

Then, by Chebyshev inequality

$$m(\text{indefinite set} = I_1 \cup I_2) \to 0 \text{ as } \epsilon \to 0^+.$$

and as $\epsilon \to 0^+$ we obtain a system of the optimality [12] for the original problem as

$$\begin{cases} A\mu + J'(\bar{u}) + q = 0, \quad q = 0 \text{ on } \{\bar{u} - \psi < 0\}, \quad \mu = 0 \text{ on } \{\bar{u} - \psi = 0\} \\ \alpha\, p^* + B^*\mu = 0 \\ A\bar{u} + \bar{\lambda} = Bp^* + f, \quad \bar{\lambda} = \max(0, \bar{\lambda} + \bar{u} - \psi). \end{cases}$$

Example 3.4 (Topological Optimization) Consider the elastic equations

$$Div\, \sigma = f$$

where σ is the stress tensor. Let ϵ is the linear strain tensor

$$\epsilon_{ij} = \frac{\partial u_i}{\partial x_j} + \frac{\partial u_j}{\partial x_i}.$$

Let p is elastic constitutive law such that $\sigma = p\epsilon$. For example the Hooke' law is written as

$$\sigma = 2\mu\epsilon + \lambda\, \mathrm{tr}\, \epsilon\, I$$

where $\mu,\ \lambda$ are the Lame' coefficients and tr denotes the trace of a matrix and I is the identity matrix. The elastic equation is equivalent to

$$\min \quad F(u, p) = \int_\Omega \frac{1}{2}\sigma : \epsilon - \mathbf{f} \cdot \mathbf{u}\, dx, \quad \sigma = p\epsilon$$

The stress maximization problem is

$$\max \quad J(u(p), p) = \int_\Omega \sqrt{\sigma M \sigma} - \frac{\alpha}{2}|\nabla p|^2.$$

where the integrant is the Von-Messe stress.

In order to illustrate the optimality condition (13.23) we consider a scalar case:

$$\min \quad J(u, p) = \frac{1}{2}\int_\Omega |u - \bar{u}|^2 + \frac{\alpha}{2}|\nabla p|^2\, dx \tag{13.27}$$

subject to $u = u(p) = \text{argmin}\, F(u, p) = \int_\Omega (\frac{1}{2}p|\nabla u|^2 - fu)\, dx$ and $0 < \underline{p} \le p \le \bar{p} < \infty$. For the existence of a minimizer (u^*, p^*) for the regularized problem, suppose $u^* \to u$ weakly in $H^1(\Omega)$ and $p^n \to p$ strongly in $L^2(\Omega)$ we have

$$\lim_{n\to\infty}((p_n, |\nabla u_n|^2) - (p, |\nabla u|^2)) = \lim_{n\to\infty}((p, |\nabla u^n|^2) - (p, |\nabla u|^2) + (p_n - p, |\nabla u|^2)) \ge 0$$

by the weakly lower semi-continuity of norms and the Lebesgue dominated convergence theorem. From (13.23)

$$\begin{cases} -\nabla\cdot(p^*\nabla\dfrac{u^* - u(p^*)}{\epsilon}) + J'(u^*) = 0 \\ ((\nabla\dfrac{u^* - u(p^*)}{\epsilon}, \nabla\dfrac{u^* + u(p^*)}{2}), p - p^*) + \alpha\nabla p^*, \nabla(p - p^*)) = 0 \text{ for all } p \in \mathcal{K}\cap H^1(\Omega) \\ -\nabla\cdot(p^*\nabla u(p^*)) - f = 0. \end{cases}$$

If we let $\bar{u} = u(p^*)$ and $\mu = \dfrac{u^* - u(p^*)}{\epsilon}$, then

$$\begin{cases} -\nabla\cdot(p^*\nabla\mu) + J'(u^*) = 0 \\ ((\nabla\mu, \nabla(\bar{u} + \dfrac{\epsilon}{2}\mu), p - p^*) + \alpha\,(\nabla p^*, \nabla(p - p^*)) = 0 \text{ for all } p \in \mathcal{K}\cap H^1(\Omega) \\ -\nabla\cdot(p^*\nabla\bar{u}) - f = 0. \end{cases}$$

From the first equation $\nabla\mu \in L^2(\Omega)$ is uniformly bounded in $\epsilon > 0$. Thus, one can prove that as $\epsilon \to 0^+$ we obtain the necessary optimality condition for $(\bar{u}, \bar{p})$ for the bi-level optimization (13.27):

$$\begin{cases} -\nabla\cdot(\bar{p}\nabla\mu) + J'(\bar{u}) = 0 \\ ((\nabla\mu, \nabla\bar{u}), p - \bar{p}) + \alpha\,\nabla p^*, \nabla(p - \bar{p})) = 0 \text{ for all } p \in \mathcal{K}\cap H^1(\Omega) \\ -\nabla\cdot(\bar{p}\nabla\bar{u}) - f = 0. \end{cases}$$

Example 3.5 (Electric Impedance Tomography) Problem of determining the conductivity σ in

$$-\nabla\cdot(\sigma\nabla u) = 0, \quad \sigma\frac{\partial u}{\partial\nu} = g \text{ at } \partial\Omega, \tag{13.28}$$

from the voltage measurement y at the boundary $\partial\Omega$ can be formulated as [9]:

$$\min \quad F(u,\sigma) = \frac{1}{2}|u-y|^2_{\partial\Omega} + \beta\,|\sigma - \underline{\sigma}|_1 + \frac{\alpha}{2}|\nabla\sigma|^2\,dx \tag{13.29}$$

subject to (13.28) and $\sigma \in Q = \{0 < \underline{\sigma} \le \sigma(x) \le \bar{\sigma} < \infty\}$. The second term is the sparsity term for $\sigma - \underline{\sigma}$ and $\beta \ge 0$ and $\alpha > 0$ are the regularization parameters [9]. Note that is equivalent to

$$\min \quad F(u,\sigma) = \frac{1}{2}\int_\Omega \sigma|\nabla u|^2\,dx - \int_{\partial\Omega} gu\,ds = 0.$$

Then, there exists a minimizer σ for the regularized bi-level optimization (13.20). If we let $\bar{u} = u(\sigma^*)$ and $\mu = \dfrac{u^* - u(p^*)}{\epsilon}$, then the necessary optimality condition for the regularized bi-level optimization is given by

$$\begin{cases} -\nabla\cdot(\sigma^*\nabla\mu)=0, \quad \sigma^*\dfrac{\partial\mu}{\partial\nu}=u^*-y \\ ((\nabla\mu, \nabla(\bar{u}+\frac{\epsilon}{2}\mu), \sigma-\sigma^*) + \alpha\,(\nabla\sigma^*, \nabla(\sigma-\sigma^*)) + \beta\,(|\sigma-\bar{\sigma}|_1 - |\sigma^*-\bar{\sigma}|_1) \ge 0 \text{ for all } \sigma \in \mathcal{K}\cap H^1(\Omega) \\ -\nabla\cdot(p^*\nabla\bar{u}) = 0, \quad \sigma^*\frac{\partial\bar{u}}{\partial\nu} = g. \end{cases}$$

Since $\mu \in H^1(\Omega)$ is uniformly bounded, letting $\epsilon \to 0^+$ we obtain the necessary optimality condition $(\bar{u}, \bar{\sigma})$ for the bi-level optimization (13.29)

$$\begin{cases} -\nabla\cdot(\bar{\sigma}\nabla\mu) = 0, \quad \sigma^*\dfrac{\partial\mu}{\partial\nu} = \bar{u}-y \\ ((\nabla\mu, (\sigma-\bar{\sigma})\nabla\bar{u}) + \alpha\,(\nabla\bar{\sigma}, \nabla(\sigma-\bar{\sigma})) + \beta\,(|\sigma-\underline{\sigma}|_1 - |\bar{\sigma}-\underline{\sigma}|_1) \ge 0 \text{ for all } \sigma \in \mathcal{K}\cap H^1(\Omega) \\ -\nabla\cdot(\bar{\sigma}\nabla\bar{u}) = 0, \quad \bar{\sigma}\frac{\partial\bar{u}}{\partial\nu} = g. \end{cases}$$

13.3.1 Constraint Bi-level Optimization

Consider the bi-level optimization with equality constraint:

$$\min J(y(p), u(p), p)$$

where $(y(p), u(p))$ is a minimizer of

$$\min \quad F(y,u,p) \text{ subject to } E(y,u,p) = 0.$$

Define the value function

$$V(p) = \inf F(y,u,p) \text{ subject to } E(y,u,p) = 0.$$

We consider the regularized problem

$$J(y,u,p)+\frac{F(y,u,p)-V(p)}{\epsilon}\ \text{subject to}\ E(y,u,p)=0$$

where (y,u) and p are independent variables. Necessary optimality condition is given by

$$\epsilon\, J_y(y,u,p)+F_y(y,u,p)+E_y(y,u)^*\lambda=0$$

$$\epsilon\, J_u(y,u,p)+F_u(y,u,p)+E_u(y,u)^*\lambda=0$$

$$J_p(y,u,p)+\frac{F_p(y,u,p)-F_p(y(p),u(p),p)}{\epsilon}=0.$$

where we use the value function calculus

$$V'(p)=F_p(y(p),u(p),p)$$

since $E(y,u)$ independent of p. Let $(\bar{y},\bar{u})=(y(p),u(p))$. Since

$$F_y(\bar{y},\bar{u},p)+E_y(\bar{y},\bar{u})^*\bar{\lambda}=0,\quad F_u(\bar{y},\bar{u},p)+E_u(\bar{y},\bar{u})^*\bar{\lambda}=0,$$

we have

$$E_y(\bar{y},\bar{u})^*(\lambda-\bar{\lambda})+(E_y(y,u)-E_y(\bar{y},\bar{u}))^*\lambda+\epsilon\, J_y(y,u,p)=0$$

$$E_u(\bar{y},\bar{u})^*(\lambda-\bar{\lambda})+(E_u(y,u)-E_u(\bar{y},\bar{u}))^*\lambda+\epsilon\, J_u(y,u,p)=0$$

$$J_p(y,u,p)+\frac{F_p(y,u,p)-F_p(\bar{y},\bar{u},p)}{\epsilon}=0$$

If we define

$$\mu=\frac{y-\bar{y}}{\epsilon},\quad q=\frac{\lambda-\bar{\lambda}}{\epsilon},\quad p=\frac{u-\bar{u}}{\epsilon},$$

$$E_y(\bar{y},\bar{u})^*q+E_{yy}(\bar{y},\bar{u})\mu+E_{yu}(\bar{y},\bar{u})p)^*\bar{\lambda}+J_y(\bar{y},\bar{u},p)=O(\epsilon)$$

$$E_u(\bar{y},\bar{u})^*q+(E_{uy}(\bar{y},\bar{u})\mu+E_{uu}(\bar{y},\bar{u})p)^*\bar{\lambda}+J_u(\bar{y},\bar{u},p)=O(\epsilon)$$

$$J_p(\bar{y},\bar{u},p)+F_{py}(\bar{y},\bar{u},p)\mu+F_{pu}(\bar{y},\bar{u})p=O(\epsilon)$$

If we let $\epsilon=0$ we obtain the necessary optimality oft the original problem.

13.4 Implicit Function Calculus

In this section we consider the implicit function case. Consider the value functional defined by

$$V(p) = F(x(p), p), \quad E(x(p), p) = 0$$

where we assume that the constraint E is C^1 and $E(x, p) = 0$ defines a continuous solution graph $p \to x(p)$ in a neighborhood of $(x(\bar{p}), \bar{p})$. Recall the classical implicit function theorem: suppose $E(\bar{x}, \bar{p}) = 0$ and E is continuously differentiable in a neighborhood of $(\bar{x}, \bar{p})$, and that $E_x(\bar{x}, \bar{p})$ is boundedly invertible. Then there exists the continuous solution mapping $p \to x(p)$ in a neighborhood of $(\bar{x}, \bar{p})$. Moreover, $x(p)$ is continuously differentiable and $x'(p) = E_x(x(p), p)^{-1}E_p(x(p), p)$ for all p in a neighborhood of $\bar{p}$. Define the Lagrangian

$$L(x, p, \lambda) = F(x, p) + (\lambda, E(x, p)) = 0$$

Thus, if we define

$$\lambda = -E_x(x(p), p)^{-*}F_x(x(p), p) = 0$$

by the chain rule

$$V'(p) = F_x(x(p), p)x'(p) + F_p(x(p), p) = L_p(x(p), p, \lambda).$$

The following is an extension, so-called Lagrange calculus [13].

Theorem 4.1 (Lagrange Calculus) *Assume F and E are C^1. Assume there exists λ that satisfies the adjoint equation*

$$E_x(x(\bar{p}), \bar{p})^*\lambda + F_x(x(\bar{p}), \bar{p}) = 0.$$

Then, $V(p)$ is G-differentiable at $\bar{p}$ if and only if the Lagrange functional

$$L(x, p, \lambda) = F(x, p) + \langle E(x, p), \lambda\rangle$$

satisfies

$$\begin{aligned} &L(x(p), \bar{p}, \lambda) - L(x(\bar{p}), \bar{p}, \lambda) - L_x(x(\bar{p}), \bar{p}, \lambda)(x(p) - x(\bar{p})) \\ &+(L_p(x(p), \bar{p}) - L_p(x(\bar{p}), \bar{p}))(p - \bar{p}) = o(|p - \bar{p}|). \end{aligned} \tag{13.30}$$

Then,

$$V'(\bar{p})\dot{p} = L_p(x(\bar{p}), \bar{p}), \lambda)\dot{p}. \tag{13.31}$$

Proof If we let

$$F(x(p),\bar{p})-F(x(\bar{p}),\bar{p})-F_x(x(\bar{p}),\bar{p})(x(p)-x(\hat{p}))+F_p((x(p),\bar{p})-F_p(x(\bar{p}),\bar{p}))(p-\bar{p})=\epsilon_1,$$

we have

$$V(p)-V(\bar{p})=F(x(p),p)-F(x(p),\bar{p})+F(x(p),\bar{p})-F(x(\bar{p}),\bar{p})$$

$$=\epsilon_1+F_x(x(\bar{p}),\bar{p})(x(p)-x(\hat{p}))+F_p(x(\bar{p}),\bar{p}))(p-\bar{p})+o(|p-\bar{p}|).$$

Note that

$$\langle E_x(x(\bar{p}),\bar{p})(x(p)-x(\bar{p})),\lambda\rangle+F_x(x(\bar{p}),\bar{p})(x(p)-x(\bar{p}))=0.$$

Let

$$\epsilon_2=\langle E(x(p),\bar{p})-E(x(\bar{p}),\bar{p})-E_x(x(\bar{p}),\bar{p})(x(p)-x(\bar{p}))+((E_p(x(p),\bar{p})-E_p(x(\bar{p}),\bar{p}))(p-\bar{p}),\lambda\rangle,$$

where

$$0=\langle E(x(p),p)-E(x(\bar{p}),\bar{p}),\lambda\rangle=\epsilon_2+\langle E_p(x(\bar{p}),\bar{p})(p-\bar{p}),\lambda\rangle+o(|p-\bar{p}|).$$

Thus, summing these up we obtain

$$V(p)-V(\bar{p})=L_p(x(\bar{p}),\bar{p})(p-\bar{p})+\epsilon_1+\epsilon_2+o(|p-\bar{p}|)$$

From assumption $\epsilon_1+\epsilon_2=o(|p-\bar{p}|)$ and thus $p\to V(x(p))$ is differentiable at $\bar{p}$ if and only if (13.30) holds and the value function derivative (13.31) holds. □

Remark Note that If we assume that E and F is C^2 in x, then it is sufficient to have the Hölder continuity

$$|x(p)-x(\bar{p})|_X\sim o(|p-\bar{p}|^{\frac{1}{2}})$$

for condition (13.30) holding.

In general if $x(\bar{p})$ is not unique, we have

Corollary 4.2 *Assume (x_n,p_n) satisfy*

$$E(x_n,p_n)=0,\quad E_x(x_n,p_n)^*\lambda_n+F_x(x_n,p_n)=0.$$

and

$$(x_n,\lambda_n)\to(\hat{x},\hat{\lambda})\ \textit{weakly}$$

for $p_n=\bar{p}+t_n\,\dot{p}$ as $t_n\to 0^+$ and the followings hold.

$$\begin{aligned} &L(x_n, \lambda_n, p_n) - L(\hat{x}, \lambda_n, p_n) + L_x(x_n, \lambda_n, p_n)(x_n - \hat{x}) \le 0 \\ &L(x_n, \hat{\lambda}, \bar{p}) - L(\hat{x}, \hat{\lambda}, \bar{p}) + L_x(\hat{x}, \hat{\lambda}, \bar{p})(x_n - \hat{x}) \ge 0 \end{aligned} \tag{13.32}$$

and

$$\begin{aligned} &\liminf_{t_n \to 0} \frac{L(x_n, \hat{\lambda}, p_n) - L(x_n, \hat{\lambda}, \bar{p})}{t_n} \ge L_p(\hat{x}, \hat{\lambda}, \bar{p})\dot{p} \\ &\limsup_{t_n \to 0} \frac{L(\hat{x}, \lambda_n, p_n) - L(\hat{x}, \lambda_n, \bar{p})}{t_n} \le L_p(\hat{x}, \hat{\lambda}, \bar{p})\dot{p}. \end{aligned} \tag{13.33}$$

Then, the (sequential) limit exists:

$$\lim_{t_n \to 0^+} \frac{V(p_n) - V(\bar{p})}{t_n} = L_p(\hat{x}, \hat{\lambda}, \bar{p})\dot{p}. \tag{13.34}$$

Moreover, if $(\hat{x}, \hat{\lambda})$ *is unique at* $\bar{p}$*, then* V *is differentiable at* $\bar{p}$ *and* (13.31) *holds.*

Proof For all solution pair $(\hat{x}, \hat{\lambda})$ of

$$E(\hat{x}, \bar{p}) = 0, \quad E_x(\hat{x}, \bar{p})^*\hat{\lambda} + F_x(\hat{x}, \bar{p}) = 0.$$

$$V(p_n) - V(\bar{p}) = F(x_n, p_n) - F(x_n, \bar{p}) + F(x_n, \bar{p}) - F(\hat{x}, \bar{p})$$

$$= L(x_n, \hat{\lambda}, \bar{p}) - L(\hat{x}, \hat{\lambda}, \bar{p}) + L_x(\hat{x}, \hat{\lambda}, \bar{p})(x_n - \hat{x}) + L(x_n, \hat{\lambda}, p_n) - L(x_n, \hat{\lambda}, \bar{p})$$

$$V(p_n) - V(\bar{p}) = F(x_n, p_n) - F(\hat{x}, p_n) + F(\hat{x}, p_n) - F(\hat{x}, \bar{p})$$

$$= L(x_n, \lambda_n, p_n) - L(\hat{x}, \lambda_n, p_n) + L_x(x_n, \lambda_n, p_n)(x_n - \hat{x}) + L(\hat{x}, \lambda_n, p_n) - L(\hat{x}, \lambda_n, \bar{p})$$

since $E(x_n, p_n) = E(\hat{x}, \bar{p}) = 0$.

Remark A sufficient condition for (13.33) is that we assume condition (13.30) for $L(x, p, \lambda))$ and the following conditions for $\langle E(x, p), \lambda\rangle$. We assume $p \to \langle E(x, p), \lambda\rangle$ is uniformly differentiable.

$$\frac{\langle E(x_n, p_n), \hat{\lambda}\rangle - \langle E(\hat{x}, \bar{p}), \hat{\lambda}\rangle}{t_n} - E_x(x_n, p_n)(\hat{x} - x_n) = \frac{\langle E(x_n, p_n), \hat{\lambda}\rangle - \langle E(x_n, \bar{p}), \hat{\lambda}\rangle}{t_n} - \langle E_p(x_n, \bar{p})\dot{p}, \hat{\lambda}\rangle.$$

and

$$\langle E_p(x_n, \bar{p})\dot{p}, \hat{\lambda}\rangle \to \langle E_p(\hat{x}, \bar{p})\dot{p}, \hat{\lambda}\rangle.$$

Example (Eigenvalue problem) Let $A(p)$ be a linear operator in a Hilbert space X. For $x = (\mu, y) \in R \times X$ consider

$$F(x, p) = \mu, \quad \text{subject to } A(p)y = \mu\, y, \quad |y|^2 = 1$$

That is, (μ, y) is an eigen pair of $A(p)$ and let μ be a simple eigenvalue. The adjoint equation for λ is given by

$$A(p)^*\lambda - \mu\,\lambda = 0, \quad (y, \lambda) = 0.$$

Since $\mu(p) \to \mu(\bar{p})$ as $p \to \bar{p}$, we have

$$\mu'(\dot{p}) = (\frac{d}{dp}A(p)\dot{p}y, \lambda).$$

13.4.1 Equality Constraint Optimization

Consider the equality constraint optimization

$$\min \quad F(x) + G(u) \text{ subject to } E(x, u) = 0 \text{ and } u \in \mathcal{K}, \tag{13.35}$$

From Theorem 4.1 we have

Theorem 4.3 (Constrained optimization) *Assume F and E are C^2 and $G(u) = G_0(u) + G_1(u)$ where G_0 is convex and G^1 is C^1 on a closed convex set $\mathcal{K}$. Let $\bar{u}$ is a minimizer of* (13.35) *and assume the assumptions in Theorem 4.1 and in addition (1) $\lambda \in Y$ satisfies the adjoint equation*

$$E_x(x(\bar{u}), \bar{u})^*\lambda + F'(x(\bar{u})) = 0. \tag{13.36}$$

(2) Conditions (13.30) *holds for $x \to L(x, \bar{u}, \lambda) = F(x) + \langle E(x, \bar{u}), \lambda\rangle$.*
(3) In a neighborhood of $(x(\bar{p}), \bar{p})$.

$$(x, u) \to L_u(x, u) \text{ is continuous.}$$

Then, a necessary optimality for (13.35) *is given by*

$$G_0(u) - G_0(\bar{u}) + (G_1'(\bar{u}) + E_u(x(\bar{u}), \bar{u})^*\lambda, u - \bar{u}) \geq 0 \text{ for all } u \in \mathcal{K}. \tag{13.37}$$

Lemma 4.4 (Sufficient optimality) *Assume* (13.36)–(13.37) *hold and*

$$\epsilon = L(x(u), u) - L(x(u^*), u^*) - L_x(x(u^*), u^*, \lambda)(x(u) - x(u^*))$$

$$-L_u(x(u^*), u^*, \lambda)(u - u^*) + G_1(u) - G_1(u^*) \geq 0 \text{ for all } u \in \mathcal{K}.$$

Then, $(x(u^*), u^*))$ is a minimizer of (13.35).

Proof It follows from

$$F(x(u)) + G(u) - (F(x(u^*)) + G(u^*)) \geq \epsilon. \qquad \square$$

13.5 Parameter-Dependent Constrained Optimization Problem

Consider the parameter-dependent constrained optimization: given $p \in \mathcal{P}$

$$V(p) = \inf_{x,u} \; F(x, u, p) \text{ subject to } E(x, u, p) = 0. \tag{13.38}$$

Define the Lagrangian functional:

$$L(x, u, , p, \lambda) = F(x, u, p) + \langle \lambda, E(x, u, p) \rangle.$$

Let $(x(p), u(p))$ is a minimizer at $p \in \mathcal{P}$.

Theorem 5.1 *Assume F and E are C^1 and assume that given $p \in \mathcal{P}$, there exists an optimal solution $y(p) = (x(p), u(p))$ for* (13.38) *and the solution graph $p \to (x(p), u(p))$ is continuous at $\bar{p}$.*
(1) $\lambda \in Y$ satisfies the adjoint equation

$$E_y(\bar{x}, \bar{u}, \bar{p})^* \lambda + F_y(\bar{x}, \bar{u}, \bar{p}) = 0. \tag{13.39}$$

(2) For $y = (x(p), u(p))$ and $\bar{y} = (x(\bar{p}), u(\bar{p}))$, $\epsilon = L(y, \bar{p}) - L(\bar{y}, \bar{p}) - L_y(\bar{y}, \bar{p})$ $(y - \bar{y}) = o(|p - \bar{p}|)$.
(3) In a neighborhood of $(x(\bar{p}), u(\bar{p}), \bar{p})$

$$(x, u, p) \to L_p(x, u, p) \textit{ is continuous.}$$

Then, the necessary optimality to (13.38) *is given by*

$$\begin{cases} E_x^*(\bar{x}, \bar{u}, \bar{p})\lambda + F_x(\bar{x}, \bar{u}, \bar{p}) = 0 \\ E_u^*(\bar{x}, \bar{u}, \bar{p})\lambda + F_u(\bar{x}, \bar{u}, \bar{p}) = 0 \\ E(\bar{x}, \bar{u}, \bar{p}) = 0 \end{cases}$$

and for $V(p) = F(x(p), u(p), p)$

$$V'(\bar{p})\dot{p} = L_p(\bar{x}, \bar{u}, \bar{p}, \lambda)\dot{p} = F_p(\bar{x}, \bar{u}, \bar{p})\dot{p} + \langle E_p(\bar{x}, \bar{u}, \bar{p})\dot{p}, \lambda \rangle.$$

Proof Let $y(p) = (x(p), u(p))$. Then,

$$F(y(p), p) - F(y(\bar{p}), \bar{p}) = F(y(p), \bar{p}) - F(y(\bar{p}), \bar{p}) + F(y(p), p) - F(y(p), \bar{p})$$

$$= F_y(y(\bar{p}), \bar{p})(y(p) - y(\bar{p})) + \epsilon_1 + F(y(\bar{p}), p) - F(y(\bar{p}), \bar{p})$$

where

$$\epsilon_1 = F(y(p), \bar{p}) - F(y(\bar{p}), \bar{p}) - F_y(y(\bar{p}), \bar{p})(y(p) - y(\bar{p}))$$

Note that

$$\langle E(y(p), p) - E(y(\bar{p}), \bar{p}), \lambda\rangle$$

$$= \langle E_y(y(\bar{p}), \bar{p})(y(p) - y(\bar{p}) + E_p((y(\bar{p}), \bar{p})(p - \bar{p}) + E(y(p), p) - E(y(p), \bar{p}), \lambda\rangle + \epsilon_2$$

where

$$\epsilon_2 = \langle E(y(p), \bar{p}) - E(y(\bar{p}), \bar{p}) - E_y(y(\bar{p}), \bar{p})(y(p) - y(\bar{p})), \lambda\rangle.$$

Since

$$\langle E_y(\bar{y}, \bar{p})(y(p) - y(\bar{p})), \lambda\rangle + F_y(\bar{x}, \bar{p})(y(p) - y(\bar{p})) = 0,$$

combining the above inequalities, we have

$$V(p) - V(\bar{p}) = L(y(p), p, \lambda) - L(y(p), \bar{p}, \lambda) + \epsilon = L_p(y(\bar{p}), \bar{p}, \lambda)\dot{p} + o(|p - \bar{p}|). \quad \square \tag{13.40}$$

In general one can have the following lemma:

Lemma 5.2 *(1) Let $y(p) = (x(p), u(p))$ is a set of minimizers of* (13.38) *at p and $y(p) \in U$, a bounded set in a neighborhood of $\bar{p}$. Assume that $y \to F(y, \bar{p})$ is weakly lower semicontinuous, $y \to E(y, \bar{p})$ is weakly continuous, and that*

$$|F(y, p) - F(y, \bar{p})| + |E(y, p) - E(y, \bar{p})| \to 0 \text{ uniformly on } U \text{ as } p \to \bar{p}$$

Then, for any weak accumulation point $\bar{y}$ of $y(p)$ as $p \to \bar{p}$, $E(\bar{y}, \bar{p}) = 0$ and $\bar{y} = (\bar{x}, \bar{u})$ is a minimizer of (13.38) *at $\bar{p}$ and $V(p) \to V(\bar{p})$ as $p \to \bar{p}$.*
(2) If $(x, u) \to F(x, u, \bar{p})$ is uniformly convex then $|y(p) - \bar{y}| \to 0$ as $p \to \bar{p}$.
(3) For $p_n = \bar{p} + t_n\,\dot{p}$ let $y_n \in y(p_n)$ as $t_n \to 0^+$, assume there exist a Lagrange multiplier λ_n for each $t_n > 0$ satisfying

$$E_y(y_n, p_n)^*\lambda_n + F_y(y_n, p_n) = 0.$$

and (y_n, λ_n) converges weakly to $(\hat{y}, \hat{\lambda})$ satisfying

$$E_y(\hat{y}, \bar{p})^*\hat{\lambda} + F_y(\hat{y}, \bar{p}) = 0.$$

Assume the second order optimality

$$\tilde{\epsilon} = L(y_n, p_n, \lambda_n) - L(\hat{y}, p_n\lambda_n) - L_y(y_n, p_n, \lambda_n)(y_n - \hat{y}) \le 0. \tag{13.41}$$

Then,

$$\limsup_{t_n \to 0^+} \frac{V(p + t_n \dot{p}) - V(\bar{p})}{t_n} \le L_p(\hat{y}, \bar{p}, \hat{\lambda})\dot{p}.$$

(4) For a minimizer $\bar{y}$ let $\bar{\lambda}$ is a corresponding Lagrange multiplier satisfying (13.39). *If we assume*

$$\epsilon = L(y_n, \bar{p}, \bar{\lambda}) - L(\bar{y}, \bar{\lambda}) - L_y(\bar{y}, \bar{p}, \bar{\lambda}))(y_n - \bar{y}) \ge 0,$$

and

$$\liminf_{t_n \to 0^+} \frac{L(y_n, p, \bar{\lambda}) - L(y_n, \bar{p}, \bar{\lambda})}{t_n} \le L_p(\bar{y}, \bar{p}, \bar{\lambda})\dot{p},$$

then

$$\liminf_{t_n \to 0^+} \frac{V(p + t_n \dot{p}) - V(\bar{p})}{t_n} \ge L_p(\bar{y}, \bar{p}, \bar{\lambda}).$$

(5) Under assumptions of (1)–(4)

$$\lim_{t \to 0^+} \frac{V(\bar{p} + t\,\dot{p}) - V(\bar{p})}{t} = \min_{(\bar{y}, \bar{\lambda})} L_p(\bar{y}, \bar{p}, \bar{\lambda})$$

$$\lim_{t \to 0^+} \frac{V(\bar{p}) - V(p - t\,\dot{p})}{t} = \max_{(\bar{y}, \bar{\lambda})} L_p(\bar{y}, \bar{p}, \bar{\lambda})$$

where $\bar{y}$ is a minimizer at $\bar{p}$ and $(\bar{y}, \bar{\lambda})$ satisfies (13.39).
(6) If $(y(\bar{p}), \lambda(\bar{p}))$ is unique, then $V'(\bar{p}) = L_p(y(\bar{p}), \bar{p}, \lambda(\bar{p}))$.

Proof For $y_n \in y(p_n)$ and $\hat{y} \in y(\bar{p})$ we have

$$F(y_n, p_n) - F(\hat{y}), \bar{p}) = F(y_n, p) - F(y_n, \bar{p}) + F(\hat{y}, p) - F(\hat{y}, \bar{p})$$

$$= F_y(y_n, p_n)(y_n - \hat{y}) + \tilde{\epsilon}_1 + F(\hat{y}, p_n) - F(\hat{y}, \bar{p})$$

and

$$0 = \langle \lambda_n, E(y_n, p_n) - E(\hat{y}, \bar{p}) \rangle$$

$$= \langle \lambda_n, E(y_n, p_n) - E(\hat{y}, p_n) + E(\hat{y}, p_n) - E(\hat{y}, \bar{p}) \rangle$$

$$+ \tilde{\epsilon}_2 + \langle \lambda_n, E_y(y_n, p_n)(y_n - \hat{y}) \rangle + \langle \lambda_n, E(y(\hat{y}, p) - E(\hat{y}, \bar{p}) \rangle,$$

where $\tilde{\epsilon} = \tilde{\epsilon}_1 + \tilde{\epsilon}_2$ and

$$\tilde{\epsilon}_1 = F(y_n, p_n) - F(\hat{y}, p_n) - F_y(y_n, p_n)(y - y(\bar{y})$$

$$\tilde{\epsilon}_2 = \langle \lambda_n, E(y_n, p_n) - E(\hat{y}, p_n) - E_y(y_n, p_n)(y_n - \hat{y}) \rangle.$$

Since as $t_n \to 0^+$

$$\langle \lambda_n - \hat{\lambda}, E_p(\hat{y}, \bar{p})\dot{p} \rangle \to 0$$

thus, from (13.44)

$$\limsup_{t_n \to 0^+} \frac{V(\bar{p} + t_n \dot{p}) - V(\bar{p})}{t_n} \leq \limsup_{t_n \to 0^+} \frac{L(\bar{y}, p_n, \bar{\lambda}) - L(\bar{y}, \bar{p}, \bar{\lambda})}{t} = L_p(\hat{y}, \bar{p}, \hat{\lambda})\dot{p}.$$

(4) Since

$$V(p_n) - V(\bar{p}) = F(y_n, p_n) - F(y_n, \bar{p}) + F(y_n, \bar{p}) - F(\bar{y}, \bar{p})$$

$$0 = \langle \bar{\lambda}, E(y_n, p_n) - E(y_n, \bar{p}) + E(y_n, \bar{p}) - E(\bar{y}, \bar{p}),$$

it follows from the assumptions we have

$$\liminf_{t_n \to 0^+} \frac{V(\bar{p} + t_n \dot{p}) - V(\bar{p})}{t} = \liminf_{t_n \to 0^+} \frac{L(y, p_n, \bar{\lambda}) - L(y_n, \bar{p}, \bar{\lambda})}{t_n} \leq L_p(\bar{y}, \bar{p}, \bar{\lambda})\dot{p}.$$

(5) From (3)–(4) we have

$$\liminf_{t \to 0^+} \frac{V(\bar{p} + t\,\dot{p}) - V(\bar{p})}{t} = \limsup_{t \to 0^+} \frac{V(\bar{p} + t\,\dot{p}) - V(\bar{p})}{t} = \min_{(\bar{y}, \bar{\lambda})} L_p(\bar{y}, \bar{p}, \bar{\lambda})\dot{p}$$

Similarly,

$$\liminf_{t \to 0^+} \frac{V(\bar{p} - V(\bar{p} - t\,\dot{p})}{t} = \limsup_{t \to 0^+} \frac{v(\bar{p} - V(\bar{p} + t\,\dot{p})}{t} = \max_{(\bar{y}, \bar{\lambda})} L_p(\bar{y}, \bar{p}, \bar{\lambda}\dot{p}$$

(6) If $(y(\bar{p}), \bar{\lambda})$ is unique, then the minimum and maximum coincide.

Remark The condition (3) holds if $y \to E(y, p)$ is afine and $y \to F(y, p)$ is convex. In general, if $y \to L(y, p, \lambda(p))$ is local convex, i.e.

$$L(y(p), p, \lambda(p)) - L(y, p, \lambda(p)) - L_y(y)(p), \lambda(p))(y(p) - y) \leq 0$$

in a neighborhood of $y(p)$, then the Lemma III holds.

Consider the nonlinear programing

$$\min \quad F(x,u,p) \text{ subject to } E(x,u,p)=0, \quad G(x,u,p)\le 0 \tag{13.42}$$

and the value functional defined by

$$V(p)=\inf_{x,u} F(x,u,p) \text{ subject to } E(x,u,p)=0, \quad G(x,u,p)\le 0. \tag{13.43}$$

Define the Lagrange functional

$$L(x,u,p,\lambda,\mu)=F(x,u,p)+\langle E(x,u,p),\lambda\rangle+\langle G(x,u,p),\mu\rangle.$$

Then, we have

Corollary 5.3 (Inequality constraint) *(1) Let $y(p)=(x(p),u(p))$ is a minimizer of* (13.42) *at p and $y(p)\in U$, a bounded set in a neighborhood $\bar{p}$. Assume that $y\to F(y,\bar{p})$ is weakly lower semicontinuous, $y\to E(y,\bar{p})$ is weakly continuous, and that $|F(y,p)-F(y,\bar{p})|+|E(y,p)-E(y,\bar{p})|+|G(y,p)-G(y,\bar{p})|\to 0$ uniformly on U as $p\to\bar{p}$. Then, for any weakly accumulation point $\bar{y}$ of $y(p)$ as $p\to\bar{p}$, $E(\bar{y},\bar{p})=0$ and $\bar{y}=(\bar{x},\bar{u})$ is a minimizer of* (13.42) *at $\bar{p}$ and $V(p)\to V(\bar{p})$ as $p\to\bar{p}$.*
(2) If $(x,u)\to F(x,u,\bar{p})$ is uniformly convex then $|y(p)-\bar{y}|\to 0$ as $p\to\bar{p}$.
(3) Assume the second order optimality

$$\begin{aligned}\tilde{\epsilon}&=L(y(p),p,\lambda(p),\mu(p))-L(y(\bar{p})),p,\lambda(p),\mu(p))\\&-L_y(y(p),p,\lambda(p),\mu(p)))(y(p)-y(\bar{p}))\le 0,\end{aligned} \tag{13.44}$$

where $(\lambda(p),\mu(p))$ satisfies

$$E_x(x(p),u(p),p)^*\lambda(p)+G_x(x(p),u(p),p)^*\mu(p)+F_x(x(p),u(p),p)=0. \tag{13.45}$$

and

$$\mu(p)\ge 0, \quad (\mu(p),G(x(p),u(p),p))=0. \tag{13.46}$$

Assume $(\lambda(p),\mu(p))$ converges weakly to $(\bar{\lambda},\bar{\mu})$ as $p\to\bar{p}$ and $(\bar{\lambda},\bar{\mu})$ satisfies (13.45)–(13.46) *at $(\bar{y},\bar{p})$. Then,*

$$L_p(x(\bar{p}),u(\bar{p}),\bar{\lambda},\bar{\mu})\in D^+V(\bar{p})$$

for all minimizers $(x(\bar{p}),u(\bar{p}))$ of (13.42) *at $\bar{p}$.*

Remark We can claim for all y

$$(\mu(p),G(y(p),p)-G(y,p)-G_y(y,p)(y(p)-y))\ge 0.$$

under appropriate conditions. For example, it holds if $y \to G(y, p)$ is coordinstewise convex in y since $\mu(p) \geq 0$. Then, if we assume $E(y, p)$ is affine and $x \to F(x, p)$ is convex, then (13.44) holds.

Value Function optimization We consider the value function optimization

$$\min \quad V(p) + \Psi(p), \quad \text{over } p \in \mathcal{K}$$

where the value function $V(p)$ is defined by (13.43). Under the second order optimality (13.44) we have the optimality

$$-\Psi'(p) \in L_p(x(\bar{p}), u(\bar{p}), \bar{\lambda}, \bar{\mu}).$$

In the following sections we present applications of our sensitivity analysis.

13.5.1 Optimal Control Problem

In this section consider the optimal control problem of Lagrange form:

$$\min \quad \int_t^T f^0(z(s), u(s))\, ds + G(z(T)) \tag{13.47}$$

subject to

$$\frac{d}{dt} z(t) = f(z(t), u(t)), \quad z(t) = x. \tag{13.48}$$

Let v is the optimal value function defined by

$$v(t, x) = \min_{u \in U_{ad}} F(z, u) = \int_t^T f^0(z(s), u(s))\, ds + g(z(T))$$

subject to

$$E(z, u) = f(z(t), u(t)) - \frac{d}{dt} z(t) = 0, \quad z(t) = x.$$

where the parameter $p = (t, x)$. Define the Lagrange functional

$$L(z, u, \lambda) = \int_t^T f^0(z(s), u(s))\, ds + g(z(T))$$

$$-(\lambda(T), z(T)) + (\lambda(t), x) + \int_t^T ((\lambda, f(z, u)) + (z(s), \frac{d}{dt}\lambda(s)))\, ds.$$

The adjoint equation for λ is given by

$$-\frac{d}{dt}\lambda(t) = f_x(z^*, u^*)^*\lambda(t) + f_x(z^*(t), u^*(t)), \quad p(T) = g'(z^*(T)).$$

The optimality condition is

$$u^*(t) = \text{argmin}_{u\in U}\{f^0(z^*(t), u) + (\lambda(t), f(z^*(t), u))\}.$$

If a minimizer is (z, u) is unique at $\bar{p} = (t, x)$ then from Theorem III we have (assuming $s \to \lambda(s)$ is differentiable at $s = t$),

$$v_x = \lambda(t), \quad v_t = f^0(x, u^*(t)) + (\lambda(t), f(x, u^*(t)).$$

Thus, the value function $v(t, x)$ satisfies the Hamilton Jacobi equation:

$$v_t + \min_{u\in U}\{f^0(x, u) + (\lambda, f(x, u))\} = 0 \text{ with } \lambda = v_x.$$

If the Hamiltonian

$$\mathcal{H}(z, u) = f^0(z, u) + (\lambda, f(z, u))$$

satisfies

$$\int_t^T \mathcal{H}(z^{t,x}(s), u) - \mathcal{H}(z^*(s), u^*(s)) - \mathcal{H}_y(y^*(s)) \cdot (z^{t,x} - z^*(s), u - u^*(s))\, ds$$

$$+g(z^{t,x}(T)) - g(z^*(T)) - g?(z^*(T)(z^{t,x}(T) - z^*(T)) \geq 0 \text{ for all } u \in U,$$

where $y = (z, u)$ and $z^{t,x}$ is the solution to (13.48) corresponding to $u \in U$, then (13.44) holds and $D^+(v)(\bar{p})$ exists and v is a supper viscosity solution, i.e.,

$$\phi_t + \min_{u\in U}\{f^0(x, u) + (\lambda, f(x, u))\} \leq 0 \text{ with } \phi(T) = g.$$

for all $\phi \in C^1$ satisfying $(\phi_t, \phi_x) \in D^+v(\bar{p})$.

Consider the parameter dependent case

$$\min \int_t^T f^0(z(s), u(s), p)\, ds + G(z(T), p)$$

subject to

$$\frac{d}{dt}z(t) = f(z(t), u(t), p), \quad z(t) = x.$$

given a fixed (x, t). Then, from Theorem III if a minimizer (z, u) is unique

$$v'(p)\dot{p} = L_p(z^*, u^*, \lambda, p)\dot{p}$$

where

$$L(z, u, \lambda, p) = \int_t^T f^0(z(s), u(s), p)\, ds + G(z(T), p)$$

$$-(\lambda(T), z(T)) + (\lambda(t), x) + \int_t^T (f(z, u, p), \lambda) + (z, \frac{d}{dt}\lambda(s)))\, ds.$$

That is,

$$v'(p)\dot{p} = \int_t^T f_p^0(z(s), u(s), p)\dot{p}\, ds + G_p(z(T), p)\dot{p} + \int_t^T (f_p(z, u, p)\dot{p}, \lambda(s))\, ds.$$

13.5.2 Shape Derivative

In this section we discuss the shape derivative [6, 13, 14] of the value function in which p corresponds to a shape of domain Ω.

$$\min_y \quad F(y, \Omega) \text{ subsect to } E(y, \Omega) = 0$$

For example, the equality constraint $E(y, \Omega) = E_0(y) + \chi_\Omega y = f$ is defined in a fixed domain Ω_0, where $E_0(x)$ is a closed nonlinear operator, and Ω is a subset of a fixed domain Ω_0 and χ_Ω is the characteristic function of potential term.

The vector filed method uses that $\Omega_t = F_t(\Omega)$ is the image of the domain Ω. In R^d under the mapping $F_t : R^d \to R^d$ defined by

$$F_t(x) = x + t\,h(x) \in R^d, \quad t \in R$$

where the deformation vector field $h \in W^{1,\infty}(R^d)$.

Given Ω, the optimal value function $v(\Omega)$ is defined by

$$v(\Omega) = \inf_x \ F(x, \Omega) \text{ subsect to } E(x, \Omega) = 0.$$

The shape derivative [6, 13] of $v(\Omega)$ is defined by

$$v'(\Omega)(x) = \lim_{t \to 0} \frac{u(\Omega_t)(x) - u(\Omega)(x)}{t}$$

Define the Lagrange functional

$$L(y,\Omega) = F(y,\Omega) + \langle E(y,\Omega), \lambda\rangle_{Y^*\times Y}$$

Then, it follow from Theorem 5.1 with $p = \Omega$

$$v'(\Omega) = L'(y,\Omega)$$

where the Lagrange multiplier $\lambda \in Y$ satisfies

$$E_y(y,\Omega)^*\lambda + F_y(y,\Omega) = 0.$$

For example, the scalar elliptic equation for $y \in H^1(\Omega_0)$

$$\int_{\Omega_0} (a(x)\nabla y, \nabla\phi)\, dx + \int_{\Omega} c(x)y, \phi)\, dx - \int_{\Omega} (f(x), \phi)\, dx - \int_{\Gamma} (g, \phi)\, ds_x$$

for all $\phi \in H^1(\Omega_0)$, where Ω and $\Gamma = \partial\Omega$ are the support of body force f and surface force g. Then, $\lambda \in H^1(\Omega_0)$ satisfy

$$\int_{\Omega_0} (a(x)\nabla\lambda, \nabla\psi) + c(x)\chi_\Omega\lambda, \psi)\, dx - \int_{\Omega_0} (F_y(y,\Omega), \psi)\, dx$$

for all $\psi \in H^1(\Omega_0)$. Then, it follows from the shape calculus [6, 13, 14] that

$$v'(\Omega)(h) = \int_\Gamma (j(y) + c(x)y\lambda - y\lambda - \frac{\partial}{\partial n}(g\lambda))(n\cdot h)\, ds$$

where $F(y,\Omega) = \int_\Omega j(y)\, dx$.

13.5.2.1 Optimal Control Problem on a Cylindrical Domain

We consider the optimal control problem:

$$\min \quad J(y,u,\Omega) = \int_0^T \int_{\Omega_0} \ell(y)\, dx + \int_{\Omega_t} \frac{1}{2}|u|^2\, dx)\, dt + g(y(T))$$

subject to

$$\frac{dy}{dt} = f(y) + \chi_\Omega(k(y) + u(t)), \quad y(0) = y_0$$

where Ω_t is the support of control action in R^d. Define the Lagrange functional

$$L(y,\Omega) = J(y,u,\Omega) + \int_0^T f(y) + \chi_{\Omega_t}(k(y)+u) - \frac{d}{dt}y, \lambda)$$

Then, the optimality condition is given by

$$u = \lambda \text{ on } \Omega_t$$

$$-\tfrac{d\lambda}{dt} = f'(y)^*\lambda + \chi_{\Omega_t}(k'(y)^* - 1)\lambda + \ell_y(y) = 0, \quad \lambda(T) = g_y(T).$$

Then, it follows from Theorem 5.1 that

$$v'(\Omega) = \int_0^T \int_{\Gamma_t} (\frac{1}{2}|\lambda|^2 + (k(y) - \lambda, \lambda)\, dsdt.$$

13.6 Conclusion

In this paper we consider the parameter-dependent constraint optimization. We derive the derivative formula of the optimality value function with respect the parameters for a general class of constrained optimization. We apply it to the bi-level optimization, the parameter design and topological optimization. We demonstrate the feasibility of the calculus for various applications. We will discuss more cases including inclusions and quasi-variational inequalities and the duality and the min-max theory in a forthcoming paper.

References

1. G. Allaire, F. Jouve, A.M. Toader, Structural optimization using sensitivity analysis and a level-set method. J. Comput. Phys. **194**, 363–393 (2004)
2. D.P. Bertsekas, *Nonlinear Programming Belmont* (Athena Scientific, Belmont, 1999)
3. J.F. Bonnans, J.C. Gilbert, C. Lemarechal, C.A. Sagastizabal, *Numerical Optimization-Theoretical and Practical Aspects* (Springer, Berlin, 2006)
4. P. Bernhard, A. Rapaport, On a theorem of Danskin with an application to a theorem of Von Neumann-Sion. Nonlinear Anal.: Theory Methods Appl. **24**, 1163–1181 (1995)
5. E. Castillo, A. Conejo, C. Castillo, R. Minguez, D. Ortigosa, A perturbation approach to sensitivity. J. Optim. Theory Appl. **128**, 49–74 (2006)
6. M.C. Delfour, J.-P. Zolesio, *Shapes and Geometries: Metrics, Analysis, Differential Calculus, and Optimization*, 2nd edn. (SIAM, Philadelphia, 2011)
7. A.V. Fiacco, *Introduction to Sensitivity and Stability Analysis in Nonlinear Programming* (Academic, New York, 1983)
8. T.H. Lee, J.S. Arora, A computational method for design sensitivity analysis of elastoplastic structures. Comput. Methods Appl. Mech. Eng. **122**, 27–50 (1995)
9. K. Ito, B. Jin, *Inverse Problems: Tikhonov Theory and Algorithms* (World Scientific, Singapore, 2014)
10. K. Ito, B. Jin, T. Takeuchi, A regularization parameter for nonsmooth Tikhonov regularization. SIAM J. Sci. Comput. **33**, 1415–1438 (2011)
11. K. Ito, K. Kunisch, On the choice of the regularization parameter in nonlinear inverse problems, (with K. Kunisch). SIAM J. Optim. **2**, 376–404 (1992)

12. K. Ito, K. Kunisch, Optimal control of elliptic variational inequalities. Appl. Math. Optim. **41**, 343–364 (2000)
13. K. Ito, K. Kunisch, *Lagrange Multiplier Approach to Variational Problems and Applications*. SIAM Advances in Design and Control (2008)
14. K. Ito, K. Kunisch, G. Peich, Variational approach to shape derivatives. ESAIM: Control Optim. Calc. Var. **14**, 517–539 (2008)
15. M. Hintermuller, K. Kunisch, Path-following methods for a class of constrained minimization problems in function space, SIAM J. Optim. **17**, 159–187 (2006)
16. A. Rothwell, *Optimization Methods in Structural Design*. Solid Mechanics and Its Applications (Springer, Berlin, 2017)

Author Index

J. Cheng et al. (eds.), *Inverse Problems and Related Topics*,
Springer Proceedings in Mathematics & Statistics 310,
https://doi.org/10.1007/978-981-15-1592-7